MODIFIED CELLULOSICS

ACADEMIC PRESS RAPID MANUSCRIPT REPRODUCTION

MODIFIED CELLULOSICS

Edited by

ROGER M. ROWELL

U.S. Department of Agriculture
Forest Products Laboratory
Madison, Wisconsin

RAYMOND A. YOUNG

Department of Forestry
University of Wisconsin
Madison, Wisconsin

ACADEMIC PRESS   New York San Francisco London   1978
A Subsidiary of Harcourt Brace Jovanovich, Publishers

COPYRIGHT © 1978, BY ACADEMIC PRESS, INC.
ALL RIGHTS RESERVED.
NO PART OF THIS PUBLICATION MAY BE REPRODUCED OR
TRANSMITTED IN ANY FORM OR BY ANY MEANS, ELECTRONIC
OR MECHANICAL, INCLUDING PHOTOCOPY, RECORDING, OR ANY
INFORMATION STORAGE AND RETRIEVAL SYSTEM, WITHOUT
PERMISSION IN WRITING FROM THE PUBLISHER.

ACADEMIC PRESS, INC.
111 Fifth Avenue, New York, New York 10003

United Kingdom Edition published by
ACADEMIC PRESS, INC. (LONDON) LTD.
24/28 Oval Road, London NW1 7DX

Library of Congress Cataloging in Publication Data

Main entry under title:

Modified cellulosics.

    Proceedings of a symposium in the Cellulose, Paper, and Textile Division of the American Chemical Society, presented at the Society's 174th meeting in Chicago, Ill., Aug. 29-Sep. 1, 1977.
    Includes index.
    1. Cellulose—Congresses.   I. Rowell, Roger M.
II. Young, Raymond Allan, Date   III. American
Chemical Society. Cellulose, Paper, and Textile Division.
TS933.C4M6   668.4'4   78-4255
ISBN 0-12-599750-7

PRINTED IN THE UNITED STATES OF AMERICA

Contents

List of Contributors	ix
Preface	xi
Anselme Payen Award 1977	xiii

PART 1   INTRODUCTION—AWARD ADDRESS

The Versatility of Cellulose   3
   Kyle Ward, Jr.

Practical Textile Finishes Based on Chemical Modification
of the Cotton Fiber   11
   J. David Reid and Robert M. Reinhardt

PART 2   OVERVIEW OF CELLULOSE SOURCES AND MODIFICATION REACTIONS

Chemical Modification of Cellulose: A Historical Review   23
   Donald F. Durso

Cotton: A World Cellulose Fiber   39
   R. S. Corkern, M. E. Carter, and B. M. Kopacz

Outlook for Wood Cellulose   65
   Necmi Sanyer

Rayon Fibers of Today   81
   G. C. Daul and F. P. Barch

CONTENTS

Modified Cellulosics—An Overview of the Future   95
    H. L. Hergert and T. E. Muller

PART 3   CELLULOSE ACCESSIBILITY AND REACTIVITY

Determination of Accessibility and Crystallinity of Cellulose   117
    Larry C. Wadsworth and John A. Cuculo

Hydroxyl Reactivity and Availability in Cellulose   147
    Stanley P. Rowland

PART 4   CELLULOSE MODIFICATION BY GRAFTING TECHNIQUES

New Methods for Graft Copolymerization
onto Cellulose and Starch   171
    Bengt Rånby

Modifications to Cellulose Using UV
Grafting Procedures   197
    Neil P. Davis and John L. Garnett

Wood Pulp Grafting with Different Monomers
by the Xanthate Method   227
    V. Hornof, C. Daneault, B. V. Kokta, and J. L. Valade

A Novel Process for the Treatment of Pulp Mill Effluents Using
the Grafted Pulp Fibers   241
    P. Lepoutre, S. H. Hui, and A. A. Robertson

PART 5   GENERAL CELLULOSE MODIFICATION REACTIONS

Antibacterial Fibers   259
    Tyrone L. Vigo

Preparation, Structure, and Properties of some Derivatives
of Cellulose and of Thermomechanical Pulp   285
    Shyam S. Bhattacharjee and Arthur S. Perlin

The Effect of Liquid Anhydrous Ammonia
on Cellulose II   303
    Luis G. Roldan

Effect of Changes in Supramolecular Structure on the Thermal
Properties and Pyrolysis of Cellulose   321
    K. E. Cabradilla and S. H. Zeronian

Recent Developments in the Industrial
Use of Hemicelluloses   341
    Roy L. Whistler and R. N. Shah

Index   *357*

List of Contributors

Numbers in parentheses indicate the pages on which the authors' contributions begin.

F. P. BARCH (81), Eastern Research Division, ITT Rayonier, Inc., Whippany, New Jersey

SHYAM S. BHATTACHARJEE (285), Department of Chemistry, McGill University, Montreal, Quebec, Canada

K. E. CABRADILLA (321), Division of Textiles and Clothing, University of California, Davis, California

MARY E. CARTER (39), USDA Agricultural Research Service, Southern Regional Laboratory, New Orleans, Louisiana

R. S. CORKERN (39), USDA Economic Research Service, Southern Regional Laboratory, New Orleans, Louisiana

JOHN A. CUCULO (117), Department of Textile Chemistry, North Carolina State University, Raleigh, North Carolina

C. DANEAULT (227), Department of Engineering, University of Quebec at Trois-Rivières, Quebec, Canada

G. C. DAUL (81), Eastern Research Division, ITT Rayonier, Inc., Whippany, New Jersey

NEIL P. DAVIS (197), School of Chemistry, University of New South Wales, Kensington, Australia

DONALD F. DURSO (23), Research Center, Johnson & Johnson, New Brunswick, New Jersey

JOHN L. GARNETT (197), School of Chemistry, University of New South Wales, Kensington, Australia

H. L. HERGERT (95), ITT Rayonier, Inc., New York, New York

V. HORNOF (227), Department of Chemical Engineering, University of Ottawa, Ontario, Canada

S. H. HUI (241), Pulp and Paper Research Institute of Canada, Pointe Claire, Quebec, Canada

B. V. KOKTA (227), Department of Engineering, University of Quebec at Trois-Rivières, Quebec, Canada

LIST OF CONTRIBUTORS

B. M. KOPACZ (39), USDA Agricultural Research Service, Southern Regional Laboratory, New Orleans, Louisiana

P. LEPOUTRE (241), Pulp and Paper Research Institute of Canada, Pointe Claire, Quebec, Canada

T. E. MULLER (95), ITT Rayonier, Inc., New York, New York

ARTHUR S. PERLIN (285), Department of Chemistry, McGill University, Montreal, Quebec, Canada

BENGT RÅNBY (171), Department of Polymer Technology, The Royal Institute of Technology, Stockholm, Sweden

J. DAVID REID (11), Consultant, New Orleans, Louisiana

ROBERT M. REINHARDT (11), USDA Agricultural Research Service, Southern Regional Laboratory, New Orleans, Louisiana

A. A. ROBERTSON (241), Pulp and Paper Research Institute of Canada, Montreal, Quebec, Canada

LUIS G. ROLDAN (303), Technical Center, J. P. Stevens & Co., Inc., Garfield, New Jersey

STANLEY P. ROWLAND (147), USDA Agricultural Research Service, Southern Regional Laboratory, New Orleans, Louisiana

NECMI SANYER (65), USDA Forest Products Laboratory, Madison, Wisconsin

R. N. SHAH (341), Department of Biochemistry, Purdue University, W. Lafayette, Indiana

J. L. VALADE (227), Department of Engineering, University of Quebec at Trois-Rivières, Quebec, Canada

TYRONE L. VIGO (259), USDA Textiles and Clothing Laboratory, Knoxville, Tennessee

LARRY C. WADSWORTH (117), Department of Textile Chemistry, North Carolina State University, Raleigh, North Carolina

KYLE WARD, JR. (3), Department of Chemistry, Institute of Paper Chemistry, Appleton, Wisconsin

ROY L. WHISTLER (341), Department of Biochemistry, Purdue University, W. Lafayette, Indiana

S. H. ZERONIAN (321), Division of Textiles and Clothing, University of California, Davis, California

Preface

Interest in cellulose modification reactions reached a peak in the early 1950s with the advent of advanced instrumental methods and prolific funding. This was followed by a lengthy period of research activity on petrochemical-based polymers. The boundaries for the synthetic polymer scientist seemed unlimited with the cheap source of monomer. New polymerization methods such as stereoregular polymerization of polypropylene solidified the optimistic attitude of the polymer chemist in the 1960s.

Significant events in the 1970s, however, instituted a dramatic change of attitude. Impending shortages of petrochemicals pronounced an end to the era of cheap petrochemicals. This, coupled with increasing environmental and health standards related to toxic monomer systems such as vinyl chloride and acrylonitrile, have forced a rethinking of important research directions in polymer science.

The research emphasis is again shifting toward the most abundant polymer in the world, cellulose. The subjects covered in "Modified Cellulosics" are based on a symposium in the Cellulose, Paper, and Textile Division of the American Chemical Society presented at the 174th National Meeting in Chicago, Illinois, August 29–September 1, 1977. The first section, an "Overview of Cellulose Sources and Modification Reactions," brings up to date the previous work on cellulose modification and outlines the inventory of cotton and wood cellulose. This is followed by a review of accessibility and reactivity of cellulose since these factors are so critical to any derivatization of the cellulose polymer. The recent discovery of a number of new cellulose solvents has broadened the possibilities for cellulose modification reactions and opened up new areas of research on cellulose regeneration to rayon.

The participants of the symposium expressed widely divergent viewpoints on the future of modification of cellulose by grafting of vinyl monomers. However, the development of grafting methods continues and more applications of grafted cellulose fibers are given in the section "Modification of Cellulose by Grafting Techniques." It is possible that graft derivatization methods will not be realized commercially until techniques are developed to avoid the large amounts of homopolymer formed in these reactions. This step would greatly alleviate present economic restraints on such processes.

Additional methods of cellulose modification and applications of modified cellulosics are presented in the last section, "General Cellulose Modification Reactions." Whether the modification reactions are applied for cotton finishes or for antibacterial fibers, the future looks bright for cellulose-derived products. Enzymatic techniques offer unique approaches to efficient derivatization of cellulosics.

A polling of the symposium participants indicated that the majority felt that the most important future area of research in the cellulose field was on cellulose modification reactions. This book should therefore serve both as a basic reference to past research and as a stepping-off point from current investigations in the field of cellulose modification reactions. With this resurgence of cellulose research, it is quite fitting that the book be dedicated to a true pioneer in cellulose research, Professor Kyle Ward, recipient of the 1977 Anselme Payen Award of the Cellulose, Paper, and Textile Division of the American Chemical Society.

The editors would like to express their appreciation to Robert F. Schwenker for chairing one of the sessions of the symposium; to Dr. Robert E. Read, Program Chairman of the Cellulose, Paper, and Textile Division, ACS, for his assistance in scheduling the symposium; and to Dr. Conrad Schuerch, Chairman of the Cellulose, Paper, and Textile Division, ACS, for his encouragement and support.

Anselme Payen Award 1977

This volume is dedicated to the 1977 Anselme Payen Award Winner, Kyle Ward, Jr.

Dr. Ward has won international recognition for his contributions to the chemistry of cellulose and wood, and particularly for his studies of the chemical modification of cellulose for both paper and textile fibers. His inspiring teaching and research guidance have made a significant impact in the field of cellulose chemistry. Dr. Ward is presently Professor Emeritus and former Chairman of the Department of Chemistry, The Institute of Paper Chemistry, Appleton, Wisconsin.

The Anselme Payen Award is named after the distinguished French scientist, discoverer of cellulose, and pioneer in the chemistry of both cellulose and lignin. The term "cellulose" was coined and introduced into the scientific literature by Payen in 1839.

The Award is presented annually by the Cellulose, Paper, and Textile Division, ACS, to honor and encourage outstanding professional contributions to the science and chemical technology of cellulose and its allied products. Previous awardees are L. E. Wise, C. B. Purves, H. M. Spurlin, C. J. Malm, W. A. Sisson, R. L. Whistler, A. J. Stamm, S. G. Mason, W. A. Reeves, T. E. Timell, Conrad Schuerch, D. A. I. Goring, V. T. Stannett, J. K. N. Jones, and R. H. Marchessault.

PART 1 Introduction—Award Address

Modified Cellulosics

THE VERSATILITY OF CELLULOSE

KYLE WARD, Jr.
1977 Anselme Payen Address

One of the great privileges of being the Payen Medalist is that of being allowed to talk, one of my favorite occupations, with a captive audience on a subject of my own choosing. The things I want to present are not new, but perhaps the juxtaposition of items may be new to some of you.

Because I want to show the chemical versatility of cellulose, I shall say little of its physical structure, nor shall I go into detail on all of its chemical reactions. I want to discuss simple etherification or esterification. I shall not discuss multiple substituents of the type Dr. Touzinsky has treated. And I am saying no more of the processes Dr. Reid mentioned - oxidation, urethane formation, grafting, cross-linking, etc., - than that they are additional evidence of the versatility of cellulose. If we rule all of these changes out, cellulose is still a remarkably versatile raw material.

We can express a simple cellulose derivative by the following formula - Figure 1. Here R represents the substituent group, x we call the D.P. and y the DS. All three are necessary criteria to characterize a derivative, but they are not sufficient. For one thing, cellulose is a polymer and x and y are average values; we must also know DP and DS distributions. Moreover, the supramolecular structure must also be known, although I shall not go into detail on this today. Mostly, I want to concentrate only on DS.

FIGURE 1

$$H[(C_6H_7O_2(OH)_{3-y}(OR)_y]_xOH$$

R = any substituent

x = DP

y = DS

Generalized Formula For a Simple Cellulose Derivative

I shall say little of x or D.P., because I am interested in cellulose as a polymer. If we start with the polymer cellulose as a raw material with a DP well over a few thousands, we can change this DP, but only downward. The production of low molecular-weight compounds from cellulose is a very important facet of its versatility, but these compounds can usually be made from simpler sources, like glucose, so I'll bypass this. But I must point out that a reduction in DP is also important in polymeric cellulose derivatives, as it changes the physical properties, especially the solution properties, like viscosity. While DP lowering usually broadens DP distribution, microcrystalline cellulose is an exception, but I'm going to ignore this important development, too.

I should like to say a great deal about the nature of the substituent R, but time prevents this. Volumes have been written about it. Obviously, the nature of the substituent determines the properties and, hence, the utility. Hydrophobic groups have different effects than hydrophilic, neutral groups than ionic, large groups than small. Color can be imparted by chromophoric groups. Functional groups with biologic activity frequently retain this activity when attached to the cellulose molecule. Since I can't cover all this in 20 minutes or even in 20 hours, I propose to pick a single substituent and show what a variety of effects can result by changing only Y, that is, DS and its distribution.

I have chosen acetyl as the substituent, a simple neutral group, yet admirable suitable to demonstrate the versatility of cellulose. Let us look at some important uses of acetylated cellulose.

"Cellulose acetate" might mean several things, but usually refers to the so-called "secondary cellulose acetate" prepared as follows. Cellulose is activated with acetic acid and then esterified with acetic anhydride and sulfuric acid. As the reaction continues, a clear solution (or dope) is formed. At this point the product is a mixed ester of sulfuric and acetic acid. The sulfuric acid groups can be removed either by simple transesterification in which magnesium oxide is added to neutralize the catalyst and continuing acetylation to give a triacetate or by hydrolysis in which dilute acetic acid is added and reaction continued to give the sulfur-free "secondary acetate" with DS about 2.5, in which each of the cellulose molecules is roughly equivalently substituted. The triacetate (DS above 2.8) is soluble in chlorinated solvents, the acetate of DS 2.2-2.8 is soluble in acetone, ketones, esters, etc. If hydrolysis is continued we find that products of DS 1.2-1.8 are soluble in methyl cellosolve and those of DS 0.6-0.8 in water. Products of lower DS are soluble only in cellulose solvents (Figure 2).

FIGURE 2

Solubility of Cellulose Acetate
of Various DS

Approximately DS	Solvent
2.8–3.0	Chloroform
2.2–2.8	Acetone
1.2–1.8	Methyl Cellosolve
0.6–0.8	Water
0.0–0.6	Cellulose Solvents

I have stressed the property of solubility, because it is the characteristic useful property of the two of these which are industrially important, the triacetate and the secondary acetate. Also, if we continue to look at solubility, it is easy to show that the degree of acetylation alone is not enough to characterize the product.

If acetylation is carried out in the presence of a non-solvent like amyl acetate or benzene, it can be carried out to the triacetate or any intermediate stage, but the products are quite different from those just described. They dissolve completely only in the usual solvents for cellulose. You all know that these solubility differences are due to differences in DS distribution, or, otherwise phrased, uniformity of reaction. But I shan't apologize for repeating what you already know; that's one way to make one remember things or even, in time, to believe them. In so-called "homogeneous acetylation," two reactions compete, acetylation and sulfation. In both reactions, the primary hydroxyl reacts more rapidly. Sulfation is more rapid than acetylation. Eventually one gets a triester containing both sulfate and acetyl groups. At this stage fibers have disintegrated and essentially all cellulose molecules are available for reaction. If one neutralizes the sulfuric acid catalyst, but permits acetylation to continue one arrives at the triacetate, which has no free hydroxyls and is soluble only in chlorinated hydrocarbons and the like. If, though, one saponifies the sulfuric acid groups from the triester, dope, one has fairly uniform substitution of the molecules and solubility is good in a number of solvents and solvent mixtures. On the other hand, fibrous acetylation is a topo-chemical reaction. The molecules are not equally accessible, the fibers are only slightly swollen and initial reaction occurs mostly on the outside of the fiber or fibril and, only after slow swelling and diffusion, is the interior reacted.

The result is a very nonuniform product, a core of poorly reacted and an outer layer of more highly reacted material, but they are not two discrete regions; methylene chloride dissolves only 8% of an acetylated fiber with 20% acetyl. Acetic acid, which swells more, dissolves twice as much. Obviously, there has been some penetration of the fiber during acetylation, but the product is distinctly different from that produced by homogeneous acetylation.

We have found use for a great number of acetylated cellulose products differing in the extent and the site of the acetylation. I should like to point some of them out as evidence of the versatility of cellulose.

FIGURE 3

Properties of Cellulose
Acetate Plastics

D.S.	2.45	2.85
R.I.	1.475	1.470
MP range, °C	230-250	290-300
Solubility		
Acetone	Sol.	Ins.
Tetrachlorethane	Sol.	Sol.
Chloroform	Ins.	Sol.

FIGURE 4

Uses of Cellulose Acetate Plastics

Secondary	Triacetate
Transparent Sheets	Transparent Sheets
Textiles	Textiles
Cigarette Filters	Photographic Film
Molded Plastics	Electrical Insulation
Tool Handles	
Toys	

First, there are the two plastics already mentioned. Some of the differences in properties and uses are shown in the Figures 3 and 4. Secondary acetate is used for molded plastics, films, sheets, and lacquers. The triacetate has found considerable use as sheeting and photographic film; it has little affinity for water and correspondingly good electrical resistance, but it is less useful for molded plastics, as its high softening temperature is not appreciably reduced by plasticizers. Both types of acetate have good clarity and toughness and both have important uses as textile fibers. Secondary cellulose acetate fibers have been important in textiles for over 50 years. Today, they have an important nontextile use in cigaret filters. The early commercial efforts with the triacetate did not make the grade, but improved technology caused a comeback 20 years ago. Secondary acetate fibers are usually prepared by dry-spinning of an acetone solution into warm air, but sometimes by wet-spinning into water. The triacetate fiber is usually dry-spun from a mixture of methylene chloride and alcohol, but a certain amount is spun from the melt, providing the dwell time at the high temperature is kept low, about 30 sec., to prevent decomposition. Cellulose acetate rayon has proved useful because it is not sensitive to moisture and gives very attractive effects in apparel fabrics. The low melting point is bad for some uses; heat-treated triacetate avoids this by partial crystallization of the regularly substituted molecules. This process is not possible with secondary acetate, where hydroxyl and acetyl groups are irregularly distributed along the chain. The cured triacetate has a higher melting point, better shrinkage and wrinkle-resistance, improved wash fastness for dyes and lower water sorption.

The subject of fibers spun from cellulose acetate is not complete without some mention of Fortisan, although this is not itself acetylated. It is related to cellulose acetate as viscose rayon is to cellulose xanthate. In other words, cellulose acetate is an intermediate form which all the acetyl groups are saponified to give a regenerated cellulose fiber. The essence of the process is that the cellulose acetate fiber is swollen (e.g., in aqueous dioxane) and highly stretched while swollen. In this state it is saponified, giving a regenerated cellulose fiber of extremely high molecular orientation, with correspondingly high tensile strength and low elongation.

Let us consider another type of acetylated cellulose textile fiber. This is formed directly by fibrous acetylation of cellulosic fibers, especially cotton. Cotton yarns were acetylated commercially in England forty years ago. In this country, the Southern Regional Research Laboratory (now Center) started work on the process during the forties. Two products were developed, one called PA (partly acetylated) and the other FA (fully acetylated) cotton. The latter is something of a misnomer, for only when the lint fiber is treated does it approach a triacetate (DS 2.7-3.0). The tighter structure of yarns and fabrics inhibits the swelling necessary for complete reaction. FA are DS 2.4-2.7 and fabrics 2.1-2.4. PA cotton has DS about 1.0. Actually, we discussed such products earlier under topochemical reaction.

DS 1 corresponds approximately for surface acetylation of cotton microfibrils of the size deduced by Scallan. For FA cotton we must have swelling and intrafibrillar penetration, but there is still too much residual hydrogen-bonded structure for solution. These acetylated cottons are very resistant to heat and rot. FA cotton has the higher tear strength and abrasion resistance.

Like cotton, paper can be acetylated - either as pulp fiber or as sheeted paper - and both have had commercial use. Paper sheets acetylated to DS 0.7-1.8 have been used for their electrical properties, dimensional stability and heat resistance. At this level of substitution, the fibers have less affinity for water and do not bond well. However, Eastman Kodak produces acetylated fiber, DS about 1, which is used for special papers. At still lower DS 0.0-0.3, moisture sorption and paper strength are better than for untreated pulp (Figure 5).

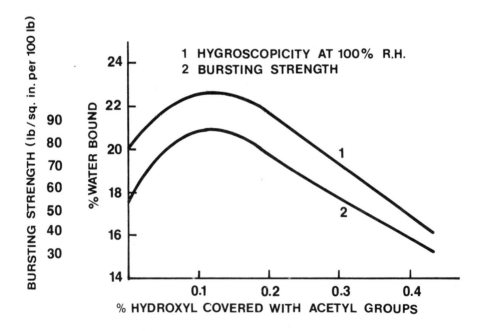

For completeness' sake, I should mention also that wood is also dimensionally stabilized by acetylation to DS 1-1.3. This is a more complex system and we are sure which components besides cellulose participate in the reaction. Probably much of the wood substance is not available for reaction. Still, the change in dimensional stability occurs as with pure cellulose; so I mention it.

To conclude, the record shows that with a single modifying substituent, cellulose properties can be changed over a wide range. Different substituents permit still wider modification. Add to this the possibility of physical changes, of other chemical changes than substitution and of multiple substitutions and we have an infinite palette. This renewable raw material for such a tremendous constellation of products deserves special consideration in our present economy.

Modified Cellulosics

PRACTICAL TEXTILE FINISHES BASED ON CHEMICAL
MODIFICATION OF THE COTTON FIBER

J. D. REID

Consultant, New Orleans, Louisiana

and

R. M. REINHARDT

USDA Agricultural Research Service
Southern Regional Laboratory, New Orleans, Louisiana

Dr. Kyle Ward, Jr., organized and initiated investigations on cotton chemical modification at the Southern Regional Research Center, which subsequently became the world's leading laboratory for cotton textile finishing research. This research profoundly influenced the utilization of modified cottons in a wide range of practical applications. Selected commercialized examples illustrate the various types of chemical modification used in the finishing of cotton, such as esterification, oxidation, etherification, and deposition within the fiber. They range from minor topochemical reactions to deep-seated changes and include treatments in which reactants are applied in the vapor, liquid, and solid phases. Illustrative developments include cotton textiles that exhibit water repellency, rot and heat resistance, durable press, flame resistance, and other functional properties. Introduction of such properties by chemical modification has kept cotton competitive with other textile fibers, both natural and synthetic.

Introduction

The coauthor of this paper, Mr. Robert M. Reinhardt, and I are delighted to have been selected to open the Anselme Payen Award Symposium that honors our good and illustrious friend, Dr. Kyle Ward, Jr.

By coincidence, it happens that I became aware of Anselme Payen and Kyle Ward in the same year, 1940. My experience with Anselme Payen was not personal, of course. However, 37 years ago, E. C. Dryden and I ran across some scientific publications written by Payen more than a hundred years earlier, and we published material describing them in a paper entitled, "Anselme Payen—Discoverer of Cellulose" (1). I think this was the first contemporary article to describe his many achievements and to acknowledge the debt owed by the modern cellulose chemist to him for discovering and naming cellulose in 1838. We pointed out that he also was the first to discover and isolate lignin, a year later, in 1839. He did this by the use of fuming hydrochloric acid in a process that was essentially the same as that proposed 90 years later in 1928 (2) by Max Phillips for analysis of lignin.

It was in 1940 in Ames, Iowa that I first met Kyle Ward. Even in those days he had a magnificent shock of crisp, iron-gray hair. Kyle was starting to assemble a group of scientists to do research on cotton. A year or two earlier it had become evident to Congress that research was needed to help the farmer establish new crops and to find expanded uses for surplus farm crops. Four great laboratories were established, one of them the Southern Regional Research Laboratory in New Orleans, Louisiana.

Although cotton was still king of the fibers in 1940, the market was flooded with an oversupply and the farmer received only a small return for his crop. Furthermore, the supremacy of King Cotton was beginning to weaken as new synthetic fibers were introduced. Obviously, cotton had to compete more vigorously with the other fibers or this important cash market of the American farmer would be lost. The major objective of research by the Southern Laboratory was to increase the utilization of cotton and thereby aid the farmer.

An obvious solution to the problem was to find methods to increase the utility of cotton by modifying it chemically to produce new and unusual properties. Kyle Ward was chosen to organize and direct research on the cotton fiber and headed the group that was called, at that time, the Cotton Fiber Division. He was the first; he set the tone for the research to follow. His was the spark that ignited the flame of a research torch that was to be a shining light in the textile world right up to the present time.

Kyle was not a bureaucrat. He believed in good research, and to that end, he avoided burdening his scientists with paper work to justify what they were doing. Our first major task was to improve cotton tire cord because cotton was losing this market to rayon. We did improve the cotton cord, but despite the improvements, it was plain that rayon and other fibers would inevitably replace cotton in this market. So we turned our attention to other chemical modification research. Because Kyle did not believe in excessive red tape, I remember that we did our first work on water repellency under the original project devoted to tire cord.

By his very nature, Kyle Ward would be the first to deny that he should be given credit for the many cotton research achievements of the Southern Laboratory. However, he cannot deny that he did get much of the work started before moving on to other challenges in 1952. He assembled a fine group of scientists, some of them still at the Southern Regional Research Center, as it is now called, actively pursuing research objectives that have their roots in the concepts he envisioned.

And now we shall discuss some of the practical finishes that have resulted from the chemical modification of cotton-- not just from the Southern Region, but from all over the world. The last 37 years of research on cotton and 10,000 or more references cannot be compressed into a short presentation. We can give neither details nor all of the credit where it is due. However, perhaps we can paint for you a word picture of a very significant era in developing textile uses of cotton.

Reactions of Cotton Cellulose

The average cotton fiber is about 1500 times as long as it is wide. It is made up of thousands of fibrils, each composed of long chains of several thousand anhydroglucose units. Some chains lie parallel to each other in a crystalline arrangement, making it almost impossible to initiate reaction with the hydroxyl groups. In other areas of the fiber, chains are jumbled together in a random arrangement that gives flexibility and accessibility to the fiber. It is in the accessible areas that reaction takes place.

The organic chemist new to cellulose chemistry must be careful that he does not consider cellulose to be merely an unusually large alcohol. The reactions of this most abundant of all of Mother Nature's polymers are unusual, and the conventional organic chemistry techniques must be altered to obtain reaction without damage to fiber strength.

Chemists who would modify cotton to improve its properties must accept the fact that they are dealing with a

very long, thin fiber that can lose strength very easily by
hydrolysis if strenuous conditions or overly active agents
are used. The reagents used to modify cotton generally react
only with hydroxyl groups on the fiber surface or in the
accessible regions where they can penetrate. More hydroxyl
groups become available for reaction if the fiber or fabric
can be swollen by water or other swelling agents, or if the
fiber is permanently swollen by mercerization.

How, then, does the textile chemist overcome the problem
of chemically modifying the cellulose of this rather intractable, fibrous, natural polymer of limited reactivity? The
problem is further compounded when the fiber is tightly woven
into the form of a fabric.

The chemist has been versatile. Treatments have been
developed that apply the desired reagent in all the familiar
phases--vapor, liquid, and solid. More explicit examples will
appear as we consider various reactions.

There are commercial processes in which fabrics are
treated with reactants applied from each of the three states.
In general, moisture content of the cotton is carefully
controlled in many vapor phase treatments, and the reagent
may be dissolved in that moisture. In some cases, the agent
is absorbed by the cellulose itself or dissolved in it.

In most practical processing, however, the reagent is
applied from an aqueous solution. There are, of course,
treatments in which application is from solvents other than
water and from emulsions. In some instances, the reaction
then proceeds at room temperature. Usually the excess liquid
is squeezed from the fabric, which is then dried, and the
agent reacts with the cotton at an elevated temperature,
perhaps 120-130°C or even higher. In some cases, because
the reagents are solid when dry, the treatment may be
considered as an example of reaction in the third phase, i.e.,
solid reacting with solid.

What types of chemical reactions have been used to modify
cotton to produce useful products? Reactions used include
oxidation, esterification, etherification, deposition or
precipitation within or on the fiber, and absorption or
adsorption. There also are other reactions such as graft
polymerization and copolymerizations but as yet, treatments of
these types with cotton have had only little commercial
acceptance. We shall restrict our discussion to finishes that
have achieved practical utilization.

In this paper we shall omit a discussion of direct dyeing
and similar procedures that depend mainly upon adsorption.
All of us have had the experience of involuntarily dyeing a
white shirt with ink from a leaky pen or the stain from certain
fruit juices. These experiences demonstrate the extraordinary adsorption power of cellulose.

Although the great majority of cotton and cotton-blend fabrics are treated with strong oxidizing agents such as sodium hypochlorite or peroxides in preparatory processing at the finishing plant, the objective is to oxidize the impurities to yield a white fabric. Oxidation of the cotton itself is to be avoided because carbonyl and carboxyl groups are formed on the cellulose chains, and subsequent scouring with alkali causes chain breaks and loss of strength.

$$-\overset{O}{\overset{\|}{C}}-\overset{O}{\overset{\|}{C}}- \qquad -\overset{O}{\overset{\|}{C}}-\overset{OH}{\underset{H}{\overset{|}{C}}}- \qquad -\overset{O}{\overset{\|}{C}}-H \qquad -\overset{O}{\overset{\|}{C}}-OH$$

The groups above illustrate typical structures resulting from the partial oxidation of cellulose. There are only a few practical examples of oxidation treatments used for cellulose modification. A particularly interesting one is the oxidized cellulose gauze used by surgeons as an absorbable hemostat. This product, made by Eastman Chemical Company, is obtained by oxidation of cotton gauze with nitrogen dioxide (3):

$$\text{Cell-OH} + N_2O_4 \longrightarrow \text{Cell-COOH} \quad (\text{Mostly 6-COOH})$$

The surgeon can use the oxidized cellulose as a surgical gauze and leave it in the body. The carboxyl groups of the oxidized cotton interact with the blood, and the gauze gradually dissolves and is absorbed by body fluids.

Esterification of cotton is an important modification for property improvement. From theoretical considerations, this would seem to be a very promising method for chemically modifying cotton. Dr. Ward will discuss the versatility of cellulose with acetylation in this symposium.

In the late 1940s and early 1950s, the Southern Laboratory did a great deal of work on the partial acetylation of cotton, and a process was developed that was used to treat cloth for sandbags for military use (4):

$$\text{Cell-OH} + O{\overset{COCH_3}{\underset{COCH_3}{\diagdown}}} \xrightarrow{HClO_4 \text{ catalyst}} \text{Cell-O-COCH}_3 + CH_3COOH$$

The treated sandbags were resistant to rotting for more than a year, even when exposed to swampy conditions. About one-third of the hydroxyl groups were acetylated, which meant that practically all of the available groups were reacted. Partially acetylated cotton had good fabric properties and resistance to rotting and heat. For a time, ironing board covers of acetylated cotton were produced commercially. They outlasted by many times those made from untreated cotton.

Acetylated cotton was also intended to be used for outdoor fabrics such as tarpaulins and awnings. Unfortunately, resistance to sunlight was only slightly improved over that of unmodified cotton, and the acetylated fabric did not succeed in penetrating this market.

Some research is presently being done at the Center on cotton cloth that has been lightly acetylated. The modern contact printing paper used in printing polyester fabric by heat transfer can be used on lightly acetylated cotton (5). Interest in acetylated cotton thus has been revived after a lapse of many years.

Undoubtedly, the most useful and versatile of all of the reactions of cotton cellulose for textile purposes is etherification:

$$Cell-OH + R-Z \longrightarrow Cell-O-R + HZ$$

Depending upon the reagents and conditions employed, reactions of this type proceed under either alkaline or acid conditions to yield many products which have been commercially successful.

As an example of an alkaline reaction, let us consider the partial carboxymethylation of cotton by reaction with chloroacetic acid:

$$Cell-OH + ClCH_2COOH \xrightarrow{NaOH} Cell-OCH_2COONa$$

Products that retain the fibrous nature and textile properties of cotton have been produced with a range of substitutions of the carboxymethyl group. Modified cottons are obtained that swell greatly with small amounts of reaction and that dissolve when the degree of substitution reaches about 1 carboxymethyl group per 3 anhydroglucose units (6).

There have been a few, small scale commercialized uses for the carboxymethylated product, such as for ion-exchange materials and for a soluble backing cloth in the manufacture of lace by the Schiffli process.

An illustration of cellulose etherification where mainly the surface hydroxyls are concerned is the English Velan process for water repellency, known in this country as the Zelan process. In 1951, we studied the reaction mechanism of this process to obtain fundamental information to guide future research. The reaction was as follows:

$$C_{17}H_{35} - CONH - CH_2 - \overset{+}{N}C_5H_5Cl^- + Cell - OH$$
$$\xrightarrow{Heat} Cell - O - CH_2 - NH - CO - C_{17}H_{35} + HCl + C_5H_5N$$

Stearamidomethylpyridinium chloride reacted with cellulose to form a long-chain ether. The molecular size of this reagent is too large for it to get into the cotton, so the reaction is confined to the fiber surface. Sodium acetate is needed as a

9. Mazzeno, L.W., Kullman, R.M.H., Reinhardt, R.M., Moore, H.B., and Reid, J.D., <u>Am. Dyest. Rep.</u> 47, 299 (1958).
10. Daul, G.C., Reinhardt, R.M., and Reid, J.D., <u>Text. Res. J.</u> 22, 792 (1952).
11. Arceneaux, R.L., Frick, J.G., Reid, J.D., and Gautreaux, G.A., <u>Am. Dyest. Rep.</u> 50, 849 (1961).
12. Reid, J.D., and Reinhardt, R.M., <u>Colourage Annual</u>, 19 (1971).
13. Schuyten, H.A., Weaver, J.W., and Reid, J.D., in "Advances in Chemistry Series," No. 9, p. 7. American Chemical Society, Washington, 1954.
14. Schuyten, H.A., Weaver, J.W., and Reid, J.D., <u>Ind. Eng. Chem.</u> 47, 1433 (1955).
15. Drake, G.L., <u>Am. Dyest. Rep.</u> 60, (5) 43 (1971).
16. Perkins, R.M., Beninate, J.V., Mazzeno, L.W., Reeves, W.A., and Hoffman, M.J., <u>Colourage</u> 20, (22) 29 (1973).

This fundamental work proved to be valuable in conjunction with the development of flame-retardant finishes for cotton textiles. Finishes based primarily on phosphonium and phosphine derivatives, pioneered in research at the New Orleans laboratory, have been utilized in several successful commercial processes. The first of these employed THPC tetrakis(hydroxymethyl)phosphonium chloride: $Cl-P(CH_2OH)_4$) which can etherify cellulose, react with itself, and react with other reagents that are included in the finishing bath (15). More recently another variation, the $THPOH-NH_3$ process has been used that makes cotton flame retardant through deposition of a highly insoluble polymer on and within the cotton fiber (16).

In summary, research on the chemical modification of cotton has been successful in expanding and maintaining the use of cotton by making it more useful and attractive to the customer. Many groups have been introduced into the cellulose molecule to impart various useful properties to cotton fabrics. Introduction of desirable functional properties has enabled cotton to remain competitive with other textile fibers, both natural and synthetic. In that competition for fiber markets, the Southern Regional Research Center's research on cotton made substantial contributions. The textile researchers there are grateful to Dr. Ward for beginning and fostering that work and for setting the tone for the research to follow. Certainly, benefits from his efforts have been realized throughout the textile industry.

We think that Anselme Payen, the father of Cellulose Chemistry, would be justifiably proud of his notable descendent, Dr. Kyle Ward, Jr.

REFERENCES

1. Reid, J.D., and Dryden, E.C., Text. Color. 62, (733), 43 (1940).
2. Phillips, M., J. Am. Chem. Soc. 50, 1986 (1928).
3. Kenyon, R.L., Hasek, R.H., Davy, L.G., and Broadbooks, K.J., Ind. Eng. Chem. 41, 2 (1949).
4. Anderson, E.V., and Cooper, A.S., Ind. Eng. Chem. 51, 608 (1959).
5. Blanchard, E.J., Bruno, J.S., and Gautreaux, G.A., Am. Dyest. Rept. 65, (7) 26 (1976).
6. Daul, G.C., Reinhardt, R.M., and Reid, J.D., Text. Res. J. 22, 787 (1952).
7. Schuyten, H.A., Weaver, J.W., Frick, J.G., and Reid, J.D., Text. Res. J. 22, 424 (1952).
8. Tootal Broadhurst Lee Co., Ltd., Foulds, R.P., Marsh, J.T., and Wood, F.C., Brit. Pat. 291,474 (Dec. 1, 1926).

dimethylolurea is shown in the following equation, which portrays a simplified, idealized representation of the reaction.

$$2 \text{ Cell-OH} + \text{HOCH}_2\text{NH-CO-NHCH}_2\text{OH} \xrightarrow[\text{cat.}]{\text{acid}} \text{Cell-OCH}_2\text{NH-CO-NHCH}_2\text{O-Cell} + 2\text{H}_2\text{O}$$

Many other nitrogenous compounds such as cyclic ureas, triazones, melamines, urons, and alkyl carbamates are employed, as their formaldehyde adducts, in durable press finishing. Many factors that bear upon the finish selected include cost, difficulty of application, formaldehyde odor, susceptibility to chlorine bleach, hydrolysis during use, yellowing with chlorine, and degradation due to difficulty in crosslinking.

The carbamate class of finishing agents, developed at Southern Regional Research Center, was first introduced to the industry in 1961 (11). Formaldehyde was reacted with a carbamate such as methyl carbamate to give the dimethylol compound. These dimethylol compounds are extremely difficult to isolate, but do exist in solution:

$$2 \text{ HCHO} + \text{CH}_3\text{OCONH}_2 \longrightarrow \text{CH}_3\text{OCON}(\text{CH}_2\text{OH})_2$$

Treatment of cellulose proceeds through a condensation reaction, as follows:

$$\text{CH}_3\text{OCON}(\text{CH}_2\text{OH})_2 + 2 \text{ Cell-OH} \longrightarrow \begin{matrix} \text{Cell-O-CH}_2 \\ \text{Cell-O-CH}_2 \end{matrix}\!\!\!\!>\!\!\text{N-COCH}_3 + 2 \text{ H}_2\text{O}$$

The finish imparts wrinkle resistance properties and is resistant both to chlorine bleach and to mild acid hydrolysis. At present, carbamate agents are used mainly on fabrics for sheets and pillowcases because the finish resists the hard laundering these items receive (12).

Only brief mention will be made of research to make cotton resistant toward burning. In our early years, Dr. Ward encouraged us to work on flame retardants, but more particularly--because he always thought fundamental work was very important--to determine why certain flame retardants prevent cotton from burning.

In 1954, Schuyten, Weaver, and Reid published their hypotheses (13,14). Briefly, they found that phosphorus-based retardants, as well as many other Lewis acids, catalyzed the decomposition of most of the cellulose carbohydrates to nonflammable carbon and water, instead of to the flammable gases which cause burning. Furthermore, the presence of certain bromine compounds, when vaporized by the heat of ignition, reduced the flammability of the gases tremendously.

buffer to prevent damage to the cotton by hydrochloric acid, a by-product from the pyridine hydrochloride. This surface treatment makes the cotton water repellent. Substitution was about 1 stearamido group per 150 anhydroglucose units. This is an addition of approximately 1.4% by weight. Only about 0.1% of the anhydroglucose units are on the surface of the fiber (7).

The most generally used modifications of cotton are those that confer wrinkle resistance, as in durable press cotton and cotton-blend garments. The easy-care garment has freed nearly every household from the former chore of ironing one day in every week.

The research leading to our present stage of development started approximately 50 years ago when Foulds, Marsh, and Woods, working for Tootal Broadhurst Lee Company in England, investigated means to make cotton as crush resistant and springy as wool (8). By 1940, the process was a mild commercial success in several areas of the world but interest in such finishes in the U. S. was largely limited to producing shrink resistance in cotton until nylon no-iron garments were introduced. These garments generated interest in no-iron cotton among textile finishers.

By 1955, the yardage of no-iron cotton fabrics being sold was considerable but the product was in difficulty in the American market because it could not withstand the prevalent rigorous laundering procedures of chlorine bleaching in the home and acid souring in commercial laundries. Many of the fabrics also had a very bad soiling problem in that they scavenged soil particles from the wash water during laundering (9).

It was obvious to us at the Southern Regional Research Center that more satisfactory no-iron cotton garments would greatly increase the use of cotton. An extensive program of research was begun to aid industry in both the practical and the theoretical aspects of the work. One of our first papers relating to crease-proofing was published in 1952 by Daul, Reinhardt, and Reid (10). The work was carried out while Dr. Ward was still head of the cotton research group.

It has been estimated that there are more than 7,000 publications and 3,000 patents relating to durable press. This volume of literature emphasizes the importance of durable press to the textile and related industries. Because not all research activities are published, it indicates the immense amount of effort expended.

To set cotton fibers in a configuration that insures the fabric will dry smooth after laundering, the long cellulose molecules are crosslinked, almost always with nitrogenous crosslinking agents. An example of such crosslinking with

PART 2 Overview of Cellulose Sources and Modification Reactions

Modified Cellulosics

CHEMICAL MODIFICATION OF CELLULOSE

A HISTORICAL REVIEW

D. F. DURSO

Research Center, Johnson & Johnson
New Brunswick, N.J.

A large variety of derivatives from cotton and wood cellulose have been prepared over the years. Some of these have become important industrial materials, while others remain research curiosities. This paper will review the reaction conditions, chemistry, products, and properties of the resulting modified cellulosics.

INTRODUCTION

    Cellulose is that ubiquitous material without which we could not have life as we know it on this planet. In nature, it is produced, consumed and destroyed in tremendous quantities each day. Its part in the carbon dioxide cycle is unmatched, and its stabilizing influence on the level of carbon dioxide in the atmosphere is of the same magnitude as that of the oceans. Cellulose provides man with both inspiration and the necessities of life; the former via unrivalled works of art (roses, orchids and other flowers) and the latter as fire, food, clothing and shelter. It is with us everywhere, serving both as the substrate for our homes and its components of comfort or as the pattern upon which man tries to improve with synthetics (fibers, films, laminates, etc.) for the same uses. When one considers the scope of the cradle provided by cellulose, it is indeed strange that, in 1977 after more than 150 years of study by eminent chemists, biologists and others, the "riddle" of its formation and structure remains as controversial a subject as in 1850 and 1900. Is this true because cellulose is complex beyond our ability to comprehend, or is it true because other research matters provide a quicker path to scientific eminence for the researcher?
    The challenge of cellulose "reactivity" has been responsible for the development of many areas of science, probably

Copyright © 1978 by Academic Press, Inc.
All rights of reproduction in any form reserved.
ISBN 0-12-599750-7

because it was the most readily available "sample" when organic, physical, polymer and other "disciplines" of chemistry became amenable to laboratory studies. Its ready availability combined with its seeming simplicity (it *is* only a simple polyglucan!) and its ability to survive the tortures of the unskilled, have been both a blessing and a curse! The continuing controversies concerning the basic structure of cellulose (raw, native or purified) had their beginnings in the gropings of the pioneers in the fields of the related sciences named above; for some reason, there has never been mounted an effort to simplify the body of "knowledge" by discarding all those publications based on faulty (or unreproducible) samples. It should be quite well-established that cellulose is not a compound; rather it is a material whose chemical and physical usefulness depends on how it was produced at biogenesis and how it has been modified by *all* steps prior to the "final" use. In other words, it has a memory which truly never forgets! Thus, each of us trying to understand our investigations upon some aspect of the material must appreciate that we are observing an *interaction* of our devices upon a *particular state* of the substrate (1, 2).

The subject for discussion in this treatise is *modified* cellulosics. The actions of chemical and physical procedures will be portrayed in terms of anomalies, mechanisms, new forms and articles of commerce by the writers of other chapters. This chapter will attempt to provide background and perspective for their contributions.

BEGINNINGS

As in any historical review of developments having large significance and complexity, it is almost impossible to determine if the technology and uses were spawned by curiosity or need. Man had been using various natural gums for centuries before the first cellulose "derivative" was prepared. It is a moot point whether this work resulted from a scientist's desire to see what could happen from a basic study of properties of derivatives or from a user's request to provide a new source, or from a need for reproducible quality or for some combination of reasons. Regardless of this chicken-egg situation, there is no argument with the fact that food, paper and textiles were modified by the addition of other natural materials for centuries before the advent of cellulose chemistry. At some early time it was discovered that solutions of cellulose had properties equal to or better than the natural additives. Thereafter, as in any new research area, there occurred explosive growth of knowledge and use as the field was explored and exploited by those having a pertinent

CHEMICAL MODIFICATION OF CELLULOSE

need. What is notable is that this was probably the first large scale program mounted by man to convert a natural material into a myriad of compositions for specific uses. With cellulose, because of its polymeric and associative properties, he was able to prepare new films and fibers in addition to soluble "gums and resins". Previously, only natural materials in a native form suited for the purpose were used for fabrics, coverings, coatings and molded (but flexible) artifacts. The revolution for custom production of raw materials and of end products, begun with cellulose, continues today but mainly with other substrates. Perhaps, as one renowned cellulose-turned-polymer chemist said "we'll return to cellulose after we gain more knowledge and technology with the simpler man-made polymers". Perhaps our current appraisal of energy and feedstocks will hasten this return. The original substrate which inspired so much useful research is now in sore need of inspired workers to simplify the methods by which it can continue to serve man in forms other than that produced by nature.

HISTORICAL REVIEW

In other publications, the reader will find discussions of modified cellulose under a multitude of headings. It is hoped that the two-category system selected for this paper will be found useful. They are:

I. Reactions Retaining the Original Morphology

II. Reactions Destroying the Original Morphology

While it is true that in many uses of the resulting modified cellulose the original structure is destroyed, the classification used herein applies only to the state of the cellulose when the modification reaction is completed.

In the remainder of this document, the reactions of commercial and theoretical importance will be listed, and their date of discovery and current status will be discussed briefly along with the basic technology of the reaction(s) involved. The order of listing (except for cellulose acetate) is chronological within each of the two categories.

Alkali Cellulose

Beginning with the work of John Mercer in 1844, the action of concentrated NaOH solution (14% or higher) on cellulose has been an important first step in the modification of cellulose. The physical changes accompanying the entry and the removal of the aqueous NaOH molecules, especially among the highly ordered regions of the cellulose matrix, are commonly called mercerization. They may be either employed per se (mercerized cotton, for example) or they are the first step for chemical reactions which proceed either by destroying or not destroying the original morphology (macro structure) of the cellulose. Where the NaOH action (without removal) is the first step in a chemical reaction sequence, the intermediate obtained is called alkali cellulose. Whether this is a mixture or a compound, it is the first attack on the cellulose structure, "opening" it for production of new forms.

REACTIONS RETAINING THE ORIGINAL MORPHOLOGY

In terms of number, the cellulose derivatives produced without a concomitant solution state are predominant. However, in terms of poundage, it appears that those requiring destruction of the original morphology are the most important.

Cellulose Nitrate

The oldest and most widely used cellulose modification is the nitrate. Since 1845 when Schönbein (3) found a practical way of producing it via reaction of cellulose fibers (cotton and related materials) with mixed nitric and sulfuric acids, this material has made many indelible marks on mankind and his history:

1. By replacing black powder, it revolutionized military strategy and provided the means, if not the incentive, for actions which have affected billions of people.

2. When a practical system was devised in 1870 for converting a plastic mixture of cellulose nitrate and camphor (celluloid), we had the beginnings of the modern technology of plastics and plasticizers (4). The knowledge that man could make his own non-metal objects suddenly opened up many avenues of research and commerce.

3. The finding that dissolution in solvents followed by evaporation produced tough films (pyroxylin) caused another revolution which is attested to today by the technical complexity and the commercial size of the coatings field. Today we live in a world of specialized functional "thin-layers" as a result of the opportunities and problems arising from the early lacquers based on a <u>synthetic</u>.

CHEMICAL MODIFICATION OF CELLULOSE

In addition to these matters of commercial importance, curiosity and product control needs inspired efforts to understand the nature of the reactions and to describe the cellulose nitrate in scientific terms. From these activities there arose large segments of the body of knowledge that we take for granted among (for example):

A. Reaction mechanisms

B. Statistical aspects of polymer composition and DP distribution

C. Heterogeneous reactions

D. Polymer degradation

From the copious literature on it, obviously cellulose nitrate was the substrate for much research and discussion. At the same time, because of its military importance, there was little published about exactly how to produce it. It seems safe to say that this material must also be considered a cornerstone in the genesis of chemical engineering. The instability of the product caused much innovation of control and sensing devices, as well as defining the need for "unit processes". The lack of scientific basics and the secrecy about this military commodity combined to produce copious amounts of "garbage" literature and to prevent the progress possible when problems are faced objectively. Thus, we see that it was 1906 before the problem of stabilization was solved and that it was 1918 before there was anyone in the U.S. capable of producing it consistently (and live to talk about it!) Further, it should be noted that not until 1943 was there a woodpulp in the U.S. of the necessary quality.

The process consists in exposing fibrous cellulose in nodular form to a mixture of nitric and sulfuric acids to obtain about 11% nitrogen content, washing and heat stabilizing the product, and drying it. It appears that equipment refinements were made throughout the years from about 1930 through 1950. Since then, other materials have been displacing cellulose nitrate in all its uses. It was announced recently that soon there will be only one producer left in the U.S. (5).

Cellulose Sulfate
―――――――――――――――――

This compound was first prepared about 1845 but there was no commercial activity until about 1950. After a few years of introductory efforts, it was withdrawn from the market. It was a specialty product prepared by reaction of cellulose with H_2SO_4 in an inert, non-reactive organic solvent.

Methyl Cellulose

The science and technology of cellulose ethers began about 1900. The field became one of commercial importance through the pioneering work of L. Lilienfeld (6). It seems that many years were spent thereafter by the Germans, British, French and Americans in carving this field into patent empires. Finally in the 1930's, the structure of cellulose ethers became known through the work of the W. N. Haworth group in England and, coincidentally, production of ethers began.

Basically all ethers are made by suspending cellulose in finely divided form in a non-reactive non-solvent, adding enough aqueous NaOH to swell the fibers and to "activate" the hydroxyl groups. Then, upon addition of the proper alkylating agent, the reaction is controlled to obtain the D.S. (degree of substitution; more properly called M.S., molar substitution) desired. By selection of raw materials and reaction conditions, it is also possible to tailor the product as to D.P. (degree of polymerization; chain length), substitution uniformity (substituents per anhydroglucose unit) and substitution distribution (reaction on secondary versus primary hydroxyls and on sequential glucose units). It is safe to say that the work by the Haworth school and others on these stable compounds made it possible to coin the terms just listed. Because of the large variety of cellulose ethers made and studied at the time when polymer chemistry was becoming a science, both the basic chemistry and the practical technology blossomed into forms that today still exist as viable entities.

The many applications for all cellulose ethers are at such a stage of growth that technological (processing) improvements and new capacity are still being introduced.

Methyl cellulose is made from any cellulosic substrate as outlined generally above. The alkali cellulose suspended (or stirred) in a proper dispersant is made to react with methyl chloride at elevated temperatures and pressures. It is recovered and sold as a fine powder, consisting of identifiable fiber residues. The remarkable property of methyl cellulose is its inverse viscosity versus temperature reaction; when the "gel point" is reached the viscosity suddenly increases tremendously. Up to that point, normal increasing fluidity is noted. The gel point temperature depends on the DS-DP parameters of the methyl cellulose.

Ethyl Cellulose

This derivative provides exceptionally durable films and is used where extreme stress is encountered, such as in bowling pin coating. In addition, this derivative can be extruded or injection molded. Other uses include paper coatings, lacquers and adhesives.

Like methyl cellulose, it is produced by suspending the cellulose raw material in a non-reactive non-solvent to prepare the alkali cellulose. Then the latter reacts with ethyl chloride at elevated temperature and pressure, with the latter higher than that for methyl cellulose.

In Europe, ethyl hydroxyethyl cellulose is produced by a two-stage reaction. It is more hydrocarbon-like than ethyl cellulose but otherwise has the same uses.

Carboxymethyl Cellulose (CMC)

Production of this derivative began in Germany in the 1920's. It has constantly grown in importance since then as it finds ever-increasing applications in food, industrial and pharmaceutical fields. The reaction of the appropriate alkali cellulose in a non-solvent suspension with chloroacetic acid proceeds easily at ambient pressure and at temperatures only slightly elevated above normal.

Unlike the other cellulose ethers, CMC finds use as a technical grade material. This can be made by simply spraying the chloroacetic acid onto a sheet or shredded nodules of alkali cellulose, reacting for a suitable time, and then drying and grinding the resultant somewhat plastic mass. The final product contains the sodium chloride and the sodium glycolate byproducts, which can be tolerated in many uses.

With constant upgrading in process control, this derivative has the largest commercial volume of all the cellulose ethers. It provides controlled properties at prices unmatched by natural and synthetic materials with similar solution properties.

Cyanoethyl Cellulose

While this compound has been known since its invention in Germany in 1938, it did not become an item of commerce until about 1953. A U.S. patent issued to General Electric in 1950 (7) describes its value as a component of condensers because of its high dielectric constant. It is a specialty product, made in rather small volume for an application ("electrical papers") where it has no equal.

As for other ethers, cellulose is first converted into alkali cellulose while suspended in a non-reactive medium. Then reaction with acrylonitrile proceeds at room temperature to obtain the rather low MS desired.

Ion-Exchange Celluloses

These are highly specialized compounds made in rather

small quantities for laboratory studies. While many anion- and cation-exchangers are possible and many have been reported on since 1930 (8), only DEAE-cellulose has found a place in our technology. The applications were first reported in 1950. This anion-exchanger is prepared from alkali cellulose (in the usual manner for cellulose ethers) by reacting it with 2-chlorotriethylamine hydrochloride to produce compounds with a nitrogen content of about 1%. Higher nitrogen-containing diethylaminoethyl celluloses can be prepared by using higher concentrations of NaOH and reacting with the amine repeatedly; in such uses the cellulose substrate must first be cross-linked in order to prevent solution.

Hydroxyethyl Cellulose

This modification of cellulose became commercial in the 1960's and is chiefly used as a component of latex paints. Because of the preparation procedure, the final product differs from all other cellulose "derivatives" in that the substituent is not a simple molecule replacing a primary or secondary hydroxyl group's hydrogen atom. The material of commercial interest does not require the preparation of alkali cellulose; NaOH is used to pre-swell the cellulose but only catalytic amounts are needed to obtain the desired polymerization of ethylene oxide in the vicinity of the cellulose chains. An MS of 2.5 as found in the usual composition would theoretically require almost complete reaction of the available hydroxyl groups. The preparation conditions used are not drastic enough for such a structure; therefore, it appears that long chains of (poly)ethylene oxide become interspersed within the cellulose matrix once a site for polymerization has been "found". Thus, the product is a mixture (or an alloy) of two polymer chains. It is remarkable that such a modification of cellulose will render it completely soluble in 7% NaOH at 0.2 MS and in water at 1.5 MS.

Microcrystalline Cellulose

This is another "derivative" introduced in the 1960's (9). O. A. Battista and coworkers discovered that hydrolysis of high-purity highly-ordered cellulose (cotton; cotton linters) to its level-off DP provided a new form which can be used in place of derivatives for some applications. The process requires 2.5 N hydrochloric acid at 105 C for 15 minutes, followed by a high shear treatment to separate the microcrystals. At 5% solids, the material becomes an opaque, white gel when the homogenization step is complete.

The material is being used as a component of pills to aid in their disintegration. It is also finding applications in the food field, especially in low calorie products.

Graft Copolymers

In the 1960's, polymer chemists learned that they could obtain materials with new properties by activating a site(s) on one polymer and then causing a second polymer to form via the initiating action of that site(s). There has been a great deal of laboratory work using cellulose and its derivatives, as witnessed by hundreds of publications. However, there is yet no commercial realization of the many processes and the cellulose-synthetic polymer combinations available. In some respects, these "new" graft copolymers are not unlike the reaction products obtained when acrylonitrile and ethylene- or propylene-oxide are caused to polymerize in the presence of cellulose. There may be true co-valent bonding at some sites, but in essence the product may be considered an intertwined system of two (or more) polymer chains. These chains would associate to some degree via hydrogen bonding and would have properties in the combined matrix like and unlike each of the constituents. While the possibilities are many, there is no commercial venture (excluding the hydroxyalkyl derivatives) probably because there has not yet been found a unique application for them which cannot be met by the more common cellulose derivatives.

Hydroxypropyl Cellulose

This close relative of hydroxyethyl cellulose became available in the late 1960's when the more complex polymerization of propylene oxide onto activated cellulose was mastered (10). It is necessary to obtain a high MS (about 4) in order to obtain this modified cellulose. It differs from all others in that the same "compound" is soluble in water, alcohols, hydrocarbons and cycloalkanes. The reaction is a "standard" etherification in that cellulose is suspended in an inert organic liquid and activated with sodium hydroxide prior to the addition of the propylene oxide. At present, mixed ethers are the main commercial evidence of this reaction where hydroxypropylation produces modified methylcelluloses for specialty applications.

Cross-Linked Celluloses

Beginning with formaldehyde, many similar compounds have been found to produce modified cellulosic materials of improved performance. The reaction is an aldol condensation followed by dehydration resulting in the joining of two hydroxyl groups with a bridge of appropriate size between them. This simple chemistry became a commercial reality in the 1960's with the appearance of "permanent-press" or "wash-wear" fabrics. In this form, the importance of a slightly modified cellulose or cellulose-synthetic blend is without peer among all the presently known cellulose reactions. The reader will find excellent discussions and reviews in a large number of papers; a condensation of the knowledge and practice has appeared recently (11).

With the inspiration of success obtained on the wrinkle-resistant projects, scientists have proceeded into two other related areas with notable success. These are flame-resistant textiles and dimensionally-stable wood where the essential process is the "... removal of hydroxyl groups from their normal reactive state ...". In the 1970's we see the basic chemistry and technology which was developed for textile forms being applied more and more to cellulosics in sheet form ("paper" is the generic term).

The present commercial successes in this field arise from an effort to "correct" the shortcomings of cellulose which are based on the weakness of the hydrogen bonding between hydroxyls of adjacent chain molecules. The large body of polymer physical chemistry was a necessary first "reaction" to make it clear that improvements in certain properties of cellulose could only be obtained by making it "less-cellulosic". In the areas discussed thus far, the practical solution provides control of the negatives (wrinkle resistance) without loss of the positives (moisture sorption = comfort) at an economical cost. Many cellulose chemists have participated in this advance for the good of all mankind; the latter has been rewarded for its support of research on cellulose.

In the late 1970's, another form of modified cellulose is seeking a place in our society. The material is best described as cellulose derivatives which would dissolve in water except for the cross-linking built into them by a second chemical reaction. While they have been known for quite some time, the practical technologies for preparation and for specific uses have only recently reached the stage where large scale utilization can be predicted.

These new materials seem especially valuable for the area of absorption of liquids because as hydrogels (12) they provide a means for retention which is several orders of magnitude greater than the original substrate. Products exploiting these materials because of this property are just now appearing in test markets.

Cellulose Acetate

There is a significant production in Belgium and in Canada of cellulose triacetate via a "fibrous process". Here the reaction is carried out in an organic inert non-solvent, otherwise the chemistry is the same as in the "dissolving" process described later. The advantages of the fibrous system have been described (13). The triacetate product is especially useful for melt extrusion into final forms such as film and fiber.

REACTIONS DESTROYING THE ORIGINAL MORPHOLOGY

While numerically inferior, the reactions producing a solution during the preparation of the derivative are the most important on a tonnage basis. The products based on them have provided major revolutions in the fiber field (artificial silks) and the molded articles area (acetate plastics). In these two product areas the current products still maintain their cost/performance position despite the competition from other synthetics.

Cellulose Xanthate

In the fiber field, this chemical reaction discovered by Cross, Bevan and Beadle (14) in 1892 provided man with his first opportunity to produce fibers with specified properties. The products, known as artificial silk and then as rayon, have made their way from textiles into industrial and other high technology fiber uses. These fibers are available as continuous filament (for V-belts, tires, other reinforcements) and as staple (for textiles, non-wovens, and paper) in a variety of deniers, cross-sections, degree of crimp, degree of luster and tenacity. In later years, the xanthation reaction became the source of polymer for film which is used for packaging, sausage casing, and membranes in artificial kidneys; it is also the means to produce artificial sponges. In short, the discovery by Cross, et al has been exploited throughout the world in the production of tremendous quantities of the necessities of life. It spawned the purified cellulose industry first with cotton linters, then with wood and now with other plant fibers such as bagasse which are too short for most of man's needs.

The process consists in the conversion of high purity cellulose into alkali cellulose of high solids content (14-15% NaOH, 33-35% cellulose), "aging" to obtain the desired average DP, reacting with CS_2 (25-30% based on cellulose weight), dissolving the xanthate crumbs in NaOH to obtain viscose solution (4-7% NaOH, 4-9% cellulose) and regenerating. In most cases, the cellulose is regenerated by extruding the viscose solution (with or without additives) at the proper DS (obtained by "ripening") through an appropriate die into a warm H_2SO_4 bath (with or without additives). In the case of sponges, regeneration is caused by heat after the addition of Na_2SO_4 crystals and of reinforcing fibers to the viscose. In all cases, the regenerated cellulose must be desulfured, desalted, and dried; in some cases, it is also bleached.

The basic technology has not been changed since the 20's when the commercial production of fiber began in a number of countries. It grew quickly, reaching the million ton annual level in 1940. Mechanical components and process efficiency have been constantly improved but a number of factors have combined to cause a levelling of annual production at about 3 million tons since the late 1950's (15).

The present state of the industry is such that one can predict a major revolution in this field; the abundance of the renewable raw material and the need for regenerated cellulose products for which there is no real substitute (the combination of esthetic, comfort, performance and economic features of rayon) contrast with production units whose costs of handling large liquid flows (10% cellulose solution seems maximum) are becoming prohibitive. This combination of need and of technical problems spotlights an opportunity for some other system to be devised and exploited; this has been the historical sequence in the past, and we are seeing the beginnings of efforts to provide new solution/regeneration schemes (16).

Cuprammonium Cellulose

Schweizer (17) discovered in 1857 another means for regenerating cellulose into filaments (Bemberg silk; Bemberg rayon). The product was a fiber of superb hand with high strength, high crimp and very fine denier. It enjoyed a special place in the textile field, especially female attire, from the 1920's until its last producer discontinued activities in the late 1950's. The inability to handle cellulose at solution concentrations above about 3% and the advent of man-made polymers such as polyester were the chief factors in its demise.

The process is quite simple. Comminuted cellulose sheet is dispersed in an aqueous solution of ammoniacal copper $Cu(NH_3)_4(OH)_2$ with or without some NaOH addition until it dissolves, and then it is extruded into a water bath through a suitable die. The regenerated form requires copper removal, washing and drying as the final steps.

Cellulose Acetate

The largest portion of the commercial uses for this cellulose derivative are based on what is called "secondary acetate". The process was first described in 1905 (18) and became commercial in the early 1920's. The first products were fiber forms where the hand and the luster were especially desired; in satin and other fabrics, the acetate fiber still has no peer today. Besides its textile uses, the acetate as fiber has a monopoly in the cigarette filter field, and as film commands the photographic, transparent tape and blister packaging fields. In plastics, as the mixed acetate-butyrate, it has a special niche where high impact resistance is required. While cellulose acetate has not grown in recent years, it also has not lost its supporters among the pulp and acetate producers, and the consumers of its end-product forms. For special filtrations and for osmotic membrane processes, the films made from acetate are unmatchable.

The process consists in activating the very high purity raw cellulose with dilute sulfuric acid (during or after comminution), acetylating with acetic anhydride in the presence of concentrated sulfuric acid as a catalyst, diluting and heating the resultant triacetate solution, and precipitating and washing the product when the DS drops to about 2.5 (secondary acetate). The products of commerce are acetate flake and acetate molding powder. Fiber is produced by dissolving the flake in acetone and extruding the solution into a hot air column; no after-treatments are required. Film is made by melt-extrusion of the flake or by solution casting from acetone. Plastic molded forms are prepared by injection molding, where it provides a reproduction of the die which is almost unmatchable by other polymers.

Its position in world commerce seems secure.

Vulcanized Fiber

This specialty cellulosic product is based on the discovery by Mercer in 1850 (19) that aqueous zinc chloride solution causes cellulose to swell and gel. If the highly swollen gel is leached with water, the gel shrinks as the zinc is removed slowly. Finally, upon drying, a very dense and very tough "fiber-reinforced plastic" is obtained. It is unequalled in its combination of physical and electrical properties. It is a component of electrical switches and is used in bolted railroad rail connections. It is a prime material for wire drawing dies. Its abrasion and fatigue resistance are the basis for its use.

It is prepared by subjecting sheets of high purity cellulose to 65-73% zinc chloride solution, piling sufficient number of such sheets together, allowing the mass to become coherent, and then gently leaching out the zinc. Drying is the final step.

SUMMARY

It is hoped that this overview of the past will provide perspective as well as inspiration for those considering cellulose as an opportunity to provide more for man in the future than it has in the past. While man has used cellulose for centuries, its chemical value has only been utilized since the classic work of Payen in 1838 (20). The writer feels strongly that far too much attention and technology have been invested in production of "pure" cellulose. In many cases (21) a useful modified cellulose can be produced by direct reaction of unbleached pulp or even whole wood. Future allocation of resources must be based on maximum value returned from investments in research, production and marketing. If necessary, a complete break with the past forms of cellulose technology must be undertaken if we are serious about continued use of the world's most abundant polymer.

REFERENCES

1. Segal, L., Part V, Chapter XVII (A. Effect of Morphology on Reactivity), in "Cellulose and Cellulose Derivatives", N. M. Bikales and L. Segal ed., Wiley-Interscience, New York, 1971.
2. Calihan, C.D., "Cellulose Derivatives; Polymers with a Future" in Cellulose Technology Research, A.F. Turbak ed., ACS Symposium Series No. 10, American Chemical Society, Washington D.C., 1975.
3. Barsha, J., in "Cellulose and Cellulose Derivatives", E. Ott, H.M. Spurlin and M.W. Grafflin, ed., Interscience, New York, 1954, p. 713.
4. Hyatt, J.W. and I.S. Hyatt, U.S. Patent 105,338 (7/12/1870).
5. Anon., Wall Street Journal, 7/20/77.
6. Savage, A.B., et al in "Cellulose and Cellulose Derivatives, 1954, loc. cit., p. 883.
7. Miller, H.F. and R. G. Flowers, U.S. Patent 2,535,690 (1950).
8. Guthrie, J.D., in "Cellulose and Cellulose Derivatives", 1971, loc. cit., p. 1277.
9. Battista, O.A. and P.A. Smith, U.S. Patent 2,978,446 (1961).
10. Klug, E.D., U.S. Patent 3,278,521 (1966).
11. Tesoro, G.C. and J.J. Willard in "Cellulose and Cellulose Derivatives", 1954, loc. cit., p. 835; H.B. Goldstein, p. 1095; and H. Tarkow, p. 1337. See also J.W. Weaver, TAPPI 4th International Dissolving Pulps Conference Preprints, Chicago, 1977.

12. Whistler, R.L., TAPPI 4th International Dissolving Pulps Conference Preprints, Chicago, 1977.
13. Geurden, J., Pure Appl. Chem., 14, 507 (1967).
14. Cross, C.F., E.J. Bevan, and C. Beadle, J. Chem. Soc., 63, 837 (1893).
15. Krässig, H., TAPPI 4th International Dissolving Pulps Conference Preprints, Chicago, 1977.
16. Hergert, H.L., R.B. Hammer and A. F. Turbak, ibid., Chicago, 1977.
17. Schweizer, E., J. prakt. Chem., 72, 109 (1857).
18. Miles, G., French Patent 358,079 (1905).
19. Mercer, J., British Patent 13,296 (1850).
20. Payen, A., Comptes rendus 7, 1052 (1838); ibid., 7, 1125 (1838).
21. Durso, D.F., Svensk Papperstdn., 79(2), 50 (1976).

Modified Cellulosics

COTTON: A WORLD CELLULOSE FIBER

R. S. CORKERN

USDA Economic Research Service
Southern Regional Laboratory, New Orleans, Louisiana

and

M. E. CARTER AND B. M. KOPACZ

USDA Agricultural Research Service
Southern Regional Laboratory, New Orleans, Louisiana

Cotton is a major world fiber and cellulose resource, contributing to the health, safety, and well being of all people. It is produced in more than 68 countries, the main producing areas being concentrated in North America, USSR, and Asia and Oceania. The three principal botanical groups of commercial importance are Gossypium hirsutum, G. barbadense, and G. herbaceum.

The processing of cotton into useful consumer and industrial products begins on the farm. Additional processing is carried out at gins, textile and pulp mills, and chemical plants. The use of both physical and chemical processes at these levels enhances the inherent properties of cotton and imparts new or modified properties desired by consumers and industrial users.

World cotton production and consumption has an upward trend. This trend has paralleled the continued development of technological innovations that are cost effective and at the same time provide users with better quality products.

INTRODUCTION

The historical development and continued use of cotton as a textile fiber and industrial cellulosic raw material was and is dependent on concurrent mechanical and chemical technologies. Some of the more important technologies that advanced the use of the cellulosic component of cotton were the high speed cotton gin, delintering machinery, modern spinning frames and weaving looms, mechanical production and harvesting equipment, and chemical treatment for cotton fibers and fabrics. Technology has also been developed for expanding the use of the major coproduct of the cotton plant--the seed. For example, cottonseed meal can now be processed for edible flour and protein, and cocoa butter-like fat can be produced from the oil for use by the confectionary industry. Further, the cellulosic content of the cotton plant (leaves, stalks, and burrs) could be utilized as a fuel, feed, paper, or as a raw material for chemical intermediates. The whole plant use concept and its annual renewability are two attributes which make cotton a major contributor to our future national resource base.

COTTON PRODUCTION, PROCESSING, AND CONSUMPTION

Cotton has been produced, processed, and consumed since about 3000 B.C. Today cotton is a major world resource, contributing to the health, safety, and well being of all people. The current status of cotton production, processing, and consumption shows that it will remain a major resource for satisfying human needs.

A. Production

Principal Cotton Groups. Cottons that are of commercial importance belong to three principal botanical groups. The first is Gossypium hirsutum, referred to as Upland cotton, and the second is Gossypium barbadense, referred to as Egyptian, American Egyptian, and Sea Island. The third group is Gossypium herbaceum, a short staple cotton grown in Asian countries.

During the 1975 season, over 72 varieties of the first group were grown in the United States (1). These varieties accounted for over 99% of total cotton production. About five varieties of the second group, and no cottons of the third group, were grown in the United States in 1975.

Fiber Development. Each fiber emerges from a cell that develops on the surface of the seed. During the growth period the fiber elongates as a thin walled tube, and as maturing progresses the wall is thickened by deposition of cellulose from within. When the growth phase ends, the living matter within the fiber dies and shrinks, leaving a collapsed wall with a central canal. On drying the matured tubular fiber twists about its own axis, forming a convolution. The different layers of the tubular wall of a single matured cotton fiber are shown in Figure 1A.

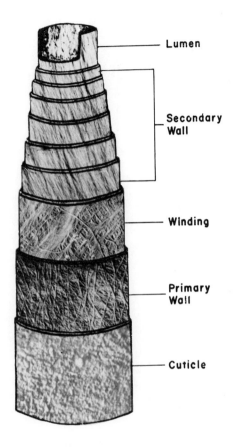

Fig. 1A. The major structural features of a typical cotton fiber before collapse.

Source: Courtesy of W. R. Goynes, Jr. Southern Regional Research Center, Agricultural Research Service, New Orleans, Louisiana.

An external view of matured cotton fibers is shown in Figure 1B.

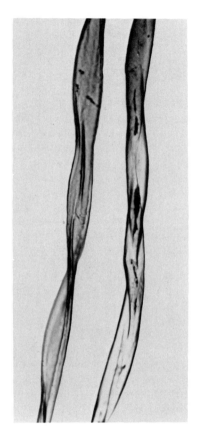

FiG. 1B. Portions of two single convoluted cotton fibers.

Source: Courtesy of W. R. Goynes, Jr. Southern Regional Research Center, Agricultural Research Service, New Orleans, Louisiana.

<u>Fiber Composition</u> The composition of a typical mature cotton fiber is shown in Table 1.

TABLE 1
Composition of typical mature cotton fiber

	Dry basis[1]		
		Range	
Constituent	Typical %	Low %	High %
Cellulose	94.0	88.0	96.0
Protein (N x 6.25)	1.3	1.1	1.9
Pectin substances	0.9	0.7	1.2
Ash	1.2	0.7	1.6
Wax	0.6	0.3	1.0
Malic, citric, and other organic acids	0.8	0.6	1.0
Total sugars	0.3	---	---
Other	0.9	---	---

1/ Moisture about 8%.

Source: John D. Guthrie in "Chemistry and Chemical Technology of Cotton," Kyle Ward, Jr., Ed. Interscience Publishers, Inc., New York, N.Y., 1955, Chapter I-A.

The cellulose component ranges from 88% to 96% of the typical fiber. The remaining components are protein, pectin substances, ash, wax, organic acids, sugars, and miscellaneous materials. These components may also vary among different individual mature fibers. In addition, a portion of these components may be removed during subsequent processing.

Geographic Production. On a world basis the most favorable zone for cotton production is between 47° north latitude and 32° south latitude. Over 68 countries have production areas within this adaptable zone.

World cotton production and its geographical distribution is shown in Table 2.

TABLE 2
Average world cotton production

	1968/69-1970/71	1974/75-1976/77
World production, 1000 bales	54,100	59,747
Percent of production accounted for by		
5 countries	66.9	69.7
10 countries	84.7	84.1
15 countries	90.3	89.7
Geographic production, percent		
North America	24.4	21.5
South America	8.6	7.2
Western Europe	1.4	1.4
Eastern Europe	0.2	0.1
USSR	17.8	21.1
Asia and Oceania	36.6	40.1
Africa	11.0	8.6

Source: *Cotton World Statistics. International Cotton Advisory Committee.* Vol. 30, No. 6 (Part II), 1977, Washington, D.C.

Average world total production increased by about 11% from 1968/69 to 1976/77. This increase was shared by the USSR and countries in Asia and Oceania, primarily the People's Republic of China. Production in the remaining areas decreased slightly or remained stable. Most of the world's cotton--67% in 1968/71 and 70% in 1974/77--is produced in only five countries. About 90% is produced in fifteen countries.

In the United States, cotton is produced in four production regions: the West, Southwest, Delta, and Southeast (Table 3).

TABLE 3
Distribution of U.S. cotton production

Producing region	1967		1976	
	Regional Production (1000 bales)	% of U.S. Production	Regional Production (1000 bales)	% of U.S. Production
West[1]	1651	22.2	3477	32.9
Southwest[2]	2958	39.7	3436	32.6
Delta[3]	2179	29.3	2871	27.2
Southeast[4]	655	8.8	773	7.3

[1] California, Arizona, New Mexico, and Nevada.
[2] Texas and Oklahoma.
[3] Arkansas, Missouri, Mississippi, Louisiana, Illinois, Kentucky, and Tennessee.
[4] North Carolina, Virginia, South Carolina, Alabama, Georgia, and Florida.

Source: "Cotton and Wool Situation," U.S. Dept. Agri. CWS-9, February 1977.

Between 1967 and 1976 some regional shifts in production have occurred. The West has gained an increasing share of production, and the Delta, Southwest and Southeast regions share of total production has decreased.

Production Cost The cost of producing cotton in the United States and selected countries for various years is shown in Tables 4 and 5. The costs given in these two tables is not directly comparable because of differences in computational procedures and different production seasons. However, the data do indicate that production cost is highly variable throughout the world. This variation is expected because production cost is influenced by climate, soils, varieties grown, cultural and management practices, and the local structure of input prices. Also, the cost shown does not indicate the competitive position of cotton relative to other farm enterprises within the respective countries.

TABLE 4
Average cost of producing U.S. cotton by regions, 1975 and 1976

	1975		1976	
	Total cost per acre (dollars)	Per lb. lint[1] (cents)	Total cost per acre (dollars)	Per lb. lint[1] (cents)
Southeast	263.28	64.1	276.95	68.0
Delta	241.25	48.6	245.65	58.3
Southern plains	143.23	48.7	159.45	45.3
Southwest	413.52	33.3	447.61	32.7
All regions	214.97	45.0	233.19	47.1

[1] Cost after credit given for cottonseed.

Source: Committee on Agriculture and Forestry United States Senate. "Cost of Producing Selected Crops in the United States, 1975, 1976, and Projections for 1977," 95th Congress--1st Session, January 21, 1977.

TABLE 5
Cost of producing cotton, selected countries

Country	Crop year	Yield Seed cotton/acre (lb)	Cost per acre (dollars)
Angola	1974/75	535-1,561	173-285
Australia	1973/74	943-1,160	429-433
Columbia	1974/75	1,266	239
Ecuador	1974	-	282-295
Egypt	1973/74	-	221
El Salvador	1974/75	-	367-494
Greece	1974	2,230	567
Honduras	1974/75	1,735	296
India	1973/74	304-2,036	33-252
Iran	1974/75	1,517-2,454	233-309
Israel	1974/75	3,569	301
Italy	1973/74	624	252
Ivory Coast	1974/75	892	181
Kenya	1974/75	714	223
Malawi	1974/75	803	80
Mexico	1974	-	197-366

Table 5 (continued)

Country	Crop year	Yield Seed cotton/acre (lb)	Cost per acre (dollars)
Morocco	1974	1,784-1249	259-328
Nicaragua	1974/75	990-1,159	349-372
Pakistan	1974/75	279	99
Paraguay	1974/75	847	77-91
Spain	1974/75	580-1,071	516-654
Sudan	1974/75	478	175
Syria	1974/75	1,784	358
Turkey	1974	795	372-411
Upper Volta	1974/75	419-1,338	61-160

Source: 34th Plenary Meeting of the International Cotton Advisory Committee, DOC 13, "Survey of Cost of Production of Raw Cotton," Abidjan, Ivory Coast, November 1975.

B. Cotton Textile Fiber Processing

The physical flow of cotton through the United States market system illustrates the principal stages where some physical processing occurs (Figure 2).

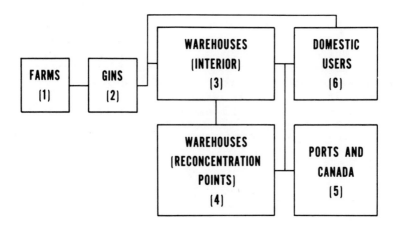

Fig. 2. Physical flow of U. S. cotton.

Cotton Farms, Stage 1. The first stage of processing raw cotton fiber is on cotton farms. At this stage, cotton producers carry out farm operations to maintain the quality of harvested seed cotton. An example of an innovative operation is on-farm storage of seed cotton in large enclosed modules and large open ricks. This innovation facilitates orderly harvesting and evens the flow of seed cotton to gins.

Gins, Stage 2. Cotton ginning is the separation of the fibers from the seed. The ginning process is usually divided into specific steps, which include feeding, drying, cleaning, extracting, conditioning, ginning, lint cleaning, sampling and packaging.

Technology developments that increase the rate of ginning minimize cost, and improve fiber quality have been readily adopted by the ginning industry. Some recent advancements are (1) the replacement of the conventional pneumatic telescopic method of unloading seed cotton with systems for trailer dumping or feeding from modules, (2) installation of universal density presses, (3) automatic sampling, (4) automatic strapping, and (5) integration of improved separators and cleaners into the processing stream for more efficient trash removal.

The United States cost for ginning a bale of Upland cotton during the 1975/76 season was slightly over $32. This cost compares with a $21 charge during the 1972/73 season (2).

Warehousing, Stages 3 through 5. Very little cotton is stored at the gins. Most of the baled cotton is shipped to interior warehouses, reconcentration points, or to ports for foreign shipment. The physical processing of cotton at warehouses is minimal but may include further sampling, segregation into more uniform lots by grade, reweighing, compressing gin-run bales to universal density bales, and rewrapping if needed.

The average United States cost of assembling cotton from gins and distributing it to domestic and foreign outlets during the 1974/75 season is shown in Table 6. Included in of 500 lb bale of cotton is used to spin yarns for weaving, knitting, sewing thread, carpeting and tufting, and miscellaneous uses. These yarns are further processed into clothing, household, and industrial textiles. The remaining 44 lb represented by 20 lb of tare and 24 lb of nonlint material is not used for textile products.

A typical cost of processing cotton from raw fiber to finished textile products usually is not available because of the large number of cotton textile end products produced and the varying degrees of processing needed to meet user requirements. However, the farm-to-retail spread of cotton denim dungarees was developed for 1974. This estimate provides an insight of the general magnitude of the cost and its distribution. For example, delivery of fibers to the mill represents slightly over 8% of the retail cost, textile mill processing almost 20%, apparel manufacturing 30% and wholesaling-retailing 42%. The farm-to-retail cost distribution would, of course, be quite different for a highly fashionable product with a much higher labor cost component and a high obsolescence factor.

TABLE 7
Cotton denim dungarees: Estimated distribution of the retail dollar by operation or service, 1974

Operation	Cost per pound of cotton (dollars)	Cost per pair produced (dollars)	Percent of retail
Farm production	0.366	0.516	6.4
Ginning	0.061	0.086	1.1
Marketing to textile mills	0.052	0.073	0.9
Textile mill processing and finishing	1.115	1.572	19.6
Apparel manufacturing	1.715	2.418	30.0
Wholesaling-retailing	2.390	3.370	42.0
Total value at retail	5.699	8.035	100.0

Source: Edward H. Glade, Jr. "Who Gets the Cotton Denim Dollar?" CWS-4, Economic Research Service, U.S. Dept. Agri. March 1976.

An average bale of U.S. cotton entering the mill processing level is distributed as shown in Figure 4. About 456 lb

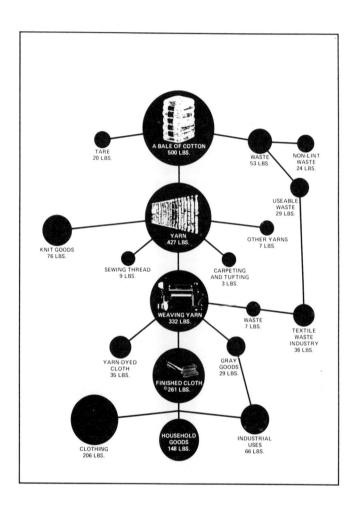

Fig. 4 Distribution of an average bale of U.S. cotton.

Source: Edward H. Glace, Jr. and Anne P. Alderman, "Distribution of Average Bale of U.S. Cotton," CWS-7 Economic Research Service, U.S. Dept. Agri., September 1976.

the fabric is bleached, dyed or printed for aesthetic appeal and it may be treated with various chemicals and resins to impart desirable consumer characteristics such as durable

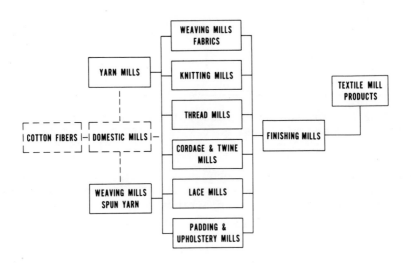

Fig. 3. Physical flow of cotton fibers through domestic textile mills.

press, flame retardance, water repellance, and static resistance. As seen in Figure 3, each of the processing steps may be performed in a single integrated mill or by mills that specialize in specific processes such as yarn spinning or finishing.

TABLE 6
Average cost per bale of assembling and distributing U.S. cotton by type of costs and outlet, 1974/75 season

	Outlet	
	Domestic	Foreign
	- - - - - dollars - - - - -	
Buying and local delivery	1.06	1.23
Storage	1.63	1.66
Compression	3.45	3.64
Other warehouse services	3.28	3.18
Transportation	7.56	36.42
Cotton insurance	.24	1.85
Financing	3.16	3.20
Selling	.87	.97
Overhead	2.40	2.53
Miscellaneous	.49	.38
Total	24.14	55.06

Source: Whitman J. Chandler, Jr. and Edward H. Glade, Jr. "Cost of Merchandising U.S. Cotton, 1974/75 Season," CWS-6, Economic Research Service, U.S. Dept. Agri. July 1976.

the cost are charges for warehouse services. The cost was slightly over $24 per bale for movement to domestic outlets, and just over $55 for cotton distributed to foreign users.

Domestic Users, Stage 6. The physical flow of cotton through domestic mills is shown in Figure 3. At this processing level, raw cotton fibers are converted into yarns and textile products. The specific processing steps depend on the final textile product output. In the case of an apparel product, likely steps would include opening-blending, cleaning, picking, carding, drawing, roving, spinning (spindle or open-end), warping, slashing, fabrication (woven, knitted or nonwoven), and fabric finishing. At the finishing step,

C. Consumption

Average world consumption of cotton expanded by 11% between 1968/71 and 1974/77, Table 8. Although all countries

TABLE 8
Average world cotton consumption

	1968/69-1970/71	1974/75-1976/77
World consumption, 1000 bales	54,680	62,224
Percent of consumption accounted for by		
5 countries	61.9	60.5
10 countries	73.7	72.3
15 countries	80.8	79.3
Geographic consumption, percent		
North America	17.7	13.3
South America	4.5	5.1
Western Europe	12.4	9.9
Eastern Europe	4.8	4.6
USSR	14.5	14.7
Asia and Oceania	42.9	48.4
Africa	3.2	4.0

Source: International Cotton Advisory Committee, Cotton World Statistics, Vol. 30, No. 6 (Part II) 1977, Washington, D.C.

consume cotton, about 60% is consumed by five countries and 79% by fifteen countries. The geographic consumption pattern has shifted since 1968-71. There was a decline in the share consumed in North America, Western Europe and Eastern Europe. The share for other regions increased, with the largest increase occurring in Asia and Oceania.

U.S. textile fiber consumption by broad product categories and types of apparel are shown in Table 9.

TABLE 9
Consumption profile for cotton and other textile fibers by product category and type of apparel, 1964 and 1975.

Category	Cotton Fiber 1964 (percent)	Cotton Fiber 1975	Other Fibers 1964 (percent)	Other Fibers 1975
Apparel	52	49	27	33
Household	31	36	27	43
Industrial	17	15	46	21
Apparel				
Men's, youths', & boys'	64	69	43	43
Women's, misses' & juniors'	21	20	48	46
Girls', children's & infants'	15	11	9	11

Source: National Cotton Council of America, *Cotton Counts its Customers*, 1964 and 1975, in press, Memphis, Tenn.

Table 9 shows the distribution of cotton and other fibers by major end-use category for the years 1964 and 1975. Apparel accounts for about half of total cotton consumption in the United States, with home furnishing and industrial uses accounting for 36% and 15%, respectively. Distribution figures for competing fibers reflect somewhat higher percentages in home furnishings and industrials with the enormous floor covering and tire cord markets for man-made fiber being major contributors to the higher level of consumption in home furnishings and industrials.

Between 1964 and 1975, the principal change in the relative importance of the three types of apparel was an increase for cotton usage in men's, youths' and boys' apparel and a decrease in the girls' children's and infants' category. For other fibers, a similar, but less pronounced trend occurred.

The apparel fiber consumption profile, by type of construction, for 1971 through 1975 is shown in Table 10.

TABLE 10
Fiber use in knit and woven apparel, 1971-1975

Construction	1971 (percent)	1972 (percent)	1973 (percent)	1974 (percent)	1975 (percent)
All fibers					
Knit	43.5	47.2	49.1	50.1	45.0
Woven	56.5	52.8	50.9	49.9	55.0
	100.0	100.0	100.0	100.0	100.0
Cotton					
Knit	31.9	31.9	31.4	31.8	34.6
Woven	68.1	68.1	68.6	68.2	65.4
	100.0	100.0	100.0	100.0	100.0
Other fibers					
Knit	52.7	58.6	60.8	61.6	51.9
Woven	47.3	41.4	39.2	38.4	48.1
	100.0	100.0	100.0	100.0	100.0

Source: National Cotton Council of America, Cotton Counts its Customers, 1971 through 1975, Memphis, Tenn.

The percentage of all fibers used for knits had an upward trend between 1971 and 1974, and a decline between 1974 and 1975. For cotton, the percentage used for knits and wovens was relatively stable up to 1974; for 1975, a slight gain was made in knits and a slight loss in wovens. The use of other fibers shows a marked shift toward knits between 1971 and 1974. Between 1974 and 1975, knits declined by about 10% of total use, but increased for wovens.

COTTON LINTER AND CELLULOSIC MATERIAL PRODUCTION, PROCESSING, AND USE

A. Output and Use

United States prices, imports, and exports of cotton linters and cellulosic raw materials for the 1972/73 and 1975/76 season are shown in **Table 11**.

TABLE 11
U.S. production, prices, exports and imports of cotton linters and cellulosic waste, 1972/73 and 1975/76 season

	Unit	1972/73	1975/76
Production			
Linters[1]			
Total	million lb	852.6	522.1
First cut	million lb	254.7	130.9
Second cut	million lb	549.0	334.9
Mill run	million lb	48.9	56.3
Gin motes mill[2]	million lb	157.5	102.9
Textile mill waste[3]	million lb	279.7	261.0
Average prices			
Linters, f.o.b			
Cottonseed oil mills, cents[4]			
Grade 2 staple 2	lb	7.95	9.30
Grade 3 staple 3	lb	7.18	8.65
Grade 4 staple 4	lb	6.00	7.96
Grade 5 staple 5	lb	5.04	6.59
Grade 6 staple 6	lb	4.34	6.05
Grade 7 staple 7	lb	3.74	5.17
Chemical 73% cellulose base	lb	3.50	4.25
Soft cotton waste imports, cents		4.39	8.29
Exports[4]			
Linters, running bales	1000	259.2	181.6
Grades 1-7	1000	42.7	55.6
Chemical grades	1000	216.5	126.0
Cotton pulp	short ton	52,696	46,694
Imports			
Linters, 500 lb. gross bale	1000	33.2	28.2
Soft cotton waste, lb.	1000	27,538	8,018

[1] Fats and Oilseed Crushings, Current Industrial Reports, Series M20J, Bureau of the Census.
[2] Anselm C. Griffin, Jr., "Fuel Value and Ash Contents of Ginning Waste," Trans ASAE, Vol 19, 1976. Values shown computed using the fiber content of lint cleaner and gin stand motes.
[3] Based on 36 lb of usable waste per 500 lb bale consumed in domestic mills, see Figure 4.
[4] Monthly Cotton Linter Review, Agricultural Marketing Service, U.S. Dept. Agri.

Between 1972/73 and 1975/76, production of linters, gin motes, and textile mill waste declined from 1,290 to 886 million pounds. This decline was due to lower cotton production and reduced consumption of cotton by textile mills. During the 1972/73 season, output was distributed as 66% linters, 12% motes, and 22% textile mill waste. Output shifted slightly in the 1975/76 season to 59%, 12%, and 29%, respectively.

Table 11 shows linter and imported soft cotton material prices for the 1972/73 and the 1975/76 seasons. The prices for each commodity increased between 1972/73 and 1975/76. This increase reflects the lower production mentioned and the continued strong domestic demand for cotton cellulosic raw material for spinning applications—particularly open-end spinning.

Linters, cotton pulp, and soft cotton material are traded in the international market. The United States usually exports linters and cotton pulp and imports linters and soft cotton materials. Both exports and imports of these raw materials declined between 1972/73 and 1975/76.

The use of linters by selected U.S. industries in 1972 is shown in Table 12. Based on the use pattern, 62% of the linters were used as batting, 36% for paper products, and 2% was used for surgical appliances and surgical supplies. The use pattern for gin motes and soft cotton waste is not available, but would probably be similar to the pattern for linters.

B. Processing

Cotton linters and waste are major sources of cellulosic raw materials for the production of consumer and industrial products. The physical flow of cotton linter and cellulosic materials through the U.S. marketing system is used to indicate the various stages at which processing occurs (Figure 5). The amount of processing that linter and cellulosic materials undergo depends on the ultimate product.

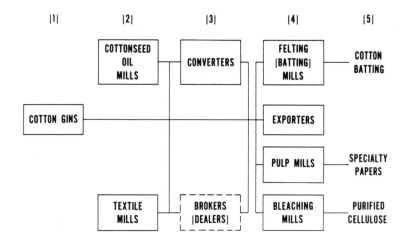

Fig. 5. The marketing system for cotton linters and waste.

Gins, Stage 1. Cotton linters are short fibers that remain on the seed after the seed cotton has been ginned to remove fibers for the textile mills. Gins usually do not remove the linters from the seed, but ship the cottonseed to a cottonseed oil mill (Stage 2). Also, cotton fibers for use in textile products are shipped to textile mills (Stage 2). During ginning, some cellulosic waste is generated. This material generally referred to as gin motes, includes short and some spinnable cotton fiber removed during the gin cleaning process. Only minimal processing of motes taken place at the gins, but cleaning and baling would be included. Gin motes are shipped to Stages 3 and 4 for further processing. During the 1975/76 season, the prices of gin motes varied between 5.5¢ to 20c/lb.

Cottonseed Oil and Textile Mills, Stage 2. Cotton linters are removed from cottonseed by the oil mills, and textile waste is generated at the textile mills.

Three methods are currently used by cottonseed oil mills to remove the linters from cottonseed received from gins--removal by saws, abrasion, and acid. The saw and abrasion methods are usually used for seeds that will be crushed for oil, meal, and hulls. The acid method is usually used to delint seeds for planting. Several processing steps are similar for each delinting method, for example, cleaning by passing the seed through a rock catcher and magnet, dust removal, and packaging. Several processing steps also differ. In the saw and abrasion methods, linters are usually separated into three grades, referred to as first cuts, second cuts, and mill-run linters. In the acid delintering method, anhydrous hydrogen chloride gas is passed through the seeds to loosen the linters. After the seeds are discharged from the acid chamber, they are neutralized by an ammonia gas treatment and the linters are scrapped from the seeds. A new acid delintering method based on aqueous dilute sulfuric acid is now being installed in commercial plants. The various processing steps in this method include seed cleaning, spraying with a dilute acid, acid removal, drying, scrapping the linters, and packaging.

The cost of delintering cottonseed by the above methods range from $4.78 to $5.85 per ton of seed processed. These costs were developed for oil mills processing a specific volume of seed; for this reason, the costs should be used with caution.

The cotton wastes generated at textile mills are further processed for their cellulose. For example, about 36 lb of usable cotton waste are generated for each 500 lb bale of cotton processed (see Figure 4). This waste is generated from the cleaning, carding, combing, and yarn spinning operations. Other than collecting and packaging only minimal physical processing takes place at the mills. However, there are a few mills that operate their own utilization and recycling plants. The selling prices of these wastes are somewhat dependent on the operation from which they were generated. During the 1975/76 season, prices ranged from 9¢ to 50¢/lb.

Convertors and Brokers, Stage 3. Convertors and brokers constitute the third stage in the marketing system for linters and waste. The convertors buy linters from Stage 1, 2 and brokers (dealers) in Stage 3. Convertors usually provide little processing other than cleaning, mixing, some storage, and packaging. Brokers and dealers do not process linters and waste. Their function is limited to contracting for linter and waste supplies and making shipping arrangements to convertors and Stage 4.

Felting, Bleaching, Pulp Mills, and Exporting, Stage 4.
Linters and cellulosic waste are converted to industrial products or exported at Stage 4.

Linters and waste are processed into batting at garnetting plants. The typical processing steps are opening, blending (usually 60% linters and 40% cotton waste), cleaning, garnetting, web-laying, and felting. If smolder-resistant batting is to be produced for the bedding, upholstery, and automotive industries, additional processing steps are needed. There are three methods for manufacturing smolder-resistant batting: immersion in a boric acid solution, dusting with a dry boric acid powder, and applying boric acid via a vapor phase process. If the immersion method is used, the application is usually on raw stock linters and waste or following the opening-cleaning-blending operation. After wetting a drying operation is required if this method is used. The second method, using dry acid powder, is performed between the garnetting and the web-laying processing steps. Application by the vapor phase method can be performed at any stage in the processing.

Purified cellulose is produced at bleaching mills. The processing steps include the removal of hull pepper, waxes, oils, and pectins; bleaching with hypochlorite or peroxide; neutralizing; and packaging. The purified cellulose may be shipped to chemical plants for conversion into chemical derivatives such as carboxymethyl-, methyl, and ethyl cellulose, or to other plants for inclusion in detergents, oil well drilling muds, plastics, sausage casings, films, and many other industrial and consumer products.

Cotton linter pulps for use in specialty paper products are produced at pulp mills. These mills use a sulfate process on second cut linters. The typical processing steps include charging through beaters, digesting, washing, bleaching, pulp beating, drying or finishing (sheet and roll), packaging (bulk, sheet pulp or roll pulp), and storing.

TECHNOLOGICAL PROSPECTS FOR INCREASED USE OF COTTON

The use of cotton as a cellulosic raw material closely parallels the development and adoption of technological innovations in production and processing. The basis for cotton's technological achievements has been the continued expansion of basic knowledge relating to the cotton plant, the chemical reactivity of cotton cellulose, and the structural characteristics of fibers. The meshing of basic knowledge and technical innovation has and will continue to open new opportunities for producing and exploiting the attributes of cotton.

A. Production Technology

The development and use of technology at the farm level has revolutionized cotton production. For example, higher lint-yielding varieties are planted and harvesting is accomplished with machines. Some examples of new technologies that will be applied in the future include the use of insect-resistant cotton varieties, minimum and automated tillage, the control of insect pests by biological methods, and the maintenance of seed cotton quality in bulk storage containers.

B. Textile Processing Technologies

Off-farm processing of cotton begins at the gins and continues through the textile mills. At each of these processing levels, new technologies are being developed to maintain and improve the quality of cotton, increase process efficiency, conserve resources, protect the environment, and evolve new cotton products that meet the needs of consumers.

Future developments applicable to the gin level will probably include the use of chemical processing aids, water jet and differential ginning, fully automated conveyor systems, incinerator and heat recycling, enclosed dust and noise control systems, and automatic moisture controls to prevent fiber breakage.

A number of new processing technologies applicable to the textile mill level will be developed and tested. These advances cover the entire spectrum of processing from basic yarn formation to textile finishing.

New methods for transforming cotton fibers into yarns are being developed. Some possible future developments include improved open-end spinning, a twistless method that uses chemicals and adhesives, core-spinning, use of electrostatic fields to effect twist, and an air vortex system for forming yarns.

The chemical reactivity and hydrophilicity of cotton cellulose, coupled with the increase in knowledge of surface properties of fibers, has opened up many new technical opportunities for conserving energy, reducing textile pollution, reducing processing costs, and imparting additional functional and aesthetic properties to consumer textile products. In the printing operation, new chemical and resin systems are being developed that increase the versatility of the heat transfer method. Solvent and gas phase systems to induce and control desirable chemical reactions are applicable to the printing as well as to the dyeing and finishing steps. The traditional method for applying surface finishes is the pad-dry-heat cure method. Several alternative methods to padding such as kiss-roll, wet-on-wet, and transfer padding are now technically feasible. Options for curing are also technically feasible. Examples are photoinitiation, ultraviolet, microwave, and chemical curing. Dyeing processes using polar solvents in conjunction with vacuum padding are being developed to increase the receptivity of cotton to dye formulations.

Cloth fabrication systems as options to weaving and knitting are technically feasible for producing many apparel and industrial textile products. These nonwoven systems may include chemical bonding, adhesions, and mechanical entanglement and are very similar to the wet and dry lay systems currently employed in the paper industry.

The prospective technical innovations for producing apparel, household, and industrial textile products can be integrated into current textile mill processing streams. A completely new integrated system referred to as the "tuft-to-yarn system," is being developed. "Tufts" refer to small clusters or bundles of fibers. In this system, fibers are moved, cleaned, and aligned by mechanical and aerodynamic forces. A number of the traditional processing steps such as picking, carding, drawing, and roving will be eliminated. In addition, the system can be completely automated and enclosed to reduce mill noise and mill dust. It is also conceptionally feasible to utilize electrostatic forces as a substitute for mechanical and aerodynamic forces to move fibers through the system.

C. Linter Technology

Currently, purified cellulose is produced from linters. Some of this purified cellulose is further converted to cellulosic derivatives that are used in formulating food products. It has been demonstrated that the food uses from cellulose derived from cotton linters or fiber can be expanded through an enzymatic hydrolysis system in which the cellulose is depolymerized before it is used in food processing.

COTTON FIBERS USE PROJECTIONS

The 1985 projection for domestic mill use of cotton, wool, and cellulosic fibers is shown in Table 13.

TABLE 13
Projected 1985 domestic per capita fiber use.

Fiber	Pounds
Cotton	14-18
Wool	1
Cellulosic	7-8

Source: George F. Dudley, "U. S. Textile Fiber Demand Price Elasticities in Major End-Use Markets," Tech. Bull. 1500, ERS, U. S. Dept. Agri.

These projections are on a "cotton equivalent" basis, which means that each fiber is adjusted for differences in processing waste and in the amount of raw fiber required to produce a pound of cloth. The analytical approaches used to make the projections include regression and nonparametric techniques, trend extrapolation, and fiber specialists' judgements.

The 1985 cotton fiber consumption is projected at 14-18 lb per capita. The predictive equation in arriving at this level included four variables: a total fiber price index deflated by an index of personal disposable income, year-to-year change in deflated personal disposable income, lagged total fiber consumption, and time. These variables were used to reflect the effects of real fiber prices, income, prior consumption, and nonquantifiable factors on the demand for fibers.

Projections for the individual types of fibers were constrained by the projected total fiber consumption. However, each fiber type was projected separately, based on the most reasonable technique. For example, the nonparametric Gompertz curve technique was used to project cotton consumption, and wool was projected by trend extrapolation.

The projection shown above for total cotton, wool, and cellulosic fiber consumption and the individual fibers could be appreciable altered if the current uncertainties relating to petroleum and petroleum-derived chemical supplies continue. In this situation, it could be expected that noncellulosic fiber consumption would be lower and consumption of cotton fiber would increase. However, disposable personal income is expected to continue its upward trend, which will moderate any future change in overall fiber consumption. In addition, another moderating element would be the adoption of the technological developments mentioned earlier.

REFERENCES

1. U.S. Department of Agriculture, Agricultural Marketing Service, Cotton Varieties Planted 1971-1975, August 1975.
2. Bounds, F., and Cole, R., "ERS Bulletin-2 (1973, 1976, revised), U.S. Department of Agriculture, Economic Research Service and Agricultural Marketing Service.

Modified Cellulosics

OUTLOOK FOR WOOD CELLULOSE

N. Sanyer

USDA Forest Products Laboratory
Madison, Wisconsin

A rapidly diminishing supply of traditional raw materials for organic chemicals has generated interest in reassessing cellulosics for manufacture of plastics, films, and fiber. Since wood is the chief source of cellulose, the development and maintenance of an accurate timber inventory is vital to future expectations and long-range planning. Also, the major economic factors governing demand and supply with respect to raw materials, energy, environment, and capital resources need to be continuously evaluated to guide research and develop improved processes and products. The quantity, quality, and distribution of timber stands in the United States, including ownership, rate of growth and removal, potential changes in growth, recovery of residues are reviewed. The manufacture of pulp for chemical conversion including projected demands, byproducts, and economics are discussed. During the next 20 to 40 years, no major changes in the traditional uses of wood are anticipated, but an increasingly higher proportion of the U.S. timber supply will be available for the manufacture of pulps including dissolving grades. Large expansion of wood cellulose production may not start, however, before 1990 due to economic and technical limitations. Substantial research and development efforts are needed to overcome these and improve processes and products.

INTRODUCTION

The future availability of wood cellulose is directly related to timber supply and projected demand situation. To provide a quick perspective here, the traditional uses of timber products will be briefly reviewed and the wood requirement for dissolving pulps compared with that of paper pulps.

Most of the data on U.S. timberland and timber products are taken from the U.S. Forest Service reports, "The Outlook for Timber in the United States" (1) the 1975 Assessment (2). These reports contain much detailed information on the geographic distribution, vegetation characteristics, ownership, productivity of forest and rangeland, and the demand and supply situation of all major products and uses of forests, including timber, outdoor recreation, fish and wildlife, and water. Several critical reviews have been published (3-6) which are also relevant to the subject of this chapter.

TIMBER INVENTORIES

The inventories of forest and rangeland renewable resources of the United States are maintained primarily by the forest surveys conducted by the Forest Service of the U.S. Department of Agriculture. Keeping these surveys current and intensifying them to provide more precise resource data for small geographic areas will greatly improve land use planning and management of forest lands including those in small private ownership (1,2). These reports recognize the need for more in-depth research and information on the growth and yield responses in stands subjected to active and intensive management. Rapid changes in economic activities, expanding international trade in products and technologies, increasing urgency for substitute products, and rising energy costs will influence the intensity of forest management necessary for providing adequate timber.

A. Forest Land Area

About 69% of the U.S. land area, 1.55 out of a total of 2.27 billion acres, is classified as forest and rangeland (Table 1). More than two-thirds of this, or over a billion

TABLE 1
Major Classes of Land Area of United States (2)

Classes	Area (million acres)
Cropland	427
Commercial timberland	499
Range and noncommercial timberland	1056
Private pasture	118
Others	166
Total	2266

acres, is classified as rangeland and noncommercial forest. Natural grasslands, savannas, shrublands, deserts, tundra, coastal marshes, and wet meadows constitute the rangelands. The noncommercial forests includes ecosystems incapable of producing crops of industrial wood because of poor site or other adverse conditions, such as pinion-juniper or high alpine forests, and productive forest lands withdrawn for parks, wildlife refuges, recreation areas, or other uses not compatible with timber production.

Nearly one-third of the forest and rangeland, or 500 million acres, is classified as commercial timberland, capable of producing more than 20 cu. ft./acre of industrial wood a year in natural stand. This is about 22% of the total land area of the United States.

About three-quarters of commercial timberlands area is in the East. Of the total forestland, hardwoods cover 53% of the area, eastern and western softwoods together equally account for 42%, and the remaining 5% includes the nonstocked or unregulated land area (Table 2).

TABLE 2
Area of U.S. Commercial Timberland by Ecosystem (2)

Ecosystem	Area (million acres)	Percent
Eastern Groups		
Softwoods		
Southern pines	71.1	14.3
Spruce-fir-jack pine	31.1	6.2
Hardwoods	253.9	50.8
Nonstocked	14.3	2.9
Total	370.4	74.2
Western Groups		
Softwoods	105.0	21.1
Hardwoods	12.8	2.5
Nonstocked or unregulated	11.1	2.2
Total	128.9	25.8
Total of all groups	499.3	100.0

B. Ownership

The future potential and efficiency of management of forests is largely influenced by the ownerships and the size of the units. About 73% of all commercial timberland is privately owned, while 27% is in federal, state, and other public holdings (Table 3). The private holdings include some of the most productive timber sites in the eastern United States which are also close to markets; these constitute 81% of the timber in the north and 91% in the south. Some 92 million acres, or 18% of the timberlands are

TABLE 3
Timberland Ownership in the United States (Million Acres) (2)

Type of ownership	East	West	Total	Percent
Federal	26.6	80.4	107.1	21
State-County	22.4	6.4	29.0	6
Forest industry	52.8	14.4	67.3	14
Farm	116.1	14.9	131.1	26
Miscellaneous - private	152.2	12.8	165.1	33
Total	370.2	128.8	499.6	100

in National Forests. These are located largely in the west in the Rocky Mountains and Pacific Coast, and made up of relatively low-quality sites at higher elevations; nevertheless, they represent a substantial part of the U.S. softwood sawtimber inventory.

There has been some shift in the designation of forest lands to noncommercial status in recent decades by establishment of parks, wilderness, and other reserved forest areas. The steady trend in land clearance for crops and pastures by the settlers ceased around 1920, and 50 million acres of these areas has since returned to commercial forest land in the East and South. On the other hand, there is a continuous loss of timberland to industrial and urban development, agriculture, and other uses. An accelerated program of reforestation of nonstocked or marginal lands by both private and public sectors could easily make up for these losses. Any discouragement caused by the increases in land values, coupled with the long-term nature of the return associated with forestry, must be overcome, however, by establishing resourceful national public policy and commitment.

C. Forest Productivity

Net annual growth in two-thirds of the commercial timberlands varies between 50 and 120 cubic feet or about 0.75 to 1.75 tons of wood per acre. The Pacific Coast region in the Northwest and the southern pine region are the areas of highest productivity. The ratio of net growth to removal varies, of course, with the age of the timber and its rate of growth, mortality rate, and remedial management practices. It is estimated that the net annual growth can be nearly doubled by intensive management, genetic improvement, improved stocking, and fertilization. The long investment period particularly discourages the several million nonindustrial private owners (who control 59% of the U.S. timberland) from managing their forests more intensively. Here lies a great opportunity for substantially increasing the timber supply by providing the incentive to these owners through educational, technical, and economic assistance and public partnership.

D. Timber Supply

Projections of supplies, based on assumptions that existing policies, practices, and trends will not materially change, are shown in Table 4. According to these projections

TABLE 4
U.S. Roundwood Supplies in 1970 with Projections to 2020 (Million Cubic Feet) (2)

Area	1970	2000	2020
East:			
North – Softwoods	579	1,109	1,113
Hardwoods	1,409	3,845	3,799
South – Softwoods	3,745	5,768	5,788
Hardwoods	1,668	3,327	3,416
West: Softwoods	4,657	4,607	4,722
Hardwoods	96	194	203
U.S. total: Softwoods	8,981	11,484	11,622
Hardwoods	3,173	7,365	7,418

softwood roundwood supply will rise 29% (from 9 billion to 11.6 billion cu. ft.) and hardwood supply will increase 134% (from 3.2 billion to 7.4 billion cu. ft.) during the next quarter century. The projected increases for the sawtimber-size material, however, are much smaller. In fact, a slight decline is indicated for both hardwoods and softwoods after 2000. This is largely due to a sharp decline in the inventory of large sawlog-size trees and harvesting the remaining old-growth timber in the West.

Without a substantial increase in prices of timber products relative to the general price levels, the projected timber demands in 10 to 20 years will be substantially above the available supplies. The potential supply problems and associated price increases are likely to be greatest for softwood sawtimber. Consequently, significant rises in price of softwood stumpage and timber products will be necessary for balancing the supply and the demand.

The projections for hardwood timber show a more favorable supply and price outlook with the possible exception of preferred species and larger sizes.

E. Pulpwood Supply

The availability of wood for dissolving pulp will be, of course, closely related to the pulpwood demand for the manufacture of pulp for paper and paperboard. Depending on the economic trends and developments, the projected consumption of paper products will continue to expand somewhat faster than solid wood products, closely following per capita disposable personal income and per capita gross national product. The trends and a somewhat conservative projection of pulpwood demand are shown in Table 5. This material from Auchter (7) assumes that by the turn of the century, the per capita paper consumption will increase from the present 600 lb to about 1,100 lb.

The wood supply situation for pulp and paper is more favorable because of the availability of relatively large-volume eastern hardwoods, and increasing primary manufacturing plant residues. Furthermore, technological advances in high-yield pulping, whole-tree chipping, bark-chip separation, and chip cleaning will significantly extend the raw material base and stabilize pulpwood prices. The use of recycled fiber will also increase, substantially reducing the woodpulp projection by as much as an additional 20% during the next two or three decades.

TABLE 5
Pulpwood Consumption with Projections to 2020
(Million Tons) (2,7,8)

Year	Total	Roundwood	Residue
1970	83	59	24
1972	85	59	26
1974	94	65	29
1976	88	56	32
1980	107	71	36
1990	122	78	44
2000	152	108	44
2020	189	145	44

The pulpwood projections were substantially lowered because of the complex economic and environmental constraints and the increasing public interest and awareness for resource conservation. They will still amount to more than 60% of the U.S. roundwood supply. In 1976 the U.S. pulp, paper, and pulpwood exports were 12 million and the imports about 16 million cords in equivalent pulpwood. These values were not represented in Table 5; they are likely to expand but the imbalance will remain essentially unchanged.

As for expansion of pulp production, it will mostly occur in areas adjacent to existing manufacturing facilities as shown in Figure 1 (9).

DISSOLVING PULP PRODUCTION

A. Historical Trends and Projection

The U.S. cellulose production has seen a rapid rise starting in the early 1950's, which leveled off in 1970, Table 6. Since a large proportion of the U.S. production of dissolving pulps is exported, demand for this commodity is severely influenced by worldwide economic fluctuations. The U.S. annual capacity is about 1.8 million tons or nearly one-half of the combined capacity of western hemisphere and Japan (10). Demand has suffered severely from the 1975 economic recession and is presently affected by the curtailed rayon fiber production, largely resulting from stiff competition of synthetic fibers which are underpriced due to much overcapacity.

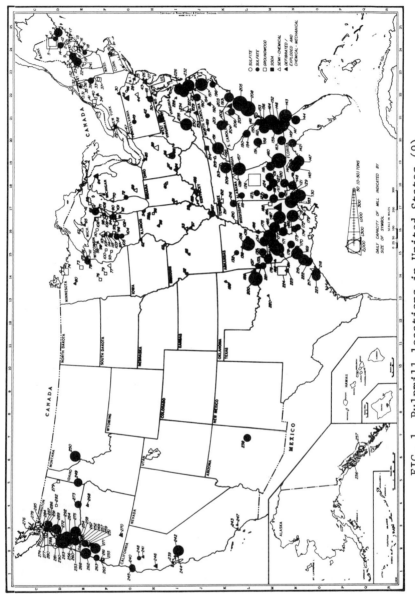

FIG. 1 Pulpmill locations in United States (9).

TABLE 6
Woodpulp Production with Projections to 2000
(Million Tons) (7,8)

		Paper and board pulps	
Year	Dissolving	Sulfite	Kraft
1935	0.19	1.39	1.47
1940	.33	2.28	3.75
1950	.48	2.37	7.40
1960	1.14	2.58	14.59
1970	1.71	2.34	29.47
1974	1.72	2.21	33.01
1975	1.37	2.03	29.36
1976	1.49	2.21	33.56
1980	1.80	2.20	40.50
1990	2.00	2.00	50.00
2000	2.80	1.80	60.00

Even if mills were operated at a full capacity, the wood required for dissolving pulp production constitutes only about 5% of total pulpwood consumption (Table 6). In 1974 and 1976, in addition to 37 million tons of chemical pulp, another 11 million tons of mechanical and semichemical pulps were produced but were not included in the table. From these projections it can be concluded that the problem is not going to be wood supply but lack of expansion in use of cellulosics. Moreover, if more use of cellulosics in place of or in conjunction with synthetic polymers can be promoted, the dissolving pulp production could be doubled or tripled during the next or the following decade without being hampered by wood supply.

The future demand for cellulosics during the next two decades will be influenced by the severe competition provided by the synthetic films and fibers of petrochemical origin. In recent years the worldwide capacity expansion in synthetic fibers, for instance, has limited the operating levels of these plants to 70 to 75% capacity, keeping the prices low and highly competitive with cotton, rayon, or wool. This situation will probably continue well into the 1990's, making it difficult for cellulosics to expand. At that time the synthetics will reach a saturation level and further shortages and price increases of petrochemicals will restrain overcapacity. Coincidentally, the demand for natural fibers, namely cotton and cellulosics, will dramatically increase and they will become indispensable for blending and designing fibers with unique properties.

Since relative costs of growing cotton are increasing, the opportunity for supplanting it with wood cellulosics will also become greater, possibly after 1990. Such demand increases could be met either by building new capacities or converting paper pulpmills into dissolving grades--for which the sulfite mills could be most likely candidates. Considering the 1980 projected worldwide textile fibers demand of 27.5 million tons (cotton 13, wool 1.7, synthetics 9.1, and cellulosics 3.7 million tons--H. Krassig, TAPPI Conference Papers, 4th International Dissolving Pulps, 1977) compared to a projected paper pulp capacity of about 156 million tons (100 million tons of chemical pulp) the wood supply should not be a serious barrier for a substantial capacity increase in dissolving pulp, an additional 5 or 6 million tons for instance. A similar case can be made for an expanded use of cellulosic in films, membranes, and plastics.

B. Processes

Cellulose is the main load-bearing polymer in wood and for fulfilling this function it is very closely associated with two other classes of polymers, lignin and hemicellulose. Therefore, the separation of pure cellulose requires complex chemical processing involving a capital intensive technology where the product yield is low. These processes are quite similar to those used in the manufacture of paper pulps, but the dissolving pulps are obtained in substantially lower yields using stringent purification steps. A thorough description of various methods used in dissolving pulp manufacture is compiled by Rydholm (11).

Acid sulfite is the oldest process first adapted in 1903 to make dissolving pulp for viscose. It is suitable for pulping northern softwoods and relatively low-density hardwoods. It is combined with a hot or cold alkali extraction step to produce various grades of celluloses of desired purity. The prehydrolysis kraft process is a newcomer which has been established since World War II. It is a more universal and broadly applicable process including species of high-density hardwoods as well as softwoods containing heartwood resins and phenolic extractives such as oaks, Douglas-fir, and southern pines. About 45% of the 1.8 million tons of dissolving pulp production capacity of the United States is in sulfite and 55% is in prehydrolysis kraft process. Depending on the end use, the residual hemicellulose, resin, metal ions, silica, and ash content of dissolving pulps are reduced to the desired levels using suitable purification methods which are usually incorporated into bleaching operations.

C. Pulp Composition and Properties

Although the dissolving pulps for manufacturing various cellulose products are often specified by their alpha cellulose content, in certain applications, such as esters for making films, there are more specific processing requirements with respect to purity and reactivity. For the manufacture of cellophane, regular tenacity rayon yarn, nitrate lacquer, and various cellulose ethers a hot alkali refined sulfite pulp with an alpha content of 91 to 92% could be adequate. Sulfite pulps for replacing cotton linters in manufacturing explosives grade cellulose nitrate are required to contain 98%, while cellulose acetate requires 95 to 97% alpha content. High tenacity or high wet modulus rayon yarns with a cotton-like property require pulps with 96 to 98% alpha content.

The yield and composition of typical pulps are shown in Table 7 which indicate that by proper manipulation of process variables pulps of similar composition can be obtained from different wood species.

TABLE 7
Yield and Composition of Dissolving Pulps (ca. 96% Alpha Content) (Percent on Wood)

Pulp	Yield	Cellulose	Glucomannan	Xylan
Spruce				
Wood	100	42	17.0	7.2
Acid sulfite	36	34	1.0	1.0
Prehydr. kraft	36	33	1.6	1.4
Birch				
Wood	100	42	2.1	31.0
Acid sulfite	36	33	1.0	2.0
Prehydr. kraft	36	33	1.0	2.0

In addition to lignin and hemicelluloses a significant amount of cellulose is also lost that contributes to low pulp yield. The amount of various wood substances dissolved in pulping liquors are included in Table 8. In making 1 ton of pulp, 2800 kg of wood is required, of which 1800 kg is degraded and solubilized.

TABLE 8
Wood Substance Dissolved in Pulping Liquors
(Kg/Ton of Pulp)

	Spruce	Birch
In pulp (96% alpha)	1000	1000
In liquors:		
Cellulose	224	250
Glucomannan	448	28
Xylan	174	800
Miscellaneous extractives	169	162
Lignin	785	560
Total wood substance	2800	2800

In sulfite process the polysaccharides are more or less hydrolyzed into simple sugars and a small fraction is further degraded. During kraft cooking polysaccharides are almost entirely converted into a variety of hydroxy aliphatic acids (12,13). The prehydrolysis liquors contain, of course, a significant fraction of partially hydrolyzed and solubilized hemicelluloses.

Low pulp yield and high capital requirement are major contributors to the relatively high manufacturing cost of dissolving pulps (Table 9). When the selling and shipping expenses are added, the pulp cost at the conversion plant would come to around $0.44/kg. This is a relatively high price for a starting material for further conversion and justifies a major research effort toward developing an improved pulping method for increasing the cellulose yield and reducing the cost of manufacturing.

The recovery of higher value byproduct chemicals from spent pulping liquors remains to be an interesting option for lowering overall cost and improving the utilization of this resource. Biological conversion of sugars in sulfite

TABLE 9
Production Cost Components of Dissolving Pulp
(1000 Tons/Day)

	$/Air-dry ton
Wood ($40/cord)	92
Chemicals and fuel	31
Labor-salaries-overhead	60
Plant cost - maintenance	86
Working capital	10
Taxes - insurance	4
	283

spent liquors to alcohol and torula yeast has long been demonstrated. More recently, however, a new and efficient process to produce protein from these liquors has been developed (14). The analysis of spent sulfite liquor components and their potential fermentation products are shown in Table 10.

TABLE 10
Carbohydrate and Acidic Components of Spruce Sulfite Spent Liquors and Their Fermentation Products (Kg/Ton of Pulp at 36% Yield) (14)

Component of the liquor	
Mannose	190
Galactose	27
Glucose	55
Rhamnose	4
Arabinose	19
Xylose	95
Total monosaccharides	390
Polysaccharides and uronic acids	88
Aldonic acids	94
Acetic acid	88
Lignosulfonates	900
Total organic	1560
Fermentation products:	
Alcohol	137
Pekilo after alcohol fermentation	147
Pekilo protein only	284

Although the questions of marketability and economics of this new protein remain to be established, the recovery of an additional 28% potential product on pulp basis would be certainly a very worthwhile development.

Finally, the manufacture of cellulosics involve complex conversion methods which are also costly in terms of capital, energy, and environmental liability. Recent efforts in discovering novel cellulose solvents (15,16) can be looked upon as glimpses of long-awaited new cellulose regeneration schemes which will do much to overcome some of these problems and open the new age of cellulosics.

IV. REFERENCES

1. U.S. Forest Service, "The Outlook for Timber in the United States." Forest Resource Report 20 (1973).

2. U.S. Forest Service, "The Nation's Renewable Resources - an Assessment, 1975." Forest Resource Report 21 (1977).

3. Cliff, E. P., "Timber: The Renewable Material." U.S. Government Printing Office, Washington, D. C., 1973.

4. Bethel, J. S. and G. F. Schreuder, Science 191, 747 (1976).

5. Spurr, S. H. and H. T. Vaux, Science 191, 752 (1976).

6. Jahn, E. C. and S. B. Preston, Science 191, 757 (1976).

7. Auchter, R. J., Tappi 59, 4, 50 (1976).

8. Lowe, K. E., Pulp & Paper 51, 21 (June 30, 1977).

9. McKeever, D. B., "Woodpulp Mills in the United States." USDA Forest Service Resource Paper FPL-1, For. Prod. Lab., Madison, Wis. (1977).

10. Nellbeck, L., Svensk Papperstid. 77, 76 (1974).

11. Rydholm, S. A., Pulping Processes, Interscience Publishers, Inc., New York, N.Y., 1965.

12. Malinen, R. and E. Sjostrom, Paperi ja Puu 57, 728 (1975).

13. Lowendahl, L., Petersson, G., and O. Samuelson, Tappi 59, 9, 118 (1976).

14. Forss, K., Paperi ja Puu 56, 3, 174 (1974).

15. Johnson, D. C., M. D. Nicholson, and F. C. Haigh, Applied Polymer Symposia No. 28, T. E. Timell, Ed., John Wiley & Sons (1976).

16. Philipp, B., M. Schleicher, and I. Laskowski, Faserforsch. Textile Tech. 23, 2, 6 (1972).

Modified Cellulosics

RAYON FIBERS OF TODAY

G. C. DAUL AND F. P. BARCH

Eastern Research Division
ITT Rayonier, Inc., Whippany, N.J.

Pending the development of a better process for production of man-made cellulosic fibers, there is a revival of interest in the development of new and improved viscose rayon fibers to meet the needs of today's consumers. This is especially true in Europe which enjoys the benefit of a more enlightened I.S.O. definition of regenerated cellulosic fibers based on superior physical properties and permits use of the new name "MODAL." In the U.S., on the other hand, any regenerated cellulosic fiber which is not chemically modified, regardless of properties, is restricted to the historically maligned name, rayon.

While rayon yarn production for industrial uses has reached a plateau at a sufficiently high level of properties to produce the very finest quality radial tires and other reinforced articles, the development of new, better rayon staple fibers for textile uses continues. Durability and launderability have been markedly improved in rayons with a sufficiently high level of wet strength and modulus. The major remaining deficiencies of rayon vis a vis cotton (hand and cover) have now been attained by hollow- and crimped HWM-versions of rayon. In 50-50 blends with polyester, it is difficult to tell the difference between textile containing these new rayons and the finest combed cotton, in hand, appearance, and physical properties.

This paper is intended to illustrate some of the kinds of superior rayon fibers in commercial production today.

INFORMAL INTRODUCTION

My first association with the Anselme Payen honoree, Dr. Kyle Ward, was in 1945 when I joined the staff of Dr. J. David Reid's Chemical Properties Section of the Cotton Fiber Division at Southern Regional Research Laboratory in New Orleans. Dr. Ward was, at that time, head of the Division.

Among the first problems I worked on was to make cotton more absorbent and higher swelling, that is, more like rayon. We carboxymethylated cotton to all degrees of substitution possible, backwards, forwards, and sideways. We succeeded in making a partially carboxymethylated cotton fabric which was indeed highly swellable and went even further to make cotton yarn which would disintegrate in water. After several patents and publications on this subject, our research led through acetylation, aminoethylation, hydroxyethylation, sulfonation, phosphorylation, grafting with β propiolactone, flame retarding with tris(disbromopropyl) phosphate, crosslinking with amino resins and on and on. You name it and if it could be done to cotton, we tried it.

Dr. Ward was a constant source of encouragement and advice, a friend and counsellor.

Dr. Ward left SRRL in 1951 and I stayed until 1954 when I accepted a position with the newly opened Courtaulds Research Lab at LeMoyne, Ala. Oddly enough, among my first assignments was to make rayon less water absorbent, less swellable, i.e., more like cotton. The result of this research was the first U.S. commercial cross-linked rayon, Corval. Because of its distinctly different properties from rayon, Courtaulds applied for a generic name for this fiber. However, the proposed FTC generic definition of rayon had been so cleverly written that it would be almost impossible to get a new name for a regenerated cellulosic fiber in the U.S., regardless of superior physical properties. This definition tied all U.S. rayon producers to a name which was, unfortunately, associated with rayon fiber made during and after World War I, which, out of necessity, was applied to end-uses where it could not possibly perform satisfactorily.

Definition of Rayon

The U.S. definition of rayon, as adopted in March 1959 is as follows:

"A *manufactured fiber composed of regenerated cellulose, as well as manufactured fibers composed of regenerated cellulose in which the substituents have replaced not more than 15% of the hydrogens of the hydroxyl groups.*"

This generic definition of rayon was probably more responsible than anything else for discouraging the development of new regenerated cellulosic textile fibers and for the reduction of the research effort on rayon in the U.S.

Technology for the production of superior rayon yarns and cords for industrial uses and tire reinforcement has existed for many years. In fact, it is possible to produce rayon tire cords which are even stronger than required for radial tires which are expected to displace more and more of the bias- and bias-belted tires used in the U.S. The very highest quality radial tires for automobiles are now made with steel belts and rayon sidewalls. Rayon's superiority over nylon in creep resistance and over polyester in longterm adhesion to rubber is well known and gives it advantages over the two synthetics in this end-use. Many prophets of doom have forecast the demise of rayon tire cord over the years but have had to extend the date with each revival.

On the other hand, as a result of the FTC definition and labelling act, only a few new textile rayons have been marketed in the U.S. in the past few years, notably the HWM types, Avril and Fiber 700. Extensive advertising campaigns and promotions of these fibers were necessary to overcome the handicap of the name rayon. They were, however, accepted by the consumer for their superior properties over regular rayon and have become an important item in rayon production today.

Recently, however, a more pragmatic approach has been taken by the International Standards Organization (ISO) in which the U.S. is represented. This organization adopted the name Modal on Jan. 1, 1975 (ISO-2076), with the following definition:

"Regenerated cellulosic fiber with wet tenacity of at least 2.5 gpd at 15% maximum elongation." The polymer is designated as Cellulose II.

While wet modulus is not directly mentioned in this definition, it is adequately covered by the wet stress strain parameters, as can be seen in Figure 1.

Only a few presently produced, regenerated cellulosic fibers are reported to be eligible for the Modal name, Courtaulds' Vincel 64, Snia Viscosa's Koplon, and Nittobo Rayon's Forte FB, to name a few. These are either HWM modifier types, or modified polynosics and are sold in direct competition with cotton for 100% and blended end-uses. How some of the commercial fibers rank with relation to the Modal definition as tested by our research laboratory is shown in Figure 2.

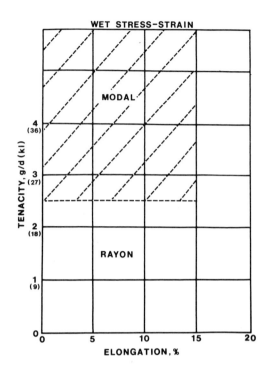

Fig. 1. *Wet Stress-Strain Areas Covered by Definitions of Rayon and Modal*

This breakthrough in terminology and acceptance of reality by ISO gives a renewed incentive to European rayon producers. Fibers with the name Modal demand a 6 to 7¢ premium over ordinary low wet modulus rayons. The Modal fibers are useful for blending with polyester or cotton and when properly processed and finished, fabrics have most of the same properties and even some advantages over their cotton counterparts. Dimensional stability to washing is no longer a problem with these fibers.

The earliest polynosic fibers had very low elongations and suffered from brittleness but these deficiencies have been largely overcome by modifications in the manufacturing process to give greater elongation and work to rupture. Toughness built into these Modal-type fibers, resists wet extensibility, thereby preventing progressive shrinkage and reducing markedly, relaxation shrinkage.

The ultimate target of scientists doing research on rayon and Modal textile fibers is obviously cotton for long-term

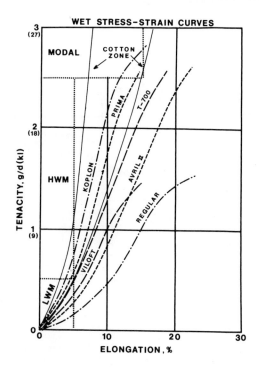

Fig. 2. *Wet Stress-Strain Curves of a Range of Cottons and Some Rayon and Modal Fibers*

economic and demographic reasons. According to United Nations statistics, (1,2) world-wide population growth is expected to reach 4.4 billion by 1980, a 400 million increase in 5 years and 4.8 billion by 1985, another 400 million, and to 6.4 billion by the year 2000. Demand will increase for cellulosic fibers in proportion to polyester as required for comfort in the temperate and tropical climates. (See Figure 3).

In 1975 per capita fiber consumption was 13.26 lbs. If per capita consumption grows at a rate of only 1.3% per year, total world fiber demand will reach 82 billion lbs. in 1985, and an astounding 135 billion in the year 2000 A.D. We expect 49 to 56% of this increase will be supplied by the noncellulosics and the rest from cotton and regenerated cellulosics. Therefore, 70 billion lbs. of synthetics and 65 billion lbs. of cellulosic fibers will be required in 2000 A.D.

The land area required for cotton to hold its present share of the market would go from 80 million acres to 164 million in just 25 years. None of the present cotton growing

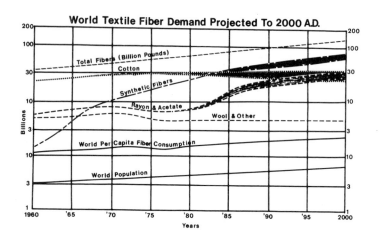

Fig. 3. *World Textile Fiber Demand and Population Growth Projected to 2000 A.D.*

nations of the world today will be able to <u>double</u> the present acreage for cotton when food will require a higher priority for the additional 2.4 billion more people. This means there will be a need for 10 billion lbs. of regenerated cellulose fibers by 1985 (double present capacity) and another 10 billion by 2000 A.D. Most of this increase will probably be in cotton-like high performance rayons.

The fiber data in this figure through 1975 are from U.N. published statistics. Projections are based on information available to ITT Rayonier's Market Development Department. All population data and projections are from U.N. reports.

While American production of HWM fibers at the present, is sold out, producers are agonizing over expansion of these fibers at the expense of regular rayon. The old standby is doing quite well in those markets which take advantage of its unique high water absorption properties, especially nonwoven disposables and commodities. On the other hand, several foreign countries are considering expansion of existing rayon plants or building new ones. Some of these are: Italy, Spain, Portugal, France, Nationalist China, South Korea,

Philippines, Egypt, Màlaysia, Japan, India and the iron curtain countries.

India, a large cotton producing country, is cutting back on cotton production in favor of food and considering production of HWM (Modal) type fibers as a direct substitute for cotton. Egypt has lost a large amount of arable land to the Aswan Dam and is looking to HWM (Modal) rayon to substitute for lost cotton production.

Rayon Variants

To meet specialized markets which have developed in the 60's and 70's, many rayon producers have introduced modifications of rayon which had been "on the shelf" waiting for markets to develop.

A. Flame Proof Rayons

As a result of the National Flammable Fabrics Act (3) several good-to-excellent flame retardant rayon fibers appeared on the market; Avisco's (Avtex) PFT, Diawabo's (Japan) HFG, Kanebo's Bell Flame, Lenzing's Flam Gard, to name a few. These fibers are made by incorporating a polymeric flame retardant agent in the viscose which, when spun, encloses the flame retardant with a cellulosic sheath which prevents it from leaching out during washing. However, as a result of the recent FDA and EPA rulings, potential rulings or even rumors of rulings against certain organic compounds, some of these fibers have been or are being withdrawn from production and new ones withheld until all decisions have been made and the dust settles. Even unfounded suspicions often expressed by spokesmen for the EPA are sufficient to deter production and to discourage research in this important area.

B. Water-Absorbent Fibers

While regular rayon is highly water absorbent, super-absorbency, increasing its capacity to hold aqueous liquids, can be built in by chemical modification, addition of water absorbent polymers to viscose prior to spinning, or other special production techniques. They have an advantage over the "super-slurpers" in that this absorbency is not always reduced by salts, such as those in urine.

Two U.S. made fibers, Avtex's P.A. and American Enka's Absorbit, are being test-marketed for the nonwoven disposables end-uses which require this super-absorbent property (diapers, sanitary pads, hospital incontinent pads, etc.).

Low substituted hydroxyethylated cellulose fibers, highly swellable in water but soluble in alkali, can be produced without the environmental problems of the viscose process. Some interest is being shown in Eastern European countries in this fiber (which oddly enough would have to be called rayon in the U.S.).

C. Others

Other rayon variants are being produced on a relatively small scale for specialty end-uses, high denier crimp as for wool blending, acid or differential dyeing, etc. One such fiber is American Enka's Enkalon, an acid-dyeable rayon suitable for differential dyeing effects or blending with other acid-dyeable fibers such as wool.

Avtex's crimped version of Fiber-40 came on the market in 1976. According to Avtex literature, the fiber would not be promoted as a cotton-like, or wool-like but emphasis would be placed on the "full, natural hand of fabrics made from Avril II". End-uses for this fiber are definitely aimed at the cotton market.

TWO NEW RAYONS

While the rayons described above have been around for some time, two brand new regenerated cellulosic fibers have been announced this year. Courtaulds' Viloft and ITT Rayonier-Snia Viscosa's Prima^R. Both fibers are aimed at direct competition with cotton.

The main properties of cotton which have eluded rayon researchers in the past have been "hand" and "cover". Whereas blending with polyester, and resin finishing techniques, have largely overcome the limpness of some rayons, up until recently, it has not been possible to build cover or opacity into a rayon fiber. The physical characteristics which nature has endowed upon cotton, the multiwalled, wound layers of cellulose, and the hollow lumen running the length of the fiber, are mainly responsible for the properties of hand and cover, crispness and opacity, commonly referred to as "cotton-like".

Two separate approaches have been taken to produce cotton-like rayon fibers.

1. VILOFT

The Courtaulds' fiber has a hollow center imitating the lumen of cotton. (Figure 4). It is less dense than ordinary rayons so there is more fiber in a pound and hence provides more cover to a fabric.

While precise details of the Viloft process are not known, a general description was given at the ACS meeting in New Orleans in March.(4) According to the authors, the fiber is made by adding a gas-forming agent, sodium carbonate, to viscose containing a modifier. The fiber physical properties

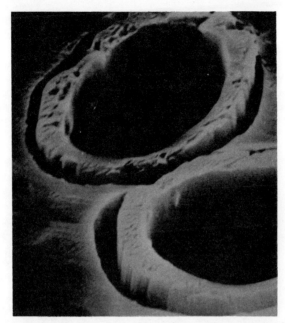

Fig. 4. SEM Cross-Section of Viloft Rayon Showing Hollow Core

are close to those of regular rayon but the wet modulus is somewhat higher and wet strength lower. For textile end-uses, this low wet strength would make blending with stronger fibers such as polyester or cotton almost mandatory.

The factors which contribute to its cotton-like hand and bulk are high torsional rigidity, low micronaire, and low effective density (more fibers per pound).

Courtaulds have sampled this fiber to over 130 spinners, knitters, converters, finishers and manufacturers or nonwovens, men's wear, ladies' wear, lingerie, etc. An impressive array of samples of fabrics, including wovens, knits, terry cloth toweling, have been shown. Viloft's 20-30% higher water retention makes it especially attractive for blends with polyester for comfort and for the absorbent nonwoven personal products market.

2. PRIMA

The fiber Prima^R, (5-8) researched by ITT Rayonier from 1970-74 and developed since then through pilot plant tests and plant production by Snia Viscosa, takes a different approach from Viloft.

Prima is a bi-lobal fiber with spiraling, which gives it a fine, low amplitude crimp, as many as 25-30 macro crimps/inch (or 10-12/cm). See Figure 5. The two lobes are seen

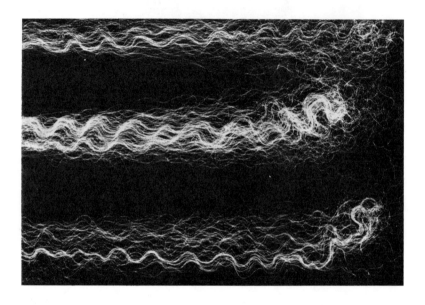

Fig. 5. Longitudinal View of Prima Showing Macro-crimp

in cross-section (Figure 6) with one having a thick skin and the other a relatively thin skin. The thin skin lobe allows easy penetration of dyes and cross-linking (durable press) resins while the thick skin lobe contributes to toughness. This tough skin has along its length, numerous microcrimps (wrinkles) which cannot be readily counted under light microscopy. The wrinkles which can be shown by SEM, resemble knee joints and probably contribute to the higher than normal conversion of fiber to yarn tensile strength which is typical of Prima. (See Figures 7 and 8).

Fig. 6. Cross-Section of Prima Showing Bilobed Thick Skin and Thin Skin Areas

Prima is a HWM fiber with minimum 0.5 gpd wet modulus at 5% elongation. The bi-lobal nature of the fiber and crimps within crimps contribute to adhesion, hand, and cover. It produces fabrics which are more luxurious than cotton in 100% and equivalent to cotton but cleaner and smoother in 50:50 blends with polyester.

Several types of fabrics - muslins, broadcloths, percale sheeting and print cloths have been made from 50:50 Prima: polyester blends which are virtually fully equivalent to similar fabrics made from combed cotton:polyester in tensile and tear strength, abrasion resistance, dimensional stability and appearance. The exception was that the Prima fabrics are cleaner, smoother and freer from neps and defects.

We do not claim that these two fibers, Viloft and Prima, will replace cotton in every market or on a tremendous scale, but they will surely do as good or better than cotton in selected end-uses. The <u>ideal</u> replacement may involve a merger of the two technologies, to produce a strong HWM rayon

Fig. 7. *SEM Longitudinal View of Prima Showing Bilobal Shape*

(Modal) with plenty of crimp, torsional rigidity, low micronaire and a simulated lumen. This should result in a <u>truly synthetic cotton</u>.

<u>REFERENCES</u>

1. United Nations Food & Agricultural Organization (1976).
2. Dykes, J.B. and Muller, T.E. World Textile Fiber Demand Projected to 2000 A.D. Presented at the SASMIRA Conference, New Delhi, India (Jan. 19-21, 1976).
3. National Flammable Fabrics Act. C.S. 191-53; Revised Dec. 14, 1967. FF 371- July 1, 1971.
4. M. Lane and J.A. McCombes. Courtaulds Challenge the Cotton Legend. 173'd American Chemical Society Meeting, March 1977.
5. Stevens, H.D. and Muller, T.E. Process for Producing High Performance Crimped Rayon Staple Fiber, U.S. Patent 3,720,743 (March 13, 1973).

Fig. 8. SEM Longitudinal View of Prima Showing Microcrimp

6. Daul, G.C. and Barch, F.P. High Crimp, High Strength Rayon Filaments and Staple Fibers and Process for Making Same, U.S. Patent 3,632,468 (January 4, 1972).
7. Daul, G.C. and Barch, F.P. High Crimp, High Strength Rayon Filaments and Staple Fibers, U.S. Patent 3,793,136 (Feb. 19, 1974).
8. Muller, T.E., Barch, F.P., and Daul, G.C. High Performance Crimped Rayon Fiber, Text. Res. J. 46, 184 (1976).

Acknowledgement

Appreciation is expressed to Eric Attle of Courtaulds N.A. for the photo of the Viloft SEM cross-section.

MODIFIED CELLULOSICS - AN OVERVIEW OF THE FUTURE

H. L. HERGERT AND T. E. MULLER

ITT Rayonier, Inc., New York, N.Y.

The forest represents a carbon source capable of providing food, clothing, energy, and shelter for mankind. It is equal in importance to petroleum, it is available to all but a few nations, and it is replenishable. The factors which influence the future development and growth of products based on cellulose vary from region to region. In a free market, of course, cost/performance is the prevailing consideration. Rayon, acetate, and other cellulosics provide comfort, aesthetics, and the desired physical properties in textiles and nonwovens, as 100% cellulose constructions and also blended with petroleum based synthetics. New, functional cellulosic fibers are being developed for both large volume and specialty applications. For the most part, these improved performance products will be made in existing plants. Hence, growth rates (with the exception of cellulose ethers) will not match those of the synthetics during the next five to ten years. Significant growth will require development of less energy- and capital-intensive processes. Major opportunities for application of modified cellulose lie ahead provided that there is a resurgence of cellulose R&D in industry and universities. There is some indication that this has begun in various parts of the world.

INTRODUCTION

The major worldwide end-use of crude cellulose is the production of paper and board. Purified cellulose (chemical cellulose or dissolving pulp) has been of major importance as a raw material for the production of man-made fibers and plastics during the present century. The objective of this overview will be to assess the likelihood of a continuation of the past growth pattern of chemical cellulose and whether the potential exists for a whole new era of opportunity for conversion of cellulose into many of the items needed by mankind in addition to paper.

In 1973-74 the world was presented with an unprecedented economic change in the form of a rapid increase in price of the basic energy source, oil. At the same time, virtually all of the world's industry was operating at full capacity to supply the needs of an overheated economy. A rapid rise in prime interest rate resulted in most major western countries reexamining inventory policy and capital additions. The net result of these three major factors was the subsequent recession in 1975-76. The production of manufactured products decreased, and with it, the concern for availability of raw material. Subsequently, a more rational economic order has emerged. The return to normal economic growth that we are currently experiencing has once again raised serious questions about the source of raw material that will be needed to accommodate a growing world population.

As the effect of OPEC decisions began to be felt around the world, there was a kind of temporary panic among planners as to the future availability of adequate raw materials for the booming fibers and plastics industry. Natural resource technologists saw this as an opportunity for converting their long held dreams into reality of a plastics and fiber industry based on a renewable resource. Judging by the number of preambles to renewable resource-based technical papers presented at scientific meetings in the last three years, this has virtually evolved into a creed. Insufficient thought has been given by renewable resource enthusiasts to the extremely modest proportion of the total oil consumption which finds its way into petrochemically-based products. For example, one tank full of gasoline in the average American car could theoretically supply the raw material requirements to manufacture sufficient man-made synthetic fiber per year for the average family of four in the U.S. The price of the tank full of gasoline before tax is roughly $5; in the form of polyester fiber, it is worth $75; in the form of finished textile garments, $500-$2000. It is difficult to visualize any type of national policy which would not ultimately protect the petrochemical industry with its high value added content.

Our projections for world chemical cellulose consumption are on the order of 4,500,000 tonnes for 1977. The major products which will be made from this cellulose are rayon staple fiber, continuous filament rayon, high tenacity rayon, cellophane, acetate fiber, mixed ester plastics, and smaller volume items such as carboxymethyl cellulose, etc. In addition to this, there are about 400,000 tonnes of cotton linters pulp produced worldwide annually, a significant portion of which also finds its way into some of the same end-use markets. To put this usage in the proper perspective, it is constructive to compare it with the worldwide use of petrochemically-derived materials.

In its newest forecast the Stanford Research Institute projects the "world" demand for petrochemically derived products as shown in Table 1.[1]

TABLE 1

World Demand Projections*
(Millions of Tonnes)

Products	Years			
	1975	1980	1985	1990
Plastics	45.0	73.0	100.0	126.0
Man-made Fibers	8.1	11.2	15.3	20.0
Synthetic Rubbers[a]	4.9	6.8	8.7	10.5
Nitrogen (fertilizer only)	45.4	60-64	75-83	92-105

* Source: Stanford Research Institute
a. Data do not include demand in the Eastern Bloc

These data do not include the demand in Eastern Europe, the U.S.S.R. and the People's Republic of China. If we add the probable needs in those countries, we foresee a world requirement of more than 80 million tonnes of plastics and 13 million tonnes of synthetic fibers by the end of 1980. In 1976, 3.47 million tonnes of man-made fiber were produced from cellulose worldwide and about 0.5 million tonnes of cellulosic plastics.[2] What are the prospects for significant inroads of cellulose-based products into the petrochemical-based markets? To answer this question, we propose to look at the major end-uses, problems associated with these uses and, particularly, the technology that will be needed to convert the hopes of the renewable resource advocates into reality.

RAW MATERIALS

The predominant raw material source for chemical cellulose is wood. Preparation involves delignification by a chemical process basically identical with that used in the preparation of chemical pulp for paper followed by removal of hemi cellulose by depolymerization and extraction, reduction of resin content, removal of heavy metal ions and careful control of the degree of polymerization (DP). Conifers are the preferred wood species since most dissolving pulp consumers have adapted the first stage of their process to use long fibers derived from coniferous species. Temperate hardwood species and plantation subtropical hardwoods, such as eucalyptus and acacia, are also usable, if appropriate steeping and shredding equipment are available, to make viscose rayon. Mixed tropical hardwoods, on the other hand, invariably would require significant adaptation of the end-use process to accommodate silica and other extraneous materials that usually accompany tropical hardwoods.

As to the current world dissolving pulp capacity, it is hard to provide an authentic figure. The principal reason for this is that virtually every dissolving pulp producer diverts some significant portion of his production into paper and specialty end-uses, such as absorptive products, automotive filters, latex saturating papers, fruit juice filtration, etc. It appears likely that 15-20% additional dissolving pulp could be produced worldwide from existing facilities without major additional equipment. If this is correct, world production capability, exclusive of cotton linters pulp, would be little over 5 million (dry) tonnes annually. Cotton linters pulp, on the other hand, could be diverted solely into chemical end-uses and this would add another 400,000 annual tonnes of productive capability.

If there were an increase of chemical cellulose usage equivalent to only 2% of the total worldwide textile fiber market per year for two years, i.e. 850,000 tonnes, it is obvious that all of the world's dissolving and cotton linters pulp capacity would be preempted for chemical end-use. One may then well ask, if there were a further sudden demand for chemical cellulose use, could it be produced? One possible answer to this question is that some wood pulp mills presently producing only paper pulp could be converted to medium alpha (a purity adequate for regular rayon) dissolving pulp production with relatively small capital investment if there were sufficient economic incentives.

In the case of a kraft paper pulp mill, this could be done by (a) installation of stainless steel digesters and circulation lines so that acidic hydrolysis could precede kraft cooking, (b) the addition of a bleaching stage to reduce cellulose intrinsic viscosity or DP, (c) provision of a suitable metallic ion reduction step, and (d) installation of roll-handling equipment. Alternatively, a separate prehydrolysis vessel might be used prior to the kraft stage, in which case the same kraft mild-steel cooking vessel could be used as is presently used for paper pulp production. However, dissolving pulp is not normally prepared in this latter fashion. In the case of a sulfite pulp mill, a major addition would have to be a pressure alkaline treatment stage for reduction of hemi cellulose. Metallic ion removal and roll-handling equipment would also be needed. In either process, the most demanding end-uses, i.e. high tenacity rayon, high wet modulus rayon staple, and certain acetate plastics also require a process step which involves extraction of pulp with "cold" mercerizing strength caustic soda solution to further reduce the hemi cellulose content of the final product. This does not appear to be an absolute necessity, especially if a rayon plant is equipped with adequate dialysis facilities. It may go without saying, but a chemical cellulose manufacturer also requires a much larger technical staff (quality control and research) than a paper pulp producer because of greater sophistication of product requirements and diversity of end-user processes.

Perhaps the greatest reason for pursuing this whole question of the role of cellulose as a raw material for fibers, plastics, etc. is the fact that there is an immense potential source of raw material more than adequate to supply all of the solid end-use products from petrochemicals. This presumes an adequate technology for conversion of the raw material to cellulose and, in turn, the cellulose into the needed final product. Such technology, of course, does not presently exist, but there is no inherent reason why it could not be developed. The raw material that we refer to is secondary fibers, i.e. waste paper and cardboard, forest residuals and agricultural byproducts such as wheat, straw, bagasse, corn stalks, etc. Total annual volume of this material in the U.S. alone is estimated to be well in excess of 300 million tonnes. Experiments recently conducted in our laboratories show that waste paper or board which contains any significant amount of groundwood (newsprint) cannot be converted to rayon or acetate by existing conventional processes unless new processes for regenerating fibers and plastics are uncovered. All of these residuals and byproducts would have to be submitted to a delignifying step before they could begin to be considered suitable for "dissolving" end-uses. Hemi cellulose (xylan and mannan) level would also have to be decreased significantly. Conversely, some new type of process to produce cellulosic fibers or plastics might be specifically developed to take advantage of these byproduct cellulosic sources.

FIBERS

World production of textile fibers is given in Table 2.[3] The data for 1976 show that synthetic fibers, mainly polyester, acrylics and nylon, now comprise 33% of the total world fiber production. Cellulose based man-made fibers are 13%. Cotton is 48%. Consumption of man-made cellulosics has declined slightly during the last seven years. Cotton consumption has been flat and man-made synthetics have grown very significantly.

TABLE 2

World Production of Textile Fibers*
(Thousands of Tonnes)

	1970	1973	1976
Rayon and Acetate[a]	3436	3661	3212
Synthetic	4700	7638	8598
Cotton	11782	13710	12502
Wool	1602	1425	1209
Silk	41	44	49

* Source: <u>Textile Organon</u>[3]
a. Excluding acetate cigarette filtration tow (256,000 tonnes in 1976).

How might these various types of fiber be expected to fare in the future? In 1974 during the peak of the raw material shortages--and oil scare--several scenarios were presented for the future textile fiber consumption in the year 2000 (Table 3).

TABLE 3

<u>1974 Scenarios</u> for Man-made Cellulose
Fiber Production in the Year 2000

Petrochemical Producers	~0
Mixed Fiber Producers	2.5 MM Tonnes
Chemical Cellulose Producers	15 MM Tonnes
<u>1977 Scenario</u> (Viscose)	3.5 MM Tonnes
(New Process)	3.5 + +

In one of these scenarios, two major worldwide petrochemical producers predicted the essential demise of the chemical cellulose industry suggesting that no man-made cellulose fibers would be produced. A European chemical company, with major productive capability for both man-made synthetics and cellulosics, forecast 2.5 million tonnes of man-made cellulosic production, i.e. 80% of the demand from the peak of the early 1970's. Optimistic representatives of a chemical cellulose producer, on the other hand, forecast 15 million tonnes. Now that three years have passed, we are beginning to cope with the future limited availability of petroleum and to see which, if any, of these forecasts might be correct. To answer this question, one must explore a number of factors which underlie forecasting future fiber needs: world population trends, further fiber properties needed, and capital availability or growth.

Projections by the United Nations show current world population at approximately four billion people. At a forecasted 1.9% per year compound rate, world population will be 6.4 billion in the year 2000 barring a catastrophe. Most of this population growth will occur in Latin America, Africa, East Asia and South Asia with year 2000 populations expected of 0.6, 0.8, 1.6 and 2.5 billion, respectively. How much fiber will that increased number of people require? Per capita fiber consumption last year was about 6 kg. Usage varies from about 2 kg. per person in third world countries to 27 kg. in the United States (Table 4). Since the current fiber usage in many developing countries is clearly inadequate by their standards and ours, a very modest doubling of that usage to 4-5 kg. per person, coupled with a correspondingly modest 1% compounded growth rate in the developed countries and the expected population growth in both sectors, results in a world fiber requirement of 62 million tonnes in 2000 A.D.

TABLE 4

Estimated Per Capita Consumption of Textile Fibers

1977

United States	27 kg.
All Developed Countries	18 kg.
Third World Nations	2 kg.
Whole World	7 kg.

Where will the 36 million tonnes of new fiber needed by 2000 A.D. come from? Production of the natural fibers, cotton, wool and silk, grew only 2.5% between 1970 and 1976, rayon and acetate declined slightly by about 0.6%, while synthetic manmade fibers grew a phenomenal 83%! Will the increased fiber requirement be supplied by the synthetics fiber industry? Or we might even ask, ought it to come from this source in light of the limited oil resource in the world? An examination of the situation in India may be instructive in this regard.

Early in 1976, a conference was convened in New Delhi to discuss the fiber needs of developing countries and how they might be supplied in the future.[4] A representative of the Indian fiber industry pointed out that the population of India, now 600 million, would be expected to grow to one billion within 25 years. Current fabric consumption is about 14 square meters per person, down 20% from consumption in the late 1960's. Even a return to the 1960's rate, grossly inadequate by Western standards, by the year 2000 would require more than twice the annual amount of fiber now produced and consumed in India. To supply this need as cotton would require more than half of the land now devoted to the growing of food crops. Obviously, this would be impossible since the food needs of this burgeoning population will probably require conversion of existing cotton growing land to food production.

According to some Western industrial representatives attending the meeting in India, an answer to the problem would be to import terephthalic acid and ethylene glycol for a large Indian polyester spinning industry which would need to be built. Since India is a net importer of oil, has balance of payment problems and imposes a high tariff on chemical imports, polyester is priced up to five times higher than cotton or rayon. In our judgment, the net result of trying to meet future fiber needs with polyester would be an even decreased availability of fiber to that segment of the population that would need it the most. Almost forgotten at this meeting until we raised the issue was another possible answer: significant expansion of the indigenous rayon industry. India has dissolving pulp and rayon spinning process and machinery know-how. The raw materials, wood, caustic soda, chlorine and sulfuric acid, although energy intensive, are not dependent upon oil. The product fits into the existing textile industry or can be optimized by blending with synthetics. Finally, there is a major underutilized subtropical wood resource which could be converted to high yield eucalyptus plantations with a new production of cellulose per hectare more than five times as high as cotton.

Indeed, why not consider rayon as the fiber to meet a significant portion of the increased fiber needs of the whole world 23 years from now? There are a number of important reasons why man-made cellulosic fibers should be considered with renewed interest. First, as already noted, oil is a resource that cannot be replenished. Some synthetic fiber intermediates are usually based on natural gas which is subject to the vagaries of politics tied in with the energy crisis. Second, fibers made from petroleum lack an important comfort factor, the ability to absorb and transmit moisture. This is especially important in warm, humid climates where the forecasted population growth is the largest. Third, the need to feed more and more people creates increasing competition for prime land over the whole world, not just in India. This results in decreased availability of cotton with concomitant increases or, at least, the potential for wild gyrations in the price of this fiber. And, fourth, significant recent advances in man-made cellulosic fiber manufacturing and application technology have opened new, potentially profitable horizons for an industry that was born before the turn of this century.

Let us take a look at these developments which have been satisfied recently or are receiving attention to assure the growth of the principal man-made cellulosic fiber, rayon.

(1) New dissolving pulp manufacturing capacity.
(2) Automated, continuous viscose/rayon production with reduced labor content.
(3) Reduction or elimination of pollution.
(4) High performance rayon staple fibers that are every bit as good as cotton, and may be preferred substitutes.
(5) Fabrics based on such rayons, either 100% or blended with synthetics.

The first point, new dissolving pulp capacity, is beyond the scope of this overview, but it can be pointed out that chemical cellulose capacity has been increased. In addition, more economical, less capital intensive processes for producing wood cellulose are under development.

The viscose rayon process used to be such a complex series of batch operations that it is not hard to see why automation, streamlining, had to come to maintain profits in the industry. Today, this process can be completely continuous and largely automated.

Significant advances have been made in cleaning up the viscose/rayon process. Not only have automated, continuous manufacturing practices tightened up the system, but efficient chemical recovery techniques have been designed and put into operations. Rayon product development, both fiber and fabric, has not lagged behind process advancements in the last few years. Fibers with suitable properties can be made into attractive fabrics with excellent wear and wash-and-wear characteristics. This allows the development of broad markets for rayon.

Second generation rayon staple fibers, call them HWM or Modal or polynosic or by any other name, are manufactured in France, Italy, Austria, the United Kingdom, Canada and the U.S. Lenzing's 333, Snia's Koplon, Courtaulds' Vincel 64, Avtex's Avril, Enka's 700 and others are second generation rayons. Today new, third generation rayon staple fibers are being introduced worldwide which yield cotton-like fabrics, either 100% rayon or blended with synthetics, such as polyester. Prima(R), manufactured commercially by Snia Viscosa, Italy,[5] Avril II of Avtex[6], Viloft made by Courtaulds[7] are examples of such fibers. Prima and Avril II are crimped high wet modulus rayon staple fibers, while Viloft is a hollow filament rayon.

When Prima rayon is substituted for cotton, the uniformity of the rayon fiber means fewer yarn breaks. Carded yarn mills can produce combed qualities at considerable savings. And for all cotton mills, the use of rayon helps eliminate harmful cotton dust and costly waste. For the making-up and retail trades, Prima offers the quality and glamour necessary for satisfactory mark-ups. And for the consumer, a third generation rayon can provide luxury, comfort, style and color in tough, low-shrinkage apparel and textile products for home and industry. Prima performs well blended with synthetics, such as polyester, and can upgrade lower quality cottons in rayon/cotton blends.

Rayon does not, in fact, compete with synthetics; in blends with synthetics, it provides the necessary comfort factor. As a cotton replacement, rayon means:
- no waste, compared with about 10% waste with cotton
- no need for scour and bleaching, an energy saving
- easier dyeability, also an energy saving
- elimination of ginning, an energy saving
- elimination of combing, a saving in energy and waste
- elimination of mercerizing (energy and chemical savings)
- wide range of deniers, yielding a broader range of goods than cotton, e.g. woolen worsted and modified spinning systems
- greater moisture pick-up than cotton
- greater ability for business planning through price stability and property uniformity

One of the major markets for rayon staple fiber is in nonwovens. Over 80,000 tonnes, or close to 50% of the U.S. rayon staple fiber production, is used in disposable nonwoven products, principally as cover stock for diapers and sanitary products. Nonwovens are expected to grow at a rate of 8-20% annually, depending on the process and application. Rayon provides the necessary moisture absorbency or, when resin finished, moisture transmission characteristics in carded webs, needle-felt fabrics, wet laid nonwovens and even as spunbonded materials. Major growth areas for nonwovens are disposable diapers, wipes, sanitary and medical products, filter materials, carpet components, coated fabric substrates, home furnishing fabrics and interlinings. While fluffed wood pulp is technically not a modified cellulose, it should be mentioned here. The market for this type of specialty cellulose, employed as the absorbent component in many disposable products, is expected to grow at comparable rates.

There are a number of interesting specialty applications for rayon or modified rayon fibers. For example, porous and hollow rayon fibers are used for insulation and as membranes for dialysis in artificial kidneys. Aldehyde groups, introduced by oxidizing rayon, bring about increased wet strengh when such fibers are used for strong, specialty papers. Rayon is one of the substances used commercially to prepare carbon fibers by pyrolysis. For textiles, an almost infinite variety of properties can be introduced, of course, by topical textile finishes.

Modified rayon and wood pulp fibers with very high moisture absorptivity have been developed recently to be used primarily in diapers, sanitary and medical applications. Carboxylated polymers added prior to spinning or grafting with acrylonitrile followed by hydrolysis can bring about this characteristic in the case of rayon while a low level of etherification or grafting can be employed to produce highly absorbent wood cellulose fibers

So far we have said little about non-textile applications of rayon fiber. High tenacity filament continues to find substantial industrial applications, such as a reinforcing agent in industrial rubber goods, i.e. belts, hoses, etc. In spite of many forecasts for total displacement by synthetics, rayon tire cord still continues to be used in many of the premium radial tires manufactured in Western Europe. It is also used in almost all types of tires produced in Eastern Europe, the U.S.S.R. and India. Rayon cord is not subject to flat-spotting or to catastrophic failure noted with certain synthetics. Unfortunately, it does not have the "glamour" appeal needed in marketing. We believe that rayon tire cord usage will continue at least at the present levels and might even grow modestly depending upon the marketing strategies of radial tire manufacturers and, of course, whether it will be available in the needed quantities.

The capital investment for a new, modern viscose/rayon staple fiber plant with all the required pollution control systems is high, about \$2.50/kg. of annual fiber capacity.[8] This means an investment which is about 40% higher than that needed to build a polyester staple fiber plant of comparable size. In addition, it can be calculated that the total energy required to manufacture rayon is 2.8 kg. of oil equivalent per kg. of rayon staple fiber, compared to 2.6 kg. of oil equivalent per kg. of polyester staple fiber. So, although rayon is not a petroleum-based fiber, the energy demand to manufacture rayon is not low. For this reason, it is imperative to develop a non-viscose system for producing regenerated cellulose fibers (Table 5). The requirements for this process should include:

- a process with lower capital intensity, equal to or lower than polyester starting from DMT and glycol,
- an automated, low labor-content system,

- a solvent which allows a simple dissolution/filtration/dry spinning sequence,
- spinning speeds of at least several hundred meters/minute,
- a closed system with simple, efficient recovery of chemicals,
- a product--rayon--with cotton-like properties.

TABLE 5

Requirements for a New Regenerated Cellulose Process

1. Capital no higher than polyester from flake plant.

2. "Dry"-spinning, hopefully. If wet-spun, water requirements must be less than half of viscose rayon system.

3. Low energy.

4. Product must be equivalent or better than current high wet modulus rayon.

Currently there is considerable activity around the world, in industrial centers of research and at universities, to develop new solvent systems for cellulose. For example, the Case Western Reserve University of Akron, Ohio, presented a paper at the American Chemical Society Meeting in New York in April, 1976, on hydrazine, which, it was claimed, can dissolve cellulose at elevated temperatures and pressures.(9) Hydrazine, of course, is not an easy chemical to work with.

Scientists from Rayonier laboratories presented several papers at the Spring 1977 American Chemical Society meeting in New Orleans which dealt with the dissolution and spinning of non-viscose rayons in solvent systems. (10) It was shown that cotton-like rayon fibers can be produced by spinning from a dimethylformamide/dinitrogen tetroxide/cellulose system. Such non-viscose rayons had a conditioned tenacity of about 3 g/d, comparable to that of regular viscose rayon staple fiber, and a wet modulus (wet tenacity at 5% elongation) of several times higher than that of regular rayon. Unfortunately, the chemical recovery of this system is very complicated, so that this part of the process still requires too high a capital investment.

Dimethylsulfoxide/paraformaldehyde/cellulose is another system under investigation. There are others, perhaps none perfect as yet. But a clear sign of vitality in this area is that new solvent systems for cellulose are being investigated at about a dozen places around the world. One should keep in mind, however, that it takes about eight years to bring any new fiber from the laboratory to the marketplace, so improvements in the viscose rayon process and product areas are certainly timely while solvent spun rayons are being developed. We feel confident that new, non-viscose rayons will eventually come into being.

Man-made cellulosic fibers are, of course, also made by esterification to the triacetate or diacetate derivative. Both can be spun to a very fine denier which is aesthetically pleasing and suitable for a variety of textile applications. Acetate fabrics are widely used for apparel, e.g. blouses, shirts, dresses. Brushed fabrics are also popular today for robes. There is a noticeable shift, especially in the U.S., from tricot-type knits to woven, more stable acetate goods. Nylon/acetate blends, using either diacetate or triacetate, have gained increased acceptance. Such fabrics also tend to be stable and have adequate abrasion resistance and strength characteristics. Another significant application for acetate fabrics is in men's suit linings. Cellulose triacetate, such as Celanese's Arnel, Courtaulds' Tricelon, Rhone Poulenc's Rhonel, Deutsche Rhodiaceta's Tri-a-faser, though more expensive to produce than diacetate, generally have better dimensional stability, improved color fastness in dyeing and increased abrasion resistance. While diacetate textile continuous filament has a number of markets in which it makes a good "fit," long term growth appears to be confined to its use as "tow" for the manufacture of cigarette filters. This end-use is expected to continue to grow with the world GNP.

There are also substantial R&D needs in the field of acetate fiber production, but most acetate producers seem to devote relatively greater effort to product and process improvement than those who manufacture rayon. Various additives have been proposed to produce cellulose acetate fibers with improved or specialized physical properties; better dye affinity, antistatic and antimicrobial properties, fibers with permanent crimp made of mixed polymer systems, etc. Stronger diacetate fiber with no shrinkage upon washing might expand usage in textiles, but there is considerable debate as to whether desirable aesthetics (elongation, softness, luxurious hand, etc.) can be retained. Flame retardant acetate and rayon fibers have been produced commercially but industry was forced to go back to the drawing board in this area because some flame retardant additives have been implicated as carcinogens.

For long-term expansion, it would be desirable to have a process for direct acetylation to the diacetate without generation of acetic byproduct. The ability to use less purified (lower priced) wood pulps without a corresponding deterioration in filterability and spinning stability is also a desirable R&D objective. Work of this type is in progress in our laboratories and will be reported later this year.

Probably the most serious immediate deterrent to rational planning for and supplying of the future needs of fiber is the economic situation prevailing throughout the fiber industry of the developed countries. All fibers are grossly underpriced and probably represent the best bargain that today's consumer has in the marketplace. Polyester, nylon and acetate filament are essentially the same price as they were in 1973, but the cost of raw materials has doubled, labor has increased 40%, and energy has tripled (Table 6).

TABLE 6

Price Index for Fiber and Raw Material 1973 - 1977

Base Year = 100

Caprolactam	212
Nylon Filament	105
DMT	145
Polyester Filament	85
Wood Pulp	195
Acetate Filament	105

The net result is a virtually profitless fiber-producing industry with little expectation for improvement during the next several years. Rayon pricing is also caught between the vise of synthetic fiber pricing and government subsidization of cotton. The low or non-existent profitability means that capital cannot be accumulated for future expansion or the necessary R&D for fiber improvement. Reasons for this situation are complex. Basically they are a consequence of gross polyester overcapacity resulting from overoptimistic projections of fiber growth demand during 1970-74 and unanticipated changes in western lifestyle. The influence of fashion has declined, as illustrated by the wide acceptance of the denim look. Textile products have an increased longevity. These factors, among others, have resulted in lower overall demand for textiles in the U.S.A., Western Europe and Japan. While a forecast published in 1972 projected per capita U.S. textile fiber consumption at 30 kg., the actual figure proved to be only 25 kg. On the other hand, world population growth is inexorable. If the developed countries can restrain their impulse to build further synthetic fiber and intermediates capacity for a few years, a return to normal growth and profitability might be anticipated by 1980-81.

FILM

Regenerated cellulose for film (cellophane) continues to show a decline in most parts of the world because of competition from non-cellulosics. Cellophane is made by the viscose process and differs from rayon fiber production primarily in the use of more "stressed" compositions, e.g. higher cellulose and lower caustic content in the viscose. Modern plants are characterized by continuous steeping and xanthation units. The final product is coated with cellulose nitrate or a whole variety of synthetic polymers to achieve specified physical and moisture barrier properties. We estimate the current world production to be about 495,000 tonnes annually.

Since cellophane is made by the viscose process, it has many of the same problems as the rayon fiber industry. Effluent and emission control are difficult and will eventually need to be upgraded in those countries which still do not have strong environmental standards. In the United States, strong competition is being exerted by polypropylene film. Polypropylene has the lowest raw material costs of all synthetics useful for film production. This is likely to be the case indefinitely even if oil prices are doubled over the next decade. If a lower cost process for cellophane production were available, the future would be much brighter. We have, for example, made regenerated cellulose films in our laboratories using $DMSO/N_2O_4$ as a solvent. Markedly superior film clarity and improved tensile properties were noted. While the capital requirements of this system are not better than the viscose process, these results demonstrate the technical feasibility of an alternative process.

Cellulose is also used in the production of diacetate and triacetate-based film and sheets. Photography is a major end-use. Triacetate is usually cast as a film from solvent, a typical mixture being a mixture of methanol and methylene chloride. Accurate world production/consumption figures are virtually impossible to estimate since much of the production is captive. We believe that total annual production of cellulose ester plastics is about 225,000 tonnes, and of this amount, about 60% is used for film and sheets. Polyester is the major competitive threat in this market.

PLASTICS

Use of cellulose in plastics, other than as a filler or reinforcing agent, requires derivatization. Only the organic esters, cellulose acetate, cellulose acetate butyrate and cellulose acetate propionate, and the inorganic ester, cellulose nitrate, have found their way into this end-use. Total world production of the organic esters is less than 1% of that of the petrochemical-based plastics.

Cellulose nitrate is used mainly for industrial lacquers and, during periods of conflict, for explosives. Cellulose acetate and the mixed esters are transparent thermoplastics which have good mechanical properties and excellent clarity. Top of the line products require cotton linters since even traces of hemi cellulose and wood resin usually present in wood pulp result in some loss of clarity and color stability during molding. The organic esters are used for laminating and coating film, powder coating, production of toys (shatter resistance is an important property in this end-use), windscreens for snowmobiles, personal household items (toothbrushes, etc.), automotive steering wheels, tools, eyeglass frames, permeable membranes (kidney dialysis, reverse osmosis for desalinization, etc.) and a host of other products.

Growth in use of organic esters of cellulose is limited by their relatively high production cost compared with hydrocarbon thermoplastics, such as polyethylene or polypropylene or even some of the acrylics. Technically, the major problem is that cellulose is a highly hydrogen-bonded crystalline solid, and all reactions involve two or more phases with very slow reaction rates. The net result is excessive byproduct formation and non-uniform substitution because the various hydroxyl groups have different reactivity and accessibility. The discovery of a convenient, homogenous reaction medium should be a major objective of future R&D. In this regard, it is important to note that the whole plastics area is a major end-use in which cellulose barely participates. This end-use will continue to grow, and major cellulose R&D ought to be directed toward invention of reactions with reduced byproduct formation, more efficient conversion to the intended product and non-aqueous systems for elimination of effluent. If we may offer an editorial comment, surely this type of research is bound to be more productive than the preoccupation many academic cellulose scientists currently have with trying to unravel the nth degree of physical structure of cellulose in its native state.

CELLULOSE ETHERS

The principal commercial cellulose ethers include sodium carboxymethyl cellulose (CMC), carboxymethyl hydroxyethyl cellulose, hydroxyethyl cellulose, ethyl cellulose, methyl cellulose and hydroxypropyl cellulose. Annual production capacity in the United States is about 80,000 tonnes. World production data are not available, but we estimate them to be about $2\frac{1}{2}$ times this figure, e.g. 200,000 tonnes. Cellulose ethers

are primarily used to impart specific physical properties to
the final product: soil suspension, antiredeposition (in
detergents), binding, film forming, thickening and sizing.
The most important ethers in terms of volume usage is CMC,
which is made by the reaction of alkali cellulose (linters for
high viscosity grades; wood pulp for lower viscosity) with
chloracetic acid. CMC is used in the manufacture of ice
cream, paint thickeners and leveling agents, hair shampoo, oil
well drilling fluids, pharmaceuticals, tobacco substitutes,
detergents, textile warp sizes and a host of other products.
Cellulose ethers were used in the past for paper coating but
have been largely displaced by modified starch.

An area of particular interest because it represents a
potentially huge volume usage is enhanced oil recovery(EOR).[11]
Polysaccharides are one of three classes of chemicals required
for EOR for microemulsion flooding (the others being sulfon-
ates and alcohols). The polysaccharides are usually xanthan
gums which are thought to be more suitable than chemicals
based on wood cellulose because of the higher molecular weight
of the gums. The fact is, however, that neither these polysaccharides nor the competitive, cheaper polyacrylamides have
the required cost/performance characteristics. Derivatives
(ethers) of cellulose could penetrate this market if a family
of such materials could be developed with the required combination of properties. The important property requirements are
high molecular weight, water solubility or extensibility, and
low sensitivity to salts and high temperature shear. Polyacrylamides cost $2.20-3.50/kg. and polysaccharides are $4.40-
5.50/kg. It is generally assumed that, without a quantum
breakthrough, this market will be equally divided between
polysaccharides and polyacrylamides. The size of the potential total market for EOR polymers is significant. It is projected that 172 MM barrels oil per year (472,000 barrels per
day) will be recovered through EOR by 1989. A ratio of 0.4-
0.5 kg. polymer per barrel of oil is generally assumed.

Our comments under plastics regarding needed R&D apply
equally to cellulose ethers. In the reaction of ethylene
oxide with alkali cellulose to form hydroxyethyl cellulose,
for example, 35% or more of the applied ethylene oxide ends up
as ethylene glycol or homopolymer which must be washed out of
the product. Similar problems exist in the preparation of
methyl, ethyl, propyl and carboxymethyl ethers. A homogenous
reaction medium and a method for retaining the starting D.P.
of cellulose would go a long way towards reducing the production costs and expanding the use of cellulose ethers.

CONCLUSION

Forecasting the future of modified cellulosic products, taking into account all the factors we have already noted, requires a crystal ball with a greater amount of clarity than the one we presently possess (perhaps it needs to be made from Grade 1 cellulose acetate plastics!).[12] Nonetheless, we offer our assessment of the growth potential of the various product lines as shown in Table 7.

TABLE 7

Growth Forecast for Modified Cellulosics

Average Annual Percent Change

1977 - 1982

Acetate Plastics	2 - 3
Acetate Fiber (West)	1
Acetate Fiber (Eastern Bloc)	7
Regular Rayon Staple (West)	1 - 2
Regular Rayon Staple (Eastern Bloc)	1 - 3
High Wet Modulus Rayon Staple	15 - 20
Cellulose Ethers	3 - 5

We believe that a good future can be anticipated for the use of cellulose in man-made fibers, films and plastics. Much of the growth will be dependent upon convincing the world's scientific establishment that more of their time ought to be allocated to this field than is presently the case. In particular, there needs to be recognition that the tree is nature's solar energy chemical factory and is, therefore, a much worthier subject for R&D exploitation than raw materials with a finite availability such as oil.

REFERENCES

1. Chem. Economics Newsletter, Stanford Res. Inst., CA (May-June, 1977).
2. Childs, E.S., Chem. Engineering, p.163 (Sept. 12, 1977).
3. Textile Organon 48, No. 6 (June, 1977).
4. Int. Conference on Man-Made Fibres for Developing Countries, New Delhi, India (Jan. 19-23, 1976).
5. Muller, T.E., Barch, F.P., and Daul, G.C., Text. Res. J., 46, 184 (March, 1976)
6. Daily News Record (March 15, 1977).

7. Lane, M. and McCombes, J.A., "Courtaulds Challenge the Cotton Legend," Paper Presented at the 173rd American Chem. Society Meeting, New Orleans, LA (March, 1977).
8. Muller, T.E., "Man-Made Cellulosic Fibers," Paper Presented at Comite International de la Rayonne et des Fibres Synthetiques, Paris, France (Oct. 14, 1976).
9. Litt, M., "A New Single Component, Volatile Solvent for Cellulose," Cell Div. Preprints, American Chem. Society Meeting, New York (April, 1976).
10. Turbak, A.F., Hammer, R.B., Davies, R.E., and Portnoy, N.A., "A Critical Review of Cellulose Solvent Systems," Paper Presented at the 173rd American Chem. Society Meeting, New Orleans, LA (March, 1977).
11. Gulf Universities Research Consortium Report No. 159, "Chemicals for Microemulsion Flooding in Enhanced Oil Recovery" (Feb. 15, 1977).
12. Hergert, H.L., Appl. Polymer Symp. No. 28, 61 (1975), John Wiley & Sons, Inc.

PART 3 Cellulose Accessibility and Reactivity

DETERMINATION OF ACCESSIBILITY AND CRYSTALLINITY OF CELLULOSE

L. C. WADSWORTH AND J. A. CUCULO

Department of Textile Chemistry
North Carolina State University
Raleigh, North Carolina

The determination of cellulose accessibility and crystallinity has been extensively reviewed. The various techniques available for estimating the proportions of ordered and disordered regions have been summarized into three general categories: physical methods, chemical swelling methods, and non-swelling chemical methods. Of the physical methods, X-ray diffraction and infrared spectroscopy have received a large proportion of the attention of previous investigators. Only crystallinities of a certain size will contribute to the X-ray diffraction maxima. The infrared crystallinity ratio, derived from the relative absorbance of bands in infrared spectra, has been used for estimates of degrees of order in cellulose. Two sets of band ratios, the a_{1429} cm^{-1}/a_{893} cm^{-1} and the a_{1429} cm^{-1}/a_{2900} cm^{-1}, have shown good correlation with the X-ray crystallinity index. Sorption studies have also proven useful for estimating cellulose order. The percent accessibility has been derived from the maximum vapor moisture regain of cellulose samples. Deuteration techniques have also been applied for estimating accessible regions in cellulose. The close correlation between moisture sorption and D_2O vapor exchange strongly indicates that the two methods measure exactly the same quantity. Iodine and bromine sorption have been utilized to varying degrees of success for the estimation of cellulose accessibility. The various techniques give quite different estimates of order and disorder in cellulose. However, a definite pattern is shown for the fibrous celluose samples and the following qualitatively indicates the degrees of crystallinity of celluloses in descending order, cotton > wood pulp > mercerized cotton > regenerated celluloses.

INTRODUCTION

In any chemical reactions involving cellulose, the accessibility of the cellulose molecules for reaction is a major consideration. Before any products can be formed, the reactants must first be brought into contact with each other. With cellulose, this process is complicated by a two-phase morphology of crystalline (ordered) and amorphous (disordered) regions. Most reactants penetrate the amorphous regions only, and it is in these areas of disorder and on the surfaces of the crystallites that the majority of reactions occur, leaving the bulk of the intracrystalline cellulose unaffected. It has been established that water molecules, for example, penetrate into the amorphous regions only, since x-ray diffractograms have shown the intramolecular spacing of the crystalline regions to be unaltered during swelling with water (1,2,3). Furthermore, since cellulose cannot normally be dissolved in the reaction media, most reactions with cellulose proceed at least initially under heterogeneous conditions, i.e. at interfaces between two phases.

Before continuing, the relationship between crystallinity and accessibility warrants further clarification. According to Warwicker, <u>et al</u>. (4) "crystallinity is concerned with the solid state and accessibility with the internal volume accessible to a given reagent; it is only under very special circumstances that these terms can be directly correlated. As a general rule, accessibility is of more importance to the chemical properties of the fiber and crystallinity to the physical properties."

METHODS OF DETERMINING CRYSTALLINITY AND/OR ACCESSIBILITY

Many techniques for determining the proportions of ordered and disordered cellulose have been developed and are compared in Table 1 (4,5). These methods can be divided into three main groups (4):

1. <u>Physical methods</u>, such as x-ray diffraction, infrared absorption, and density which generally measure some function of the order or disorder of the sample.

2. <u>Chemical swelling methods</u> which measure the portion of the polymer that is accessible to a given reagent, of which there are two subgroups: absorption methods such as moisture sorption and iodine sorption which do not react chemically with cellulose, and secondly, the chemical reaction methods such as

Table 1
Average ordered fraction measured by various techniques (4,5)

Technique	Cotton	Mercerized cotton	Wood pulps	Regenerated celluloses
Physical				
X-ray diffraction	0.73	0.51	0.60	0.35
Density	0.64	0.36	0.50	0.35
Absorption and Chemical Swelling				
Deuteration or moisture regain	0.58	0.41	0.45	0.25
Acid hydrolysis	0.90	0.80	0.83	0.70
Periodate oxidation	0.92	0.90	0.92	0.80
Iodine sorption	0.87	0.68	0.85	0.60
Formylation	0.79	0.65	0.75	0.35
Non-swelling Chemical methods				
Chromic acid oxidation	0.997	*0.66–0.60	—	—
Thallation	0.996	*0.69–0.42	—	—

*Mercerization followed by solvent exchange

formylation and acid hydrolysis. Since the reagents swell cellulose, disrupt the interchain bonding, and to some extent penetrate the molecular structure, these chemical swelling methods give in effect a "semi-quantitative" measure of the degree of disorder.

3. <u>Non-swelling chemical methods</u> which utilize agents with substantially no ability to disrupt hydrogen bonds or thereby penetrate between the cellulose chains, but instead interact with cellulose at existing surfaces. Thus non-swelling methods primarily measure the amount of accessible surfaces such as external surfaces of crystallites, plus surfaces of internal voids, capillaries, and fibrillar structures, and do not measure any function of the degree of disorder of the cellulose polymer. This group can likewise be divided into two sub-groups: absorption methods and chemical reactivity methods with nitrogen sorption and the thallation technique being examples of each, respectively.

Warwicker, <u>et al</u>. (4) further noted that in considering these methods, four factors must be kept in mind. First, the values have limited utility in that they separate the polymer into two distinct components, i.e. crystalline/amorphous, inaccessible/accessible, or ordered/disordered. This is now generally recognized to be an over-simplification in that a range in degrees of order is thought to exist in the structure of cellulose. Secondly, it was noted that there is a clear distinction between disorder and accessibility in that the fraction of the cotton accessible to a given reagent depends not only upon the state of order of the cellulose chains, but also on the properties of the reagent and on the reaction conditions. Thus a considerable fraction of the cellulose structure which can be defined as disordered in some respects may not be penetrated and may be termed as inaccessible to a particular reagent. Generally, the greater the swelling ability of a chemical reagent, the greater the fraction of disordered material available to it, and furthermore, if the swelling power is sufficiently strong as in mercerization, even the highly ordered regions become accessible. Third, a wide variation of results is obtained from the literature because a given technique, as applied by different workers does not always give a similar value on the same type of cellulose due to the fact that different workers often employ different definitions, interpretations, and modifications of a technique, and these variations are coupled with normal experimental uncertainties. Fourth, it was noted that many modern workers attempt to characterize the order/disorder of the cotton in terms of a lateral order distribution diagram in preference to single value method. The characterization of cellulose by the lateral order distribution method will be covered in greater detail in the next section.

From the values of the average ordered fraction (crystalline fraction) in Table 1, it is obvious with few exceptions that the physical, chemical, and sorption methods yield results that rank cotton as being the most ordered followed by wood pulp, mercerized cotton, and lastly regenerated cellulose. However, there is considerable variation in the average ordered fraction of each type of cellulose. Generally, deuteration and moisture regain resulted in the lowest determinations of crystalline fraction with all of the cellulose types; however, water can be absorbed on the surfaces of crystallites, and can penetrate further into the more ordered regions than corresponding liquid systems. Tripp (5) noted that the size of the volume element as "seen" by one technique may differ considerably from those observed by other methods. For example, x-ray diffraction will be sensitive only to well-ordered regions above a minimum size, whereas, density methods will be responsive to elements of intermediate order as well.

The two non-swelling chemical methods both indicated crystallinity values of greater than 99 percent for cotton. Thus these values are not measures of order, but instead are related to the total surface area of cotton. On untreated cotton, the accessible surface to non-swelling agents is clearly very small, but certain treatments such as mercerization with sodium hydroxide followed by solvent exchange has been shown to drastically increase the surface area. For example after alkali swelling/solvent exchange, thallation and chromic acid oxidation techniques indicate that the percent accessibility increased from 0.4 to 0.3 up to 58 and 40 percent, respectively (4).

Forziati, et al. (6) found that by mercerizing cotton, the specific surface area as measured by nitrogen absorption decreased from 0.4 m^2/g for mature Locklett cotton and 0.8 for immature Memphis cotton to 0.2 m^2/g with the two cottons.

However, when both the untreated and mercerized cottons were swollen in water followed by solvent exchange with anhydrous methanol and pentane, and drying the area available to nitrogen increased to 25 and 34 m^2/g for the unmercerized mature and immature cottons, and increased considerably to 100 and 148 m^2/g, respectively, with the mercerized cottons. Since mercerization has been shown to result in increased sorption of moisture and alkalis, as well as reduce the crystallinity of cellulose (4), the greatly reduced sorption of nitrogen after mercerization alone was quite puzzling. In additional experiments, the specific surface of cellulose hydrolyzed with hydrochloric acid (hydrocellulose) was found to increase from 0.6 m^2/g in the air-dried state to 126 m^2/g when it was swollen in water before solvent exchange and drying. When the hydrocellulose was subjected to solvent exchange and drying without pre-swelling with water, however, the specific surface was only 6 m^2/g. Thus the water treatment was thought to play an important role in the development of the larger surface area by separating the close-fitting crystallites for most of their surfaces to be available for nitrogen sorption.

A. Lateral-Order Distribution

The lateral-order distribution of the ordered regions of cellulose has been investigated by gradually increasing the temperature or the concentration of the swelling agent and measuring the effect of the swelling treatment on the ordered region. Aqueous sodium hydroxide in a range of concentrations has been found suitable for this type of study. The transition range is determined in terms of such cellulose properties as the proportion of cellulose II, the LODP, or the amount of disordered cellulose, usually measured as a function of sodium hydroxide concentration. The amount of cellulose being modified at a particular concentration is proportioned to the slope of the curve relating the cellulose property to the alkali concentration. A curve which is related to the lateral order distribution of the sample is obtained by plotting this slope against the alkali concentration. It is interpreted that the higher the alkali concentration at the peak of this lateral order distribution curve, the greater the perfection of the crystalline material in the sample, and the narrower the later order distribution curve, the narrower the range in degrees of perfection of the crystalline material (4).

Similarly, lateral order distribution curves have been derived from moisture regain measurements, x-ray diffraction studies comparing the relative proportion of the cellulose I to cellulose II lattice, the decrease in the "levelling-off" degree of polymerization (LODP) which occurs in the transition range, methods based upon acid degradation followed by solubility in sodium hydroxide solutions, the formylation of cellulose, the oxidation of cellulose with nitrogen dioxide, and other methods based on swelling in which ethylamine and nitrogen dioxide are utilized. In assessing the parameters obtained from any of the lateral order distribution curves, Warwicker, et al. (4) cautioned that it should be realized that the position of the transition range, and thus the lateral order distribution curve derived from it, depends, to an extent, on the particular property of cellulose measured, and that the methods measure the lateral order distribution of the ordered regions only, while providing no information about the order of the less ordered regions.

B. X-ray Diffraction

In his excellent review, Tripp (5) noted that whether or not the classical two-phase (crystalline-amorphous) system or the concept of a wide range of degrees of order, or the continuum of order with defects is accepted as the true representation of cellulose morphology, the determination of crystallinity by x-ray diffraction devolves on the means of apportioning the observed diffractogram into two complementary parts. The shape of the diffraction curve for completely crystalline cellulose is now known, but diffracting bodies of estimated dimensions in celluloses account for the intensity distributions usually observed. Only crystallites above a certain size will contribute to the diffraction maxima, and as Viswanathan (7) points out, metallic crystalline aggregates as small as 64 unit cells will not produce discrete reflections, but will result in diffuse scattering. Hosemann (8) in his treatment of the paracrystalline state, attributed the broading of reflections by small crystallites and the diffuse scattering caused by lattice defects as major factors complicating the solution of the problem. Also the effects of a "sloppy lattice" are recognized by many workers who attempt to take the possible effects into account in their interpretations.

Hermans and Weidinger (9,10) were among the first to propose quantitative methods for measuring the percentage crystallinity of cellulose by x-ray diffraction. As shown in Figure 1 (11) the peaks are attributed to the crystalline scattering and the more diffuse scattering to the disordered regions. The percent crystallinity was determined by comparing the integrated intensities of the crystalline peaks and that of the diffuse background.

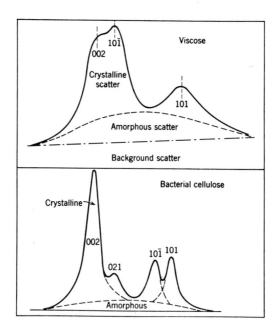

Figure 1. Determination of proportions of crystalline and amorphous material by the x-ray method (11).

The projection onto the a-c plane of the monoclinic unit cell of native and mercerized cellulose is depicted in Figure 2 (12). The a-b or 002 lattice plans coincides with the plane of the chain rings and therefore results in the highest scattering band in native cellulose. Hydrogen bonding is the primary source of cohesion between chains in the a-c plane, and the 101 plane in cellulose I is denser in hydroxyl groups than the 10$\bar{1}$ plane. In mercerized or regenerated cellulose, however, the majority of the hydrogen bonding is in the 10$\bar{1}$ plane (13). This association of the regenerated cellulose toward the 10$\bar{1}$ plane is reflected by the greatly increased intensity of the 10$\bar{1}$ bond in the

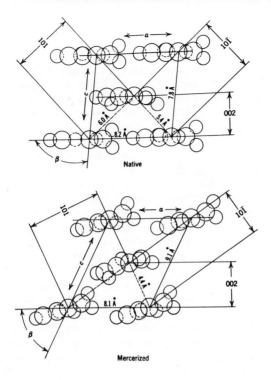

Figure 2. Projection onto a-c plane of the unit cell of native and mercerized cellulose (12).

diffractogram of cellulose II. Thus the use of x-ray diffraction to determine the crystallinity obviously looses its quantitative significance when mixed lattices are present.

1. External Reference Methods. All methods for determining crystallinity by x-ray diffraction can be conveniently divided into two groups, depending upon whether external or internal reference standards are used (14). The external methods are often referred to as relative methods in that the scattering of the sample is compared with a reference material at a single point or over a scattering range. Refinements in the technique were made by Goppel (15) who devised effective means of correcting the observed

intensities for absorption of the x-ray beam by the sample, air scatter, and instrumental variation. Ellefsen and co-workers (16) utilized the diffraction-intensity curve of ball-milled cellulose as the amorphous standard which is considered to represent the limiting state of disorder. In comparing the "amorphous" curve with that derived from a randomly oriented test specimen, the assumption is made that scattering minimum between the $10\bar{1}$ and 002 planes in cellulose I (or between the 101 and $10\bar{1}$ planes in cellulose II) results entirely from scattering of x-ray radiation by the disordered portion. The degree of amorphousness is obtained from the ratio of the intensities at these minima to those of the corresponding standard "amorphous" curve. Since there is likely some contribution to the minima intensities from the adjacent peaks, these values probably reflect an upper limit for the disordered fraction. The crystalline fractions are obtained merely by substracting the amorphous fraction from 1. Typical degrees of crystallinity obtained included the following values: hydrolyzed surgical cotton, 0.62; cotton linters, 0.57; mercerized acetate-grade pulp, 0.49; acetate-grade wood pulp, 0.52; viscose-rayon staple fiber, 0.35 (5).

Using the x-ray diffractometer technique with Geiger counter detection and strictly monochromatized radiation, Hermans and Weidinger (17) made a large number of radial and azimuthal recordings at different angles, reexamining, in part the results of their earlier work. Their main conclusions, bearing in the work of Ellefsen and co-workers described above was that the diffuse background underlying the main paratropic interference range was at no point equal to the scattering by the amorphous component. Due to the complicated anisotropic intensity of the background, no absolute determination of crystallinity could be made. By relating the intensities of the seven principle interferences to a crystalline fraction of 0.40 for "common textile rayon," they nevertheless found that the crystalline values for several regenerated celluloses did not differ much from their earlier determinations. Fiber G, an experimental viscose rayon supplied by du Pont de Nemours, mercerized Fiber G, and cuprammonium rayon had crystalline fractions of 0.42, 0.48, and 0.42 by the new methods while values of 0.48, 0.52, and 0.40, respectively, were obtained by the old method. The crystalline fraction of rayon tire cord was determined by the improved techniques to be 0.28, and was indeterminable by the old method owing to a high content of cellulose IV. In the same work, they found a correlation between crystallinity and line width in that rayon tire cord showed the widest equatorial lines, indicating smaller crystallite sizes for "all skin" types of rayon. Comparing the cuprammonium rayons, mercerization did not change the crystallite size although the crystallinity increased.

Wakelin and co-workers (18) developed a correlation method in which the diffraction intensity of the sample is compared to both an amorphous standard (ball-milled cellulose) and to a crystalline standard (acid-hydrolyzed cellulose). The corrected intensity for the amorphous standard was subtracted from the cotton sample and from the crystalline standard at the same scattering angle. Point-by-point differences were thereby obtained at each scattering angle on the spectrometer from 5° to 50°. As shown in Figure 3 (18), cotton-amorphous standard differences were then correlated with the crystalline standard-amorphous differences from which the slope of the regression line provided an estimate of crystallinity and the correlation coefficient an estimate of the error associated with the slope. By this technique the average crystalline index for six native cottons was 68.3 percent which is in good agreement with earlier findings of Hermans and Weidinger (9) at 69.0 $\pm$ 2 percent and by Ant-Wuorinen (19) at 70.0 $\pm$ 2.5 percent. Nelson and O'Connor (20) extended the method to include cellulose II specimens and determined that Fortisan fibers had an index of 74 percent and mercerized cotton, 51 percent.

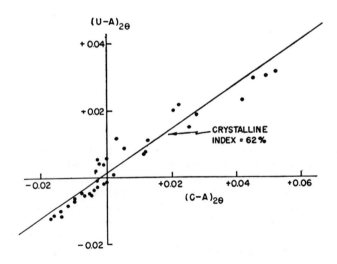

Figure 3. Correlation of intensity differences of cotton-amorphous standard with those for crystalline-amorphous standards -- SXP cotton (18).

Patil, Dweltz, and Radhakrishnan (21,22,23) further adapted the Wakelin method to obtain crystallinity of samples containing mixed lattice forms. Wakelin's technique and similar methods, however, have been found to be strongly dependent on crystallite dimensions. Also, Patil and co-workers considered ball-milled cellulose an adequate zero-crystallinity standard, but the crystallinities of the crystalline standards were found to be less than 90 percent. They suggested that the intensities of a sample displaying moderate line broadening be corrected to values corresponding to 100 percent crystallinity in the Wakelin correlation. Typical crystalline fractions reported by this modification of the Wakelin technique include: native cotton, 80 percent cellulose I; cotton exposed to 24 percent sodium hydroxide, 6 percent cellulose I, 50 percent cellulose II; cotton swelled in 78 percent ethylenediamine, 46 percent cellulose I; and rayon tire-cord, 25 percent cellulose II (5).

2. Internal Reference Methods. Internal reference methods (also known as absolute methods) are based on the demonstrable principle that the total radiation scattered by a given mass is independent of its physical state, i.e. whether the material is in the form of gas, liquid, or crystal. Thus, in principle all the information required for a crystallinity determination is contained within each diffraction pattern. The internal methods differ only in the manner in which they account for the fraction of the total radiation detected during the experiment and in the application of various correction factors to the x-ray data. The x-ray crystallinity by the internal method, X_i, is given by equation 1,

$$X_i = 0_c/(0_c + K0_a) \tag{1}$$

where 0_c and 0_a are measures of the crystalline and amorphous intensities, respectively. The constant K includes geometrical correction factors and a factor to take into account the relative efficiency of amorphous and crystalline scattering over the diffraction range measured (14).

In extracting the fraction of scattering associated with each state from the pattern, reference is usually made to model substances, and in this sense the internal reference methods do not approach "absolute" values any closer than to the external reference indices. Gjonnes, Viervoll, and Norman (24) noted that the corrected intensities for the diffractogram peaks of cellulose are not symmetrical, but are similar to the long-tailed Cauchy distributions observed in cold-worked metals. By comparing the synthesized curves with the observed line shapes for typical celluloses, they were able to show excellent fit at the critical minima, which in the past have been attributed to background scatter from the amorphous regions. Their conclusion from this new approach to crystallinity was that such samples "consisted of aggregates of distorted crystallites and that only small quantities of highly disordered material were present" (calculated 5 percent). The fit was poorer for acid-swollen and mercerized celluloses, indicating that the fraction of disordered polymer in cellulose II was probably greater. Tripp (5) notes that although this study was less quantitative in terms of the order-disorder estimates obtained, it does support the concept of cellulose being a paracrystalline polymer, and defines crystallinity to include the spectrum of order present in crystallites as well as the order of the specimen.

In contrast low crystallinity values with the narrow spread of 36-44 percent were obtained for native and regenerated celluloses by Bonart and co-workers (25) using the "so-called difference method". In this method a background line is drawn tangent to the minima at 10, 2 and 1.5 $\mathring{A}$, and the area below this is considered to arise from the non-crystalline fraction. All corrections for sample absorption, air scatter, Compton, and polarization effects are taken into account. The diffractogram is obtained on an unoriented specimen, a second scan with the sample in a nondiffracting position is made as the reference, and differences in corrected intensities are obtained. Using similar techniques, Viswanathan and Venkatakrishnan (26) examined a representative series of raw and mercerized cottons, polynosic and high-tenacity rayons, and ramie fiber, obtaining only 30-32 percent crystallinity on all cottons and rayons, and 36 percent for ramie. The rather small differences in crystallinities obtained in both of these studies invoke serious doubts about the precise value of x-ray diffraction for such determinations. The extension of the scattering range as far as 1.5 $\mathring{A}$, however, adds an area over which a strong background persists and in which scattering from small crystallites would be seen as amorphous scattering, thereby compressing the range of values obtained (5).

An elaborate extension of the "difference" method was made by Ruland (27) who introduced a "disorder" function which corrects for the loss of intensity of crystalline peaks due to deviations of the atoms from their ideal lattice positions, as in paracrystalline distortions (14). Thus the rather uncompromising crystallinity values obtained by the above methods indicate that most celluloses have about the same fraction of crystallinity, differing only in the character of the disorder. Viswanathan and Venkatakrishnan (28) adopted Ruland's treatment to cellulose. As can be seen in Table 2, mercerization increases the disorder of native cellulose, and hydrolysis increases the crystalline component, but does not change the characteristic disorder. Ball-milled cellulose and rayon have lower crystallinities, but the lower disorder parameters indicate that the fewer crystalline regions are more ordered than native cellulose. The crystallinity of ball-milled cellulose increased to 36 percent in wetting and drying, but with an accompanying increase in disorder parameter to 1.5. Elaborate calculations are involved in the treatment of this data which is indicative of the complexities involved in calculating paracrystallinity, but the method provides useful insight into the origin of response of cellulose materials to lateral order distribution studies (5).

Other internal reference methods that have been proposed do not offer a quantitative estimate of the degrees of crystallinity, but provide indices of crystallinity. They are attractively simple and are usually based on functions relating the intensity of the principle interferences with that of a simultaneously varying minimum in the diffractogram. They include the methods of Clark and Terford (29), Segal, et al. (30), and Ant-Wuorinen and Visapaa (31). Segal and co-workers (30) suggested as a "crystallinity index" (CrI) the ratio of the intensity of the 002 reflectance to the minimum intensity at $2\theta = 18°$, as expressed by equation 2:

$$CrI = \frac{(I_{002} - I_{AM})}{I_{002}} \times 100 = (1 - \frac{I_{AM}}{I_{002}}) \times 100 \qquad (2)$$

By this method, the crystallinity index of cotton yarn was found to be 79.2 percent while the CrI of cotton swollen with ethylamine was 63.3 percent, but increased to 76.0 percent after boiling in water for 2 hours. Ant-Wuorinen and Visapaa (31) proposed as a measure of the state or order of cellulose a crystalline ratio (Cr.R.) in the form of equation 3:

$$Cr.R. = 1 - \frac{Am.h.}{Cr.h.} = 1 - \frac{Am.h.}{Tot.h. - Am.h.} \qquad (3)$$

As shown in Figure 4, the total height (Tot.h.) is the intensity of the maximum interference between 22° and 23°, 2θ, in the case of cellulose I and between 18° and 22°, 2θ with cellulose II, Figure 5. The amorphous height (Am.h.) is the intensity of the minimum between 18° and 19°, 2θ, for cellulose I and between 13° and 15°, 2θ, in the case of cellulose II. Crystallinity ratios reported by this method are 0.83

Table 2
Degree of crystallinity and disorder parameter of celluloses according to Ruland's method (5,27)

Sample	Degree of crystallinity	Disorder parameter
Cotton	47	1.5
Mercerized cotton	42	2.0
Hydrolyzed cotton	51	1.5
Ramie	44	1.5
Mercerized ramie	45	2.0
Hydrolyzed ramie	56	1.5
Polynosic rayon	36	1.0
Ball-Milled cellulose	28	1.0

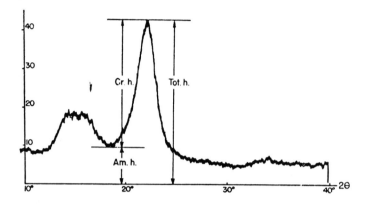

Figure 4. Diffractogram of bleached sulfite pulp (31).

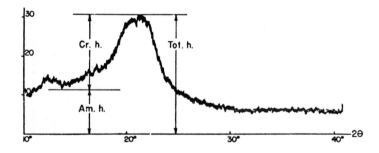

Figure 5. Diffractogram of viscose staple fiber (31).

for bleached cotton linters, 0.70 for bleached sulfite pulp, and 0.41 for viscose staple fiber.

C. Infrared Crystallinity Ratio

Nelson and O'Connor (32) observed that some peaks in the infrared spectrum of cellulose increased while others decreased or remained constant with changes in lateral order as measured by x-ray diffraction. They proposed as a "crystallinity index" for native cellulose the ratio of the absorbance at 1429 cm^{-1} (CH$_2$ scissoring) to the absorbance at 893 cm^{-1} (C$_1$ group vibration), and found that the intensities of these bands decreased and increased, respectively, with a decrease in x-ray crystallinity. Crystallinity index (a_{1429}/a_{893}) values reported by this method range from 2.80 for untreated cotton to 0.78 for cotton decrystallized with ethylamine and 0.11 for ball-milled cellulose (5). In the same study, Nelson and O'Connor noted, however, that this technique could not be utilized with cellulose samples having mixed lattices in that amorphous cellulose and cellulose II were found to have similar intensities at both absorption bands. In subsequent work (20), they proposed that the ratio of absorbance at 1372 cm^{-1} (C-H bending) to the absorbance at 2900 cm^{-1} (C-H stretching) be utilized as a "crystallinity index", and found that the latter technique could be used with samples containing a mixed lattice. The 2900 cm^{-1} band appeared to be independent of changes in crystallinity and served as an internal standard correcting for differences in sample presentation while the band at 1372 cm^{-1} was believed not to be affected by moisture absorption and decreased in intensity with decreases in crystallinity.

The infrared crystallinity ratios, a_{1429} cm^{-1}/a_{893} cm^{-1}, are plotted against the corresponding empirical and Wakelin crystallinity values in Figure 6 showing that there is a complete separation of points obtained from cellulose I and cellulose II. The infrared ratio, a_{1372} cm^{-1}/a_{2900} cm^{-1}, are plotted against x-ray crystallinity index in Figure 7 and against crystallinity by density measurements and against accessibility in Figure 8. In all four plots, cellulose I, cellulose II, and partially mercerized cellulose follow the same trends with changes in crystallinity and there is no discernible difference in the behavior of the different types of cellulose. The data in Figures 7 and 8, however, showed considerable scatter about the regression lines, as reflected in correlation coefficients of 0.86 for the empirical x-ray data, 0.88 for the density data, and -0.92 for accessibility.

The "empirical" crystallinity values were crystallinity indexes obtained by the previously described method of Segal

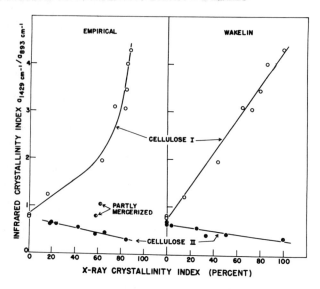

Figure 6. Relations between infrared crystallinity (or lateral order) index, a_{1429} cm^{-1}/a_{893} cm^{-1}, and empirical (left) and Wakelin (right) x-ray crystallinities (20).

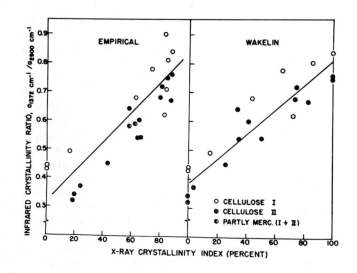

Figure 7. Relations between proposed infrared crystallinity ratio, a_{1372} cm^{-1}/a_{2900} cm^{-1}, and x-ray crystallinity indexes (20).

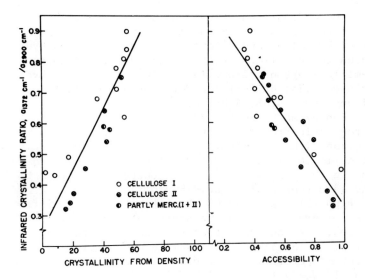

Figure 8. Relations between proposed infrared crystallinity ratio, a_{1372} cm^{-1}/a_{2900} cm^{-1}, and crystallinity from density (left) and accessibility (right) (20).

et al. (30); whereas, the Wakelin correlation method (18) provides more of a quantitative estimate of the percent crystalline component. The closer correlation of infrared crystallinity ratio to accessibility determined by moisture sorption as compared to crystallinity determinations by x-ray and density methods was explained by the fact that theoretically both infrared absorption and moisture sorption are dependent on the environment of atomic groups in the molecules. Moisture sorption depends on the number of hydroxyl groups that are free from interchain packing and can absorb water molecules while the intensity of infrared bands can be affected by the closeness of packing to adjacent molecules which could interfere with the vibrations of atomic groups. While both infrared and accessibility data provide a measure of order at the molecular level, density and x-ray methods provide a measure of the quantity and size of larger subunits, the crystallites. The crystallites must be larger than certain minimum dimensions to be detected by x-ray diffraction, and defects in the ordered regions as well as crystallite dimensions will affect the results. Density measurements are likewise dependent on the degree of order of larger volume elements, since the immersion liquids cannot penetrate into some of the regions of intermediate order. These

suppositions were born out by a correlation coefficient of
0.99 between density and x-ray crystallinity values.

Typical infrared crystallinity ratios (a_{1372}/a_{2900})
obtained by Nelson and O'Connor included: cotton hydrocellulose, 0.84; lint cotton, 0.71; mercerized cotton, 0.54;
ball-milled cotton hydrocellulose, 0.44; Fortisan hydrocellulose, 0.75; ball-milled Fortisan hydrocellulose, 0.34;
Fortisan rayon, 0.68; Fiber 40 rayon, 0.60; and XL rayon,
0.54. Pandey and Iyengar (33) report a reduction in the
infrared ratio from 0.86 for cotton to 0.69 on treatment of
the cotton with 30 percent sodium hydroxide and to 0.60 after
swelling in ethylamine. Berry, et al. (34) reported an increase in this index from 0.55 in cotton printcloth to 0.84
in the same fabric after acid hydrolysis.

Tripp (5) acknowledged that the use of such infrared
indices offered a convenient approach to crystallinity
estimates, but cautioned that the results were subject to
poor reproducibility unless sample preparation and instrumental operating conditions were carefully standardized.
An example was noted in which films of "valonia ventricosa"
exhibited a very high index, approaching 1.00, in the procedure of Nelson and O'Connor, but by cutting the film and
pressing into a potassium bromide pellet, the infrared index
was lowered to 0.78 without affecting the crystallinity as
determined by x-ray diffraction.

D. Sorption Studies

Garments made from cellulosic fibers primarily owe their
great aesthetic appeal to the moisture absorbing ability of
cellulose. Yet, x-ray studies have shown that water molecules do not penetrate the crystalline regions (1,2,3). Thus
the moisture must be absorbed in the less ordered regions and
on the surfaces of crystallites. Moisture absorption increases with increasing disorder of the cellulose structure
in that the moisture regain of cotton at 65 percent relative
humidity and 20°C is 7-8 percent, mercerized cotton up to
12 percent, and viscose rayon 12-14 percent regain (35). The
moisture regain of completely accessible cellulose was found
by indirect means to be 17.0 percent by Gibbons (36) at 65
percent relative humidity and 25°C. This value was obtained
from extrapolating the straight line portions of the plots of
moisture absorption versus degree of substitution of homogeneously reacted methyl cellulose and cellulose acetate back
to zero DS, and represented the average of the two extrapolations. This treatment with the two cellulose derivatives
yielded remarkably close maximum moisture regains with 16.7
percent being obtained with methyl cellulose and 17.3 percent

being obtained with cellulose acetate. The methyl and acetate groups served to prop the structure open and thereby facilitate penetration by water molecules, but conversely these groups blocked the cellulose hydroxyls reducing the number of hydroxyls available for hydrogen bonding with water. With both derivatives the effect of increasing DS was to increase moisture absorption up to a maximum point after which the moisture regain decreased, but by extrapolating back to zero DS the adverse effect of the more hydrophobic groups was, in effect, eliminated thereby allowing the moisture regain of completely accessible cellulose to be determined. Some fifteen years later, Jeffries (37) found that the moisture regain of almost completely amorphous cellulose (98 percent amorphous as determined by infrared deuteration) was 16.75 percent at 57 percent relative humidity and 20°C. That particular cellulose sample was regenerated into a non-aqueous bath by saponifying secondary cellulose acetate in a 1.0 percent solution of sodium hydroxide in ethanol for 1 day at room temperature, followed by a prolonged wash in ethanol. Thus the following formula for calculating the accessibility of a cellulose sample is derived assuming the maximum moisture regain (MR) for completely accessible cellulose at 65 percent relative humidity to be 17.0 percent (38)

$$\% \text{ Accessibility } (A_{H_2O} \text{ vap}) = \frac{100 \text{ MR}}{17.0} \qquad (4)$$

where MR is expressed in moles of water per anhydroglucose unit (38).

Correlations have been made between the amount of water vapor sorbed by a polymer and the amount of disordered material in the polymer. Howsmon (39) showed that there was a direct relationship between moisture sorption and the accessibility to liquid deuterium oxide. Mann and Marrinan (40,41,42) determined from infrared studies that deuterium vapor exchanges with the cellulose hydroxyls in the amorphous regions only, whereas, deuteration from the liquid phase results in partial exchange with the hydroxyls on the surfaces of crystallites as well as in the disordered regions. They found after 1 hour of vapor phase deuteration of viscose film, that the broad OH absorption at 3600-3000 cm^{-1} disappeared completely leaving four well-resolved OH bands of much lower intensity with those four bands resulting from hydrogen-bonded OH groups in the crystalline regions. This reduction in absorption intensity at the 3600-3000 cm^{-1} region was accompanied concurrently by an absorption peak at 2700-2400 cm^{-1} due to OD groups which approached a maximum intensity after 1 hour of deuteration. Assuming that vapor phase deuteration affects only the amorphous regions the fractional crystallinities of cellulose samples were determined from the ratio of the peak optical density of the OD band at 2530 cm^{-1} to that of the OH band at 3360 cm^{-1} utilizing the formula,

$$\frac{\log_{10}(I_o/I)_{OD}}{\log_{10}(I_o/I)_{OH}} = \frac{k_{OD} \, c_{OD}}{k_{OH} \, c_{OH}} \qquad (5)$$

where I_o is the intensity of the incident radiation on the film, I is the intensity of the transmitted radiation, k is the extinction coefficient per unit mole fraction, and C is the concentration of absorbing material expressed as a mole fraction such that $C_{OH} + C_{OD} = 1$ (42). The ratio k_{OD}/k_{OH} was determined to be 1.11 (41). The percent amorphous components of several celluloses determined by the method described above are given in the first column of Table 3.

To determine the percentage of crystalline regions affected by liquid phase deuteration, the infrared results after exposure to liquid D_2O for 4 hours were compared to the values obtained by vapor phase deuteration. From this comparison the percentage of the crystalline regions of three samples that were deuterated in the liquid phase are shown in the second column of Table 3.

Since only the disordered regions are accessible to both D_2O vapor and H_2O vapor, a close correlation between the two accessibility methods should be expected. This is, in fact, the case as shown in Figure 9 (43). The fractions of amorphous material were those measured by Mann and Marrinan and the sorption ratios (S.R.) were calculated from the ratio of the moisture sorption of the particular cellulose sample to that of cotton under the same conditions, and the values should be independent of the relative humidity (R.H.) over the R.H. range of 20-70 percent (43).

The fact that the best line through the points also passes through the origin was taken as strong evidence in favor of the essential correctness of the correlation. Thus from the sorption ratio at any relative humidity the amorphous fraction, Fam, as would be determined from infrared deuteration can be obtained, the exact ratio between the two being:

$$S.R. = 2.60 \, Fam \qquad (6)$$

The sorption ratio of completely amorphous cellulose is 2.60 which corresponds to a moisture regain of 17.4 percent at 58 percent R.H. or 19.3 percent at 65 percent R.H.

Table 3
The percent amorphous component and the percent crystalline regions deuterated (42)

Sample	Percent Amorphous OH	Percent Crystalline Regions Deuterated
Saponified acetate	75	--
Viscose oriented	74	--
Viscose unoriented	73.5	33
Viscose hydrolyzed in H_2SO_4	68.5	--
Precipitated cellulose	68	--
Mercerized cellulose	67	--
Viscose treated with NaOH	66.5	23
Fortisan micelles	46.5 $\pm$ 5%	--
Bacterial cellulose	30	13
Cotton micelles	31 $\pm$ 5%	--

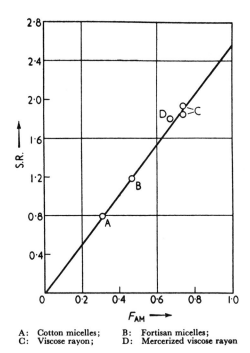

A: Cotton micelles; B: Fortisan micelles;
C: Viscose rayon;   D: Mercerized viscose rayon

Figure 9. The relation between sorption ratio (S.R.) and the fraction of amorphous OH groups (F_{AM}) for various types of cellulose (43).

The close correlation between moisture sorption and the exchange with D_2O vapor strongly indicates that the two methods are measuring exactly the same quantity. The determination of accessibility of cellulose by moisture sorption is much simpler than the measurement of the absorption of deuterated samples and can be applied to fibers whereas the infrared technique is conveniently applied only to films (43). Knight and co-workers (44) extended the infrared deuteration technique to include fibers but encountered problems with the rehydrogenation of the deuterated samples. Guthrie and Heinzelman (45,46) determined the deuterium-hydrogen-exchange accessibility by the use of D_2O^{18} and mass spectroscopy and found that the method gave higher values than other techniques based on D_2O and infrared spectroscopy when applied to the same or similar samples. Considering the inherent complications of the newer techniques, the determination of accessibility by moisture sorption has a great deal of merit.

Iodine sorption has been widely used to measure the accessibility of cellulose (38,47,48,49,50,51,52,53), but this method is not without its problems. Jeffries, et al. (38) observed that the absorption of iodine decreased with increasing sample size since with the larger samples there is a decrease in the concentration of iodine solution in equilibrium with the sample, which makes the values obtained from iodine sorption doubtful. Nelson and co-workers (52) noted that there were large experimental errors and a great variability in results associated with the iodine test. Minhas and Robertson (54) pointed out that the iodine sorption method, itself, causes swelling of the celluloses. X-ray studies have shown that the iodine even penetrates the crystalline regions when the absorption exceeds 11-12 percent, and the potassium iodide in which the iodine is usually dissolved is also thought to act as a swelling agent. Furthermore the adsorption of iodine is believed to occur in concentrated patches penetrating into the cellulose gel as opposed to monolayer adsorption (51). On the other hand, the sorption of bromine is said to proceed in a monomolecular layer, not to cause swelling and to be highly reproducible (55).

Bromine molecules are believed to form an addition compound at right angles to the glycosidic bond, as represented by the structure below:

```
         R
         |
         O .... Br ____ Br                    (A)
         |
         R
```

With an ether oxygen, a weak charge transfer can form resulting in both halogen atoms being involved in a "halogen-molecular bridge" between two ether oxygens:

```
         R                    R
         |                    |
         O .... Br ____ Br .... O             (B)
         |                    |
         R                    R
```

Obviously, the formation of such bromine bridges requires that two glycosidic oxygens from two adjacent chains in the less ordered regions be in close proximity. But as Lewin, et al. (55) point out the process may be very slow and depends on the flexibility and mobility of the chain segments and on their internal rate of diffusion. The possibility also exists for a reversible interchange between two A-compounds forming a B-compound and liberating a molecule of bromine:

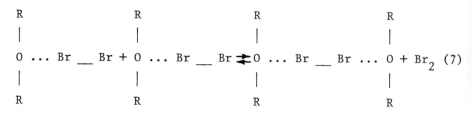

An alignment of the chain occurs in the formation of bromine bridges, and consequently upon removal of the bromine through washing with water or through reduction, the chains may crystallize. Thus the crystallization caused by the bromine treatment may result in significant error in determining the accessibility. However, the authors showed that the error in accessibility values was only 1-2 percent when the bromine concentrations were below 0.02 mole/ℓ at a cellulose concentration of 10 g/ℓ and with sorption times of 1-2 hours.

After treating the samples in bromine water the concentration of bromine in the treatment solution was determined by adding an aliquot to a measured excess volume of arsenite solution and back titating with iodine solution, using starch as an indicator. Similarily the concentration of bromine in the portion containing the fibers was determined. The concentrations of bromine in the sorption experiments were expressed as C_f (the concentration of bromine in mmoles/ℓ in the solution), C_B (the concentration of bromine in mmoles/kg in the fibers), and as $C_{B(s)}$ (the saturation concentration of bromine in cellulose). From the Langmuir plot of $1/C_B$ against $1/C_f$, the intercept $1/C_{B(s)}$ is obtained when $1/C_f = 0$, which coincides to the saturation point. Having obtained the value of $C_{B(s)}$, the ratio, n, of the number of moles of anhydroglucose units per mole of bromine at saturation is determined from the relation:

$$n = \frac{m}{C_{B(s)}} \tag{8}$$

where m = 1000/162 = 6.17 is the number of moles of anydroglucose units per kg of cellulose. The percent accessibility, A, of the cellulose is obtained from the value 100/n. Sixteen samples of cellulose were analyzed and some of the accessibility values obtained were, Pay Master cotton fibers, 22 percent; Pima cotton fiber, 20 percent; Pima cotton fabric, 32 percent; hydrolyzed Pima cotton fabric, 14 percent; ramie fabric, 20 percent; hydrolyzed ramie fabric, 6.5 percent; high modulus rayon fibers, 22.4 percent; Vincel 66 fibers 30.7; and cellulose triacetate, 70 percent. A correlation coefficient of 0.855 was obtained from the plot of accessibility by moisture regain versus bromine absorption. Better correlations were obtained between the bromine method and the x-ray and infrared crystallinity indices. The correlation coefficient between bromine accessibility and the x-ray crystallinity of Segal, et al. (30) was 0.95 while correlation coefficients of -0.984 and -0.92 were obtained between bromine accessibility and the infrared crystallinity indexes (20), a_{1429} cm^{-1}/a_{893} cm^{-1} and a_{1372} cm^{-1}/a_{2900} cm^{-1}, respectively.

REFERENCES

1. Trotman, E. R., Cotton and the Chemistry of Cellulose in "Dyeing and Chemical Technology of Textile Fibers", Charles Griffin and Company Ltd., London, 3rd Ed., 33-61 (1964).
2. Hermans, P. H. and A. Weidinger, J. Polym. Sci. 4, 135 (1949).
3. Valentine, L., J. Polym. Sci., 27, 313 (1958).
4. Warwicker, J. O., R. Jeffries, R. L. Colbran, and R. N. Robinson, "A Review of the Literature on the Effect of Caustic Soda and Other Swelling Agents on the Fine Structure of Cotton," the Cotton Silk and Man-made Fibers Research Association, Shirley, Institute, Didsbury, Manchester, England, 1966.
5. Tripp, V. W., Measurements of Crystallinity, "Cellulose and Cellulose Derivatives," Vol. 5, M. Bikales and L. Segal, eds., Wiley-Interscience, New York, 305-323, 1971.
6. Forziati, F. H., R. M. Brownell, and C. M. Hunt, J. Res. Nat. Bur. Stand., 50, 139 (1953).
7. Viswanathan, A., Fiber Structures. J. Appl. Polym. Sci., 11, 1027 (1967).
8. Hosemann, R., Polymer 3, 349 (1962).
9. Hermans, P. H. and A. Weidinger, J. Appl. Phys., 19, 491 (1948).

10. Hermans, P. H. and A. Weidinger, J. Polym. Sci., 4, 709 (1949).
11. Marrinan, H. F., The Fine Structure of Cellulose in "Recent Advances in the Chemistry of Cellulose and Starch," J. Honeyman, ed., Interscience Publishers, Inc., New York, 147-187, 1959.
12. Howsmon, J. A. and W. A. Sisson, Submicroscopic Structure in "Cellulose and Cellulose Derivatives," E. Ott, H. M. Spurlin, and M. W. Grafflin, eds., Vol. V., 2nd ed., Interscience Publishers, Inc., New York, 231, 1954.
13. Nissan, A. H. and G. K. Hunger, Cellulose in "Encyclopedia of Polymer Science and Technology," Vol. 3, H. F. Mark, N. G. Gaylord, and N. M. Bikales, eds., Interscience Publishers, Inc., New York, 131-226, 1965.
14. Miller, R. L., Crystallinity in "Encyclopedia of Polymer Science," Vol. 4, H. F. Mark, N. G. Gaylord, and N. M. Bikales, eds., Interscience Publishers, Inc., New York, 449, 1966.
15. Goppel, J. M., Appl. Sci. Res., A1, 3 (1947).
16. Ellefsen, O., K. Kringstad, and B. A. Tonnesen, Norsk Skogind. 18, 419 (1964); from Tripp, 1971, op. cit.
17. Hermans, P. H. and A. Weidinger, Text. Res. J., 31, 558 (1961).
18. Wakelin, J. H., H. S. Virgin, and E. Crystal, J. Appl. Phys., 30, 1654 (1959).
19. Ant-Wuorinen, O., Paperi ja Puu, 37, 335 (1955).
20. Nelson, M. L. and R. T. O'Connor, J. Appl. Polym. Sci., 8, 1325 (1964).
21. Patil, N. B., N. E. Dweltz, and T. Radhakrishnan, Text. Res. J., 32, 460 (1962).
22. Patil, N. B. and T. Radhakrishnan, Text. Res. J., 36, 746 (1966).
23. Patil, N. B. and T. Radhakrishnan, Text. Res. J., 36, 1043 (1966).
24. Gjonnes, J., N. Norman, and H. Viervoll, Acta Chem. Scand., 12, 489 (1958); from Tripp, 1971, op, cit.
25. Bonart, R., R. Hosemann, F. Motzkus and H. Ruck, Norelco Reptr., 7, 81 (1960); from Tripp, 1971, op. cit.
26. Viswanathan, A. and V. Venkatakrishnan, "Proceedings of the 7th Technological Conference," Section B, Ahmedabad Textile Industry Research Association, 41, 1965.
27. Ruland, W., Acta Cryst., 14, 1180 (1961); from Tripp, 1971, op. cit.
28. Viswanathan, A. and V. Venkatakrishnan, J. Appl. Polym. Sci., 13, 785 (1969).

29. Clark, G. L. and H. C. Terford, Anal. Chem., 27, 888 (1955).
30. Segal, L., J. J. Creeley, A. E. Martin, Jr., and C. M. Conrad, Text. Res. J., 29, 786 (1959).
31. Ant-Wuorinen, O. and A. Visapaa, Norelco Reptr., 9, 48 (1962).
32. Nelson, M. L. and R. T. O'Connor, J. Appl. Polym. Sci., 8, 1311 (1964).
33. Pandey, S. N. and R. L. N. Iyengar, Text. Res. J., 38, 675 (1968).
34. Berry, G. M., S. P. Hersh, P. A. Tucker, and W. K. Walsh, Evaluation of Modern Textile Resins as Reinforcing Agents for Textile Materials, Advances in Chemistry, American Chemical Society, In Press (1977).
35. Morton, W. E. and J. W. S. Hearle, "Physical Properties of Textile Fibers," John Wiley and Sons, New York, 170, 1975.
36. Gibbons, G. C., J. Text. Inst. Trans., 44, 201 (1953).
37. Jeffries, R., J. Appl. Polym. Sci., 12, 425 (1968).
38. Jeffries, R., T. G. Roberts and R. N. Robinson, Text. Res. J., 38, 234 (1968).
39. Howsmon, J. A., Text. Res. J., 19, 152 (1949).
40. Mann, J. and H. J. Marrinan, Trans. Fara. Soc., 52, 481 (1956).
41. Mann, J. and H. J. Marrinan, Trans. Fara. Soc., 52, 487 (1956).
42. Mann, J. and H. J. Marrinan, Trans. Fara. Soc. 52, 492 (1956).
43. Valentine, L., Chem. and Ind., 1279 (1956).
44. Knight, J. A., H. L. Hicks, and K. W. Stephens, Text. Res. J., 39, 324 (1969).
45. Guthrie, J. D. and D. C. Heinzelman, Text. Res. J., 40, 1133 (1970).
46. Guthrie, J. D. and D. C. Heinzelman, Text. Res. J., 44, 981 (1974).
47. Schwertassek, K., Melliand Textilber., 13, 536 (1932); from Chem. Abstr. Vol. 28, No. 6571 (In German).
48. Hessler, L. G. and R. E. Power, Text. Res. J., 24, 822 (1954).
49. Bailey, A. V., E. Honold, and E. L. Skau, Text. Res. J., 28, 861 (1958).
50. Modi, J. R., S. S. Trivedi, and P. C. Mehta, J. Appl. Polym. Sci., 7, 15 (1963).
51. Doppert, H. L., J. Polym. Sci., Part A-2, 5, 263 (1967).
52. Nelson, M. L., M. A. Rousselle, S. J. Cangemi, and P. Trouard, Text. Res. J., 40, 872 (1970).
53. Rousselle, M. A., M. L. Nelson, C. B. Hassenboehler, Jr., and D. C. Legendre, Text. Res. J., 46, 304 (1976).

54. Minhas, P. S. and A. A. Robertson, <u>Text. Res. J.</u>, 37, 400 (1969).
55. Lewin, M., H. Guttman, and N. Saar, New Aspects of the Accessibility of Cellulose, "Proceedings of the Eight Cellulose Conference. II. Complete Tree Utilization" and Biosynthesis Structure of Celluoose, T. E. Timell, ed., John Wiley Interscience, New York, 791, 1976.

Modified Cellulosics

HYDROXYL REACTIVITY AND AVAILABILITY IN CELLULOSE

S. P. ROWLAND

USDA Agricultural Research Service
Southern Regional Laboratory
New Orleans, Louisiana

Generally, reagents must first penetrate channels and pores to accomplish the purpose for which a chemical modification is designed. Thereafter, relative reactivities of $O(2)H$, $O(3)H$, and $O(6)H$ toward various types of reagents and relative availabilities of these hydroxyl groups for reaction become critical factors in achieving specific physical performance properties from chemical modifications of cellulose. The marvelous simplicity and complexity of microstructural features of cellulosic fibers and the amenability of cellulose to alteration of physical properties by chemical modification argue convincingly that much remains to be achieved in tailor-making chemical modifications of cellulose by reactions that are conducted with full cognizance of the roles played by chemistry and microstructure. An attempt is made here to put into perspective the chemistry of cellulose and its relationship to the microstructure of cellulose.

Cellulose is chemically modified for numerous reasons, some of which are: (a) to solubilize it for regeneration into fibers and films, (b) to make it thermoplastic for moldings and extrusions, and soluble for coatings, (c) to modify its bulk properties (e.g., resilience, dimensional stability) without change in physical form, and (d) to modify its chemical properties (e.g., flame retardancy, resistance to microorganisms). The chemistry of these modifications is superficially simple, primarily the esterification or etherification of cellulosic hydroxyl groups. Yet the overall situation is complex; improvements depend upon abilities to facilitate and control reactions at hydroxyl groups (or even at a specific type of hydroxyl group) while minimizing or eliminating other reactions such as oxidation and molecular degradation. Complications are increased because cellulose in its natural forms has high molecular weight and crystallinity, and because all reactions of cellulose begin as heterogeneous reactions, some taking place with little or no visible alteration in cellulose fibers and others resulting in dissolution during the reaction. It is for these reasons that relative chemical reactivities of O(2)H, O(3)H, and O(6)H of cellulose are important and that relative physical availabilities of these hydroxyl groups for reaction are equally important.

It is the objective of this review to present a <u>concise</u> overview and perspective on the titled subject; consequently some details and original references are sacrificed. Details are available in selected, major references. Where possible and appropriate, reference is made to compilations of information on a particular subject rather than to isolated publications. In particular, attention is called to "Cellulose: Distribution of Substituents" (1), "Cellulose: Pores, Internal Surfaces and the Water Interface" (2), and "Solid-Liquid Interactions: Inter- and Intracrystalline Reactions in Cellulose Fibres" (3).

PRE-REACTION STAGE: PENETRATION AND SWELLING

It is important to note that reagent molecules must enter cellulose fibers or substrates to perform their function of chemically modifying the bulk of the material (4). Organic solvents are generally poor penetrating and swelling agents for cellulose, but when an organic solvent-soluble derivative of cellulose is formed, organic solvents or excess of organic reagent can effectively facilitate penetration and reaction. In the case of aqueous reagent systems, small reagent molecules commonly penetrate channels and pores of cellulose to reach as few as 10-15% of the total hydroxyl groups of a highly crystalline cellulose or as many as 85-95% of a decrystallized cellulose (5). Pores of cellulose discriminate by a sieving action among solute molecules on the basis of molecular size (6); as solute molecules in aqueous media approach the size of the water molecule, potentiality that they will reach all of the internal surfaces that are wet by water increases (2).

Considerable evidence indicates that water at the water-cellulose interface exhibits properties that are different from the remainder of the bulk water (2). This water at the interface has commonly been called bound water and has often been equated to nonsolvent water, although much data intended to prove the existence of nonsolvent water are inconsistent and essentially self-defeating. Not all attempts to identify bound water have been successful, nor have successful investigations measured discrete layers of bound water or even similar amounts of bound water. New light was thrown on this seeming dilemma by a gel permeation chromatographic study of interactions of water-soluble solutes with cellulose (7). The outcome of the study of a series of mono-to-tetrasaccharides, which are capable of strong donor and acceptor hydrogen bonding comparable to that of the D-glucopyranosyl units of cellulose, showed no evidence for bound water. As the saccharides decreased in molecular weight to approach that of water, extrapolation showed that total internal water became available as solvent water at molecular weight = 18; thus, all water in an accessible pore is apparently available as solute water to these solutes. On the other hand, in the case of a series of polyethylene glycols, which, except at low molecular weight, are limited primarily to acceptor hydrogen bonding, there is reduced penetration at a given level of molecular weight or size. The difference in penetration between these two types of solutes is attributed to inability of the polyethylene glycols to make use of bound water as solvent water. The ability of mono-, di-, tri-, and tetraethylene glycols to penetrate cotton is further reduced by methylation of the hydroxyl groups; the methylated glycols at a given molecular weight evidently sense a thicker layer of bound water as a result of their reduced hydrogen-bonding capacities. It appears, therefore, that the capability of a water-soluble solute to penetrate pores of cellulosic fibers depends not only upon molecular weight or size, but also upon the ability of the solute to participate in strong donor and acceptor hydrogen bonding in order to compete at the water-cellulose interface for use of bound water as solvent water.

HYDROXYL REACTIVITY: IRREVERSIBLE REACTIONS (1)

Distributions of substituents provide direct measures of relative reactivities of $O(2)H$, $O(3)H$, and $O(6)H$ for irreversible reactions carried to low degrees of substitution. The distributions resulting from eight different etherification reactions of cellulose in mercerizing strength NaOH have been described (1). The reagents were methyl chloride, ethyl chloride, dimethyl sulfate, N,N-diethyl aziridinium chloride, sodium chloroacetate, sodium allyl sulfate, ethylene oxide, and sodium 2-aminoethyl sulfate. The 2-O-:6-O- and 3-O-:6-O- distributions of substituents ranged progressively downwards from 2.79 and 0.46, respectively, for the methyl substituents from methyl chloride, to 0.64 and 0.14, respectively, for the 2-aminoethyl substituents from sodium 2-aminoethyl sulfate. Relative substitution in the 2-O position decreases as the size of the incoming substituent increases, and perhaps also as the size of the leaving group (Cl^-, etc.) increases. A limiting case of the steric effect is illustrated by reactions of trityl chloride, in which case only 5-12% of substituent enters the 2-O-position, the major portion of reaction taking place at O(6)H.

Of all the data available from reactions of cellulose, the above results best measure changes in relative reactivities of O(2)H, O(3)H, and O(6)H as a function of changes in reagent molecules. However, data cited above must be treated with caution. The justification for making comparisons and drawing conclusions from data published by various investigators is twofold: (a) in all cases, due to use of various mercerizing media, O(2)H, O(3)H, and O(6)H are expected to be equally available for reaction, and (b) relative reactivities are probably not sensitive to small changes in concentration of NaOH in the mercerizing range of concentration. Although the reactivities of O(2)H, O(3)H, and O(6)H are not very sensitive to small changes in reaction conditions, distributions of substituents are altered by some types of changes. For example, the 2-O-:6-O- and 3-O-:6-O- distributions of 2-diethylaminoethyl groups from reaction of N,N-diethylaziridinium chloride (the actual reagent in basic aqueous solution from 2-chloroethyl diethylamine hydrochloride) are 1.25 and 0.35, respectively, as a result of reaction in 20% NaOH at room temperature, but they are 2.50 and 0.50, respectively, from reactions involving KOH in dioxane at 100°C; and the distributions are 0.40 and 0.43, respectively, from thermal reactions at 140°C for 5 minutes, and 2.88 and 0.60, respectively, from thermal reaction involving 1 $\underline{N}$ NaOH (1). It is evident from these examples that the distribution of a substituent is not an invariable characteristic of a reagent; however, these examples confound effects from changes in reaction conditions with effects from changes in availabilities of O(2)H, O(3)H, and O(6)H for reaction. Only in the case of the reaction in 20% NaOH is there basis to believe that the three different hydroxyl groups in these examples are equally available for reaction; this phenomenon is discussed in the section concerning crystalline celluloses.

HYDROXYL REACTIVITY: REVERSIBLE REACTIONS (1)

The distribution of substituents from reversible reactions most often approximates an equilibrium-controlled rather than rate-controlled situation. The distribution of acetyl groups from homogeneous reaction of partially acetylated cellulose with acetyl chloride and pyridine is the same as that from reaction with acetic anhydride and catalyst, but it is different from that resulting from the heterogeneous reaction with these reagents. In the first case, the ratio 2-O- + 3-O-substitution to 6-O-substitution was 0.2-0.4, and in the second case it was 0.54 (substitutions at 2-O- and 3-O-positions were not resolved). The difference has been attributed to the existence in fibrous cellulose of hydrogen bonds involving O(6)H that are more stable than those at O(2)H and O(3)H.

Additional reagents that undergo reversible reactions and that have been studied with cellulose are formic acid, acrylonitrile, acrylamide, methyl vinyl sulfone, trimethylolmelamine, and formaldehyde. For all of these, the 2-O-:6-O-distributions of substituents are 0.2-0.5 and 3-O-:6-O-distributions are 0.03-0.2. Results of distributions of linkages from methyl vinyl sulfone and from formaldehyde are considered most reliable, since these are measured by direct and semidirect methods, respectively (1).

Reversible reactions favor the 6-O-position, as compared to the 2-O-position for irreversible reactions. In most of the studies of reactions with cellulose, however, the actual state of equilibrium has not been identified. The best illustration of the effect of state of equilibrium on the distribution of substituents exists for the reaction of methyl vinyl sulfone with cotton fabric (1). Interrupted rapid and moderately fast reactions at 140°C, and slow and extended reactions at 25°C, resulted in 2-O-:6-O- and 3-O-:6-O-distributions as follows: 0.75 and 0.00; 0.28 and 0.10; 0.22 and 0.05; and 0.14 and 0.05, respectively. This type of change from initially higher to subsequently lower 2-O-:6-O-distribution with increasing duration of reaction was observed also during the course of a homogeneous reaction of the same reagent with cellulose (8). The final or equilibrium distribution of substituents is representative of the relative thermodynamic stability of the substituent-oxygen linkages, which is generally the controlling factor for distribution of substituents resulting from reversible reactions. The higher 2-O-:6-O-distribution observed from interrupted rapid reaction is a reflection of a partially rate-controlled distribution and is indicative that the relative reactivities of O(2)H, O(3)H, and O(6)H may actually be similar (other factors being equal) toward reagents undergoing etherification reactions by irreversible or reversible reactions. Thus, rate-controlled distribution (i.e., high distribution at the 2-O-position) may be approached in extremely short reaction periods, whereas equilibrium-controlled distribution or near equilibrium distribution is the normal result to be expected from a reagent that undergoes reversible reaction.

EFFECT OF REACTION MEDIUM ON HYDROXYL REACTIVITIES (1)

Mention has already been made that relative reactivities of the three hydroxyl groups in a D-glucopyranosyl unit are affected by substantial changes in reaction conditions. It was not realized until recently (1968) that reactivities of cellulosic hydroxyl groups in base-catalyzed etherifications are dependent upon the concentration of NaOH. For both N,N-diethylaziridinium chloride (DAC) (9) and ethylene oxide (10) in reaction with crystalline cellulose, the ratios of substitution at the 2-O-:6-O- and 3-O-:6-O-positions decreased progressively as the concentration of base increased from 0.5 N to 4 N; the distributions leveled off for reactions in 4 N and 6 N base. For fibrous cellulose, such a change immediately brings to mind questions concerning swelling and microstructural effects. However, these same changes in distribution of substituents have been shown to occur in reactions of DAC with decrystallized cellulose (11) and in reactions of ethylene oxide with cellulose in solution (12). A similar pattern of results also characterized the reaction of DAC with methyl β-D-glucopyranoside in homogeneous solution (13). Thus, it is indisputible that the effect is chemical rather than physical. It has been proposed that preferential reaction at O(2)H in dilute base occurs due to an accentuation of the reactivity of O(2)H by an inductive effect from the anomeric carbon and by intramolecular bonding from O(3)H to O(2)$^-$, and due to a repression of reactivity of O(6)H by a strong solvation sheath around that anion (13). It was further proposed that with increasing concentration of NaOH in the reaction medium, adduct formation at vicinal hydroxyl groups increases progressively, and the stability of the adducts is sufficient to repress oxyanion formation at O(2)H and O(3)H. The isolated O(6)H is not subject to this effect.

The foregoing observations are significant in two ways: (a) they indicate that changes in aqueous media can have substantial effect on reactivities of the hydroxyl groups of cellulose, changing O(2)H:O(3)H:O(6)H reactivities in the case of DAC and decrystallized cellulose from about 2.2:0.8:1.0 in 0.5 N NaOH to 1.25:0.3:1.0 in 4 and 6 N base; and (b) the formation of a NaOH adduct at vicinal O(2)H and O(3)H, while O(6)H exists as an alkoxide, bears on the mechanism of mercerization, i.e., conversion of lattice I cellulose to lattice II cellulose. Thus, the energy for the change in attitude of the cellulose chains relative to one another in the unit cell is provided by the strong interaction and solvating effects of base on the two sides of each of each of the D-glucopyranosyl units in the cellulose chain.

AVAILABILITIES OF HYDROXYL GROUPS ON SURFACES OF CRYSTALLINE CELLULOSES (1)

The reaction of DAC with cellulose has served as a "chemical microscope" to clarify the availability of O(2)H, O(3)H, and O(6)H for reaction. This particular reagent is well suited for the purpose because it is water soluble and undergoes reaction with cellulose in dilute and concentrated base and under a broad range of conditions, and because the substituted cellulose is readily amenable to analyses of the distribution of 2-diethylaminoethyl substituents. When DAC is reacted with methyl β-D-glucopyranoside or various forms of cellulose in 0.5 N NaOH at 25°C for 45 minutes, different distributions of substituents result. These are shown in Table 1. Each distribution characterizes the

TABLE 1
Distributions of 2-Diethylaminoethyl Substituents in Various β-D-Glucopyranosyl Units

Substrates	2-O-:6-O-	3-O-:6-O-	4-O-:6-O-	Ref.
Methyl β-D-glucopyranoside	1.48	0.78	0.52	13
Decrystallized cellulose	2.15	0.80	-	11
Hydrocellulose I[a]	3.80	0.41	-	11
Hydrocellulose II[a]	2.80	0.24	-	14
Native cotton fiber	2.55	0.25	-	15
Mercerized fiber	2.41	0.72	-	16

a. High crystallinity samples, termed exemplar hydrocellulose (11, 14); I and II refer to lattice I and II structures, respectively.

particular features of the β-D-glucopyranosyl unit under examination. The relative distributions of substituents for methyl β-D-glucopyranoside represent relative reactivities of the hydroxyl groups, since this reaction occurs in homogeneous solution and all hydroxyl groups are available for reaction. Although the reaction of decrystallized cellulose occurs in a heterogeneous system, the results also represent relative reactivities of the hydroxyl groups because the hydroxyl groups are highly accessible. Again, O(2)H, O(3)H, and O(6)H are believed to be equally available for reaction (11). Differences in relative reactivities between corresponding hydroxyl groups in methyl β-D-glucopyranoside and the β-D-glucopyranosyl unit in decrystallized cellulose are attributed to the difference in activity of each specific hydroxyl group in a dissolved, solvated unit versus that on a surface of a dispersed solid.

Differences between distributions of substituents resulting from reaction of disordered cellulose and from reaction of crystalline celluloses must be due to factors other than the real chemical reactivities of the hydroxyl groups. Two factors emerge for consideration: conformation of glucosidic units and hydroxyl groups, which could affect relative reactivities of hydroxyl groups, and intra- and intermolecular hydrogen bonding of hydroxyl groups that would tie up specific types of hydroxyls and prevent them from being available for reaction in the absence of hydrogen-bond breaking solutes. Although it has been established that relative reactivities of hydroxyl groups change with structural alterations such as from the β-D-glucopyranoside to the α-D-glucopyranoside ring (i.e., from the β- to the α-glucoside, and from cellulose to starch) (13,17), the effect of small changes in conformation such as might be caused by molecular packing or stress in segments of molecular chains of cellulose, is unknown and perhaps nonexistent. On the other hand, intra- and intermolecular hydrogen bonding of hydroxyl groups is real and is recognized as an important aspect of the crystalline structure of cellulose. It is reasonable, therefore, to look for an explanation of the data in Table 1 in terms of availabilities of hydroxyl groups (i.e., freedom from intra- and intermolecular hydrogen bonding) on accessible surfaces of crystalline cellulose.

The basis for calculating availabilities of hydroxyl groups lies in the following equations (1). Equation $\underline{1}$ is the rate expression for substitution (S_3) at O(3)H,

$$dS_3/dt = k_3[O(3)H]_a[R] \qquad \underline{1}$$

where k_3 denotes the specific rate constant, R denotes the reagent, and the quantities in brackets are activities of reagent in solution and hydroxyl groups on solid surfaces, the subscript $\underline{a}$ indicating that only the hydroxyl groups that are available for reaction are included. Equations similar to 1 describe the rates of reactions of other hydroxyl groups. The result of dividing equation 1 by the corresponding equation for reaction at O(2)H is shown in equation 2. When this equation is applied to reaction of a re-

$$\frac{dS_3}{dS_2} = \frac{k_3[O(3)H]_a}{k_2[O(2)H]_a} \qquad 2$$

agent (e.g., DAC) with decrystallized cellulose, for which $[O(3)H]_a = [O(2)H]_a$, the activities of the two hydroxyl groups cancel, and the ratio of substitution is a measure of the ratio of rate constants. When equation 2 is applied to the reaction of a crystalline cellulose, it is appropriate to rearrange it to the form of equation 3, which states that the

$$\frac{[O(3)H]_a}{[O(2)H]_a} = \frac{k_2(dS_3)}{k_3(dS_2)} \qquad 3$$

relative availabilities of these two hydroxyl groups are calculated from the reciprocal of the ratio of their rate constants and the ratio of substitution measured for the crystalline cellulose. The ratio of rate constants is obtained from the reaction of decrystallized cellulose (equation 2).

Availabilities of hydroxyl groups for crystalline and whole fiber celluloses, calculated from experimental data based on DAC as the reagent (as listed in Table 1), are summarized in Table 2. In these calculations, availabilities of O(3)H and O(6)H were indexed against the availability of O(2)H for two reasons: (a) O(2)H was found to be most available among the three hydroxyl groups, and (b) there was reason to believe, at least at the time of initial publication of these results (vide infra), that O(2)H did not denote a proton in hydrogen bonding in crystalline celluloses. The results in Table 2 show that O(3)H and O(6)H are available for reaction only to fractions of the extent of O(2)H. The validity of these results is verified by very similar values from reactions conducted in 1 and 2 N NaOH, although actual distributions of substituents resulting from reactions in these media were quite different from those listed in Table 1 (for reasons already discussed in the section concerning reaction medium). The indication is that the hydrogen bonding systems on accessible crystalline surfaces are stable in 0.5, 1, and 2 N NaOH, just as they are stable in cellulose crystals. In 4 and 6 N NaOH, selective availabilities of hydroxyl groups of native cotton fibers are no longer evident, because these strong caustic solutions disrupt hydrogen bonds on surfaces and interiors of cellulose crystals, penetrate the

TABLE 2

Selective Availabilities of Hydroxyl Groups in Cellulose

Cellulose	$\dfrac{[O(3)H]_a}{[O(2)H]_a}$	$\dfrac{[O(6)H]_a}{[O(2)H]_a}$	Ref.
Decrystallized	1.0^a	1.0^a	11
Hydrocellulose I^b	0.28	0.55	11
Hydrocellulose II^b	0.23	0.74	14
Native cotton	0.27	0.82	11
Mercerized cotton	0.79	0.86	16

a. Assumed values.
b. Refer to footnote a of Table 1.

crystalline regions, and cause $[O(2)H]_a = [O(3)H]_a = [O(6)H]_a$. However, a low level of selective availability of $O(6)H$ still remains in 4 and 6 $\underline{N}$ NaOH in the case of hydrocellulose I, which is evidently more resistant to penetration and lattice transformation than the whole cotton fiber (1).

In attempts to relate the above results to structures of crystalline celluloses (11), it is evident that there is no experimental evidence to distinguish whether hydrogen bonding on surfaces of cellulose crystals (elementary fibrils) is the same as or different from that in the interior of the crystals. The common separation of cellulose microstructural units into elementary units (widths of 20-50 A) suggests that hydrogen bonding between surfaces of elementary fibrils may be different from that within the fibrils; but this difference may be a simple reflection of perfection of interfibrillar bonding between units that are laid down as discrete units during biosynthesis. X-ray evidence points toward a larger crystallite size for the elementary fibril than that measured by microscopy (18), indicating that elementary fibrils may be bound together by a continuation of the same hydrogen bonding that characterizes the interior of the crystal but perhaps with a lower degree of perfection. It appears most reasonable, therefore, to look for a correspondence between hydrogen bonding on surfaces of elementary fibrils, as indicated by selective availabilities of hydroxyl groups, and specific hydrogen bonding systems proposed within cellulose crystals.

HYDROXYL REACTIVITY AND AVAILABILITY

Surfaces of elementary fibrils are constituted of the molecular chains that lie in the $1\bar{1}0$ and 110 planes (101 and $10\bar{1}$ planes by original cellulose crystallographic terminology) of the unit cell (19). Hydroxylic sides of D-glucopyranosyl units in cellulose chains protrude from these surfaces at angles in the range of 45° in the case of native cellulose (lattice I) and at angles in the range of 75° on two opposite sides and 25° on the other sides in the case of mercerized (lattice II) cellulose. Lattice I cellulose is illustrated in Figure 1 for surfaces on the upper left and lower right corners of a cross section of an idealized elementary fibril based on the antiparallel arrangement of cellulose chains preferred by Meyer and Misch (20), on the intramolecular O(3)H··O(5') hydrogen bond of Hermans et al. (21), and on the intermolecular O(6)H··O(1") hydrogen bond proposed by Frey-Wyssling (19) and apparently confirmed with infrared measurements (22). Although this antiparallel arrangement of molecular chains and the hydrogen bonding systems are at variance with results of recent studies (vide infra), this structure has historical interest and was used previously for demonstration of availability of hydroxyl groups. The O(2)H and O(3)H protrude from the plane of the surface at angles that are fixed by the plane of the D-glucopyranosyl unit, but O(6)H protrudes with greater freedom due to rotation about C(6). However, O(6) may be tied back into the plane of the surface by hydrogen bonding as on the left and upper surfaces of Figure 1, as well as be free to rotate in the vicinity of the locations shown on the right and bottom surfaces of this figure. Thus, it is envisioned that in the vicinity of the locations shown on the right bottom surfaces of this figure. Thus, it is envisioned that O(2)H, O(3)H, and O(6)H that are unencumbered by intra- and intermolecular hydrogen bonds to tie their protons back into the surfaces of the elementary fibrils are available for participation in chemical reactions; those hydroxyl groups that are tied back into the surface are unavailable for reaction in the absence of hydrogen-bond breaking solvents (i.e., swelling agents for cellulose).

As water or aqueous solutions penetrate cellulose fibers and cause accessible surfaces to be formed by separation of elementary fibrils (2,3), certain hydrogen bonds that exist within cellulose crystals would be expected to survive on surfaces; these are intramolecular bonds and the intermolecular bonds that lie in $1\bar{1}0$ or 110 planes. Hydrogen bonds along 020 planes (diagonals in Figure 1) would not survive on a surface since they would necessarily have to break for a surface to be formed. On this basis, then, it is possible to estimate availabilities of hydroxyl groups that might occur on surfaces of crystals (elementary fibrils) and to compare

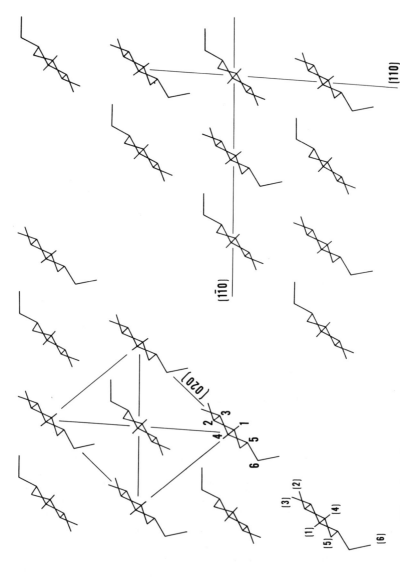

Fig. 1. Cross section of upper left and lower right corners of idealized elementary fibril of lattice I cellulose. Numbers refer to carbon (open) and oxygen (brackets).

these with values measured by chemical analyses and summarized above. The comparison is made in Table 3. Results of chemical analyses are consistent with the widely-accepted O(3)H⋯O(5') bond, but they differ from results of current X-ray studies that call for an O(2')H⋯O(6) bond. Chemical analyses show no evidence that O(2)H is restrained from availability for reaction; the results indicate that, relative to O(2)H, half (in the case of lattice I) or a quarter (in the case of lattice II) of the O(6)H are restrained from reaction by hydrogen bonds. It is logical, then, that the closest agreement with chemical analyses should occur for those hydrogen bond systems that do not involve O(2')H⋯O(6) bonds but do place O(6) in a conformation suitable for hydrogen bonds in 1$\bar{1}$0 and/or 110 planes that would tie up some of the O(6)H on the surface as well as in the interior of a crystal (entries 2 and 8, Table 3). However, these hydrogen bond systems resulted from studies (14,19-22) that were much less comprehensive and definitive than subsequent studies (24-29), so their pertinence here may be primarily for the suggestion that crystalline surfaces are best described by absence of O(2')H⋯O(6) bonds and location of O(6)H in a position that permits hydrogen bonding in 1$\bar{1}$0 and/or 110 planes on surfaces (in contrast to presence of O(2')H⋯O(6) bonds and O(6)H donor hydrogen bonding predominantly in 020 planes within the crystals). The freedom of O(6)H to assume alternate conformations on peripheries of crystallites was indicated in a recent study of mercerized cotton, which excludes only the tg and gg conformations for the "up" and "down" chains, respectively (28).

AVAILABILITIES OF HYDROXYL GROUPS ON VARIOUS ACCESSIBLE SURFACES OF CELLULOSE

The elementary fibril of cotton fibers is considered completely crystalline (30). However, there must be allowance for different degrees of order in the packing of molecular chains along the longitudinal dimension of the elementary fibril, due to concentrations of stress from distortion, and along the lateral dimension, from ordered interior to a surface that may be accessible simply because it is distorted or incapable of perfect packing with adjacent elementary fibrillar surfaces. These features are depicted in Figure 2. From data summarized in Table 2, it is concluded that the surfaces of hydrocellulose segments (B, Figure 2) are much more highly ordered in O(6)H hydrogen bonding (i.e., lower selective availability of O(6)H) than surfaces of whole fibers.

TABLE 3

Comparison of Chemically Measured Availabilities of Hydroxyl Groups to Those Consistent with Proposed Structures for Crystalline Celluloses from X-ray Studies

Cellulose	Measurement	Relative Availabilities			Hydrogen bonds	O(6)H Conformation	Ref.
		O(2)H	O(3)H	O(6)H			
Lattice I:							
Hydrocellulose I	Chemical	1.0	0.28	0.55	—	—	11
Ramie	X-ray	1.0	0.0	0.50	O(3)H··O(5') O(6)H··O(1")	gt	19, 20-22
Ramie	"	0.0(1.0[a])	0.0	1.0	O(2')H··O(6) O(3)H··O(5')	tg	23
Valonia	"	0.0(1.0[a])	0.0	1.0	O(2')H··O(6) O(3)H··O(5') O(5)H··O(3")[c]	tg	23, 24
Valonia	"	0.0(1.0[a])	0.0	0.5[b]	O(2')H··O(6) O(3)H··O(5') O(6)H··O(3")[c] O(6)H··O(5")	tg	25
Ramie	"	0.0(1.0[a])	0.0	1.0	O(2')H··O(6) O(3)H··O(5') O(6)H··O(3")[c]	tg	26
Lattice II:							
Hydrocellulose II	Chemical	1.0	0.23	0.74	—	—	14

TABLE 3 (continued)

Cellulose	Measurement	Relative Availabilities O(2)H	O(3)H	O(6)H	Hydrogen bonds	O(6)H Conformation	Ref.
Fortisan	X-ray	1.0	0.0	0.75	O(2)H··O(2") O(3)H··O(5') O(6")H··O(3)	tg	14
Fortisan, merc. cotton	"	0.25(0.75[a])	0.0	1.0	O(2')H··O(6) O(2)H··O(2") O(3)H··O(5') O(6)H··O(3")[c] O(6)H O(2")[c]	tg, gt	27, 28
Fortisan	"	0.25(0.75[a])	0.0	0.75	C(2')H··O(6) O(2)H··O(2") O(3)H··O(5') O(3)H··O(6") O(6")H··O(3) [d]	tg, gt	29

a. These are the values if O(2')H··O(6) is considered too weak to restrain O(2)H from reaction.
b. Assuming the O(6)H··O(5") bond on a surface is sufficiently strong to restrain O(6)H from reaction.
c. These hydrogen bonds are in the 020 plane and cannot survive as surfaces are formed.
d. Additional O(6)H··O(3") and O(6)H··O(2") bonds are indicated in the 020 planes.

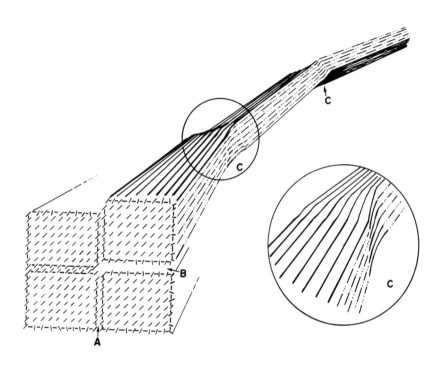

Fig. 2. Schematic representation of the crystalline elementary fibril of cotton to illustrate A, surfaces of high order; B, accessible, slightly disordered surfaces on highly ordered segments; and C, readily accessible surfaces on strain-distorted, tilt and twist segments.

The nature of accessible surfaces in the native cotton fiber was studied by tagging accessible surfaces with a low level of 2-diethylaminoethyl substitution, subjecting the substituted cotton cellulose to progressive acid hydrolysis with removal of soluble and insoluble components, and analyzing these products for distribution of substituents (31). Since the degree of substitution was very low, the crystalline structure of native cotton was little altered, as was also the case for the acid hydrolysis. Like the study described in the preceding section, distribution of substituents on a particular surface reflects the availability of hydroxyl groups for reaction and the regularity of hydrogen bonding on the surfaces.

Two types of accessible surfaces were evident from the study. One type of surface (B, Figure 2) corresponded to that on the highly crystalline hydrocellulose I. The availabilities of hydroxyl groups were: $[O(3)H]_a/[O(2)H]_a =$

0.13, and $[O(6)H]_2/[O(2)H]_2 = 0.57$. On the assumption that all of the O(3)H and half of the O(6)H are tied up in hydrogen bonds on the surfaces of a perfectly ordered, completely crystalline elementary fibril (or crystal), the percentage of disorder in this type of surface is 13% based on disruption of O(3)H bonds and 14% based on disruption of O(6)H hydrogen bonds. These results are in generally good agreement with those for highly crystalline hydrocellulose I (Table 2), for which disruptions of these two types of hydrogen bonds are 28% and 10%, respectively.

The second type of accessible surface is located on fibrillar segments that are rapidly hydrolyzed from the chemically modified cotton fibers (C, Figure 2). The availabilities of hydroxyl groups for these segments were $[O(3)H]_a/[O(2)H]_a = 0.93$ and $[O(6)H]_a/[O(2)H]_a = 0.80$, representing surfaces with 93% and 60% disruptions of O(3)H and O(6)H hydrogen bonds, respectively. In this study it was observed that removal of the substituted glucose units representative of the highly disordered surfaces of the elementary fibrils was not complete at the end of the rapid hydrolysis phase (9.5% cellulose removed at this stage), but was complete after an additional 6% of cellulose was hydrolyzed away. This latter hydrolyzate is estimated to have $[O(3)H]_a/[O(2)H]_a = 0.88$ and $[O(6)H]_a/[O(2)H]_a = 0.78$, representing surfaces with 88% and 56% disruptions of the two types of hydrogen bonds, respectively. The two types of surfaces noted in this paragraph are obviously similar, differing only in that they are two portions of the same type of surface. The latter portion of this surface, which hydrolyzed at the slower rate, underwent crystallization before hydrolysis of this segment was complete.

The B and C types of surfaces were estimated to constitute 64% and 36% of the total accessible surfaces. The relatively ordered, accessible surfaces on highly crystalline segments (B) constitute the 64%, and the highly disordered accessible surfaces on segments under stress constitute the 36%. When the higher availability of hydroxyl groups in the latter type of surface is taken into consideration, the total number of available hydroxyl groups on the two types of surfaces is essentially equal.

HYDROXYL AVAILABILITY FOR TERMINAL REACTION OF A DIFUNCTIONAL REAGENT

The introduction of crosslinkages into cellulose is dependent upon reactions of both functionalities of the crosslinking agent with cellulose hydroxyl groups. Presumably, the two hydroxyl groups are in different elementary fibrils to contribute the desired stabilization and resilience to cotton fibers. The question of spatial availability of hydroxyl groups for reaction of the second functionality of a difunctional reagent has received little consideration.

The first functionality of divinyl sulfone (DVS) reacts at O(6)H (similarly to methyl vinyl sulfone, see section on reversible reactions) but without evidence for reaction at O(2)H and O(3)H (32). It might be anticipated, then, that the second functionality would undergo reaction only with an available O(6)H. An alternative to this is reaction with water followed by chain extension (i.e., DVS reaction with this new hydroxyl group on a DVS substituent). The efficiency with which DVS forms simple crosslinks (i.e., single units of reagent in a crosslink) was observed to vary with the swelling capacity of the reaction medium, but in all cases only 6,6'-O-crosslinks were found. In 0.1 $\underline{N}$ NaOH, which served as catalyst as well as mild interfibrillar swelling agent, 82% of the DVS residues were estimated to be involved in simple crosslink formation, and 18% in simple substituent formation (i.e., incompleted crosslinks). No chain extension was observed in this case. However, in 0.5, 1.0, and 2.0 $\underline{N}$ NaOH, which exhibit increasing interfibrillar swelling capacity, the simple crosslinks decreased to 60%, 18%, and 25%, respectively. When the reaction medium was 4.0 $\underline{N}$ NaOH, an intracrystalline swelling agent, simple crosslinkages constituted 72% and extended chain structures only 8% of the reagent residue. Thus, the result of increasing concentration of NaOH from 0.1 $\underline{N}$ to 2.0 $\underline{N}$, increasing interfibrillar swelling, and concomitant increasing distance between O(6)H's on vicinal surfaces of elementary fibrils is a decreasing efficiency of formation of simple crosslinkages as lateral distance between vicinal elementary fibrillar surfaces is extended beyond the locus of the second functionality of the difunctional reagent. The high percentage of simple crosslinks resulting from reaction in 4 $\underline{N}$ NaOH constitutes evidence that sheets of cellulose molecules, which are displaced from immediate contact with one another as a result of intracrystalline swelling, are about as widely separated as accessible elementary fibrillar surfaces in 0.1 N NaOH.

Crosslinking of cotton cellulose with a reagent such as $C_2H_5N(CH_2CH_2Cl)_2$, which undergoes irreversible reaction with cellulosic hydroxyl groups and shows more balanced reactivities than DVS toward O(2)H and O(6)H, illustrates another aspect of the dependency of the terminal functionality upon availability of hydroxyl groups (33). The fractions of simple substituents introduced into the 2-O-, 3-O-, and 6-O-positions (0.707, 0.027, and 0.260, respectively) are measures of the probabilities of reaction of the first functionality of

this difunctional reagent with O(2)H, O(3)H, and O(6)H. The
probabilities for reaction of the terminal functionality
of the crosslinking agent can be estimated on the assumption
that this reaction is subject to the same limitations that
applied to the first functionality. These calculated and
normalized probabilities, together with actual experimental
results, are listed below:

	2,2'-O-	2,6'-O-	6,6'-O-
Calculated	0.525	0.399	0.075
Experimental	0.342	0.552	0.105

In actual fact, the 2,2'-O-crosslinks were 35% below expectations, and the 2,6'-O- and 6,6'-O-crosslinks were 35-40% above
expectations. The deviations from the calculated populations
appear to indicate that reagent residues with initial linkages
at 2-O-positions found a pronounced deficiency of O(2)H's
within locus for reaction with the terminal functionality and
that residues with initial linkages at the 6-O-position found
relative abundance of O(6)H within range for reaction. The
deviation is attributed to constraint applied to the locus of
reaction of the terminal functionality.

CONCLUDING COMMENTS

There is little question but that chemical factors are
controlling in reactions of cellulosic hydroxyl groups and
that reactions of O(2)H, O(3)H, and O(6)H proceed in manners
that are normal with regard to the electronic and steric
factors that apply to each of the hydroxyl groups. The
features that make reactions of cellulosic hydroxyl groups
challenging are the special effects that arise from the
fact that these reactions are initiated or occur in heterogeneous phase, that the three types of hydroxyl groups are
not equally available for reaction in crystalline celluloses,
that reactions of difunctional reagents are critically
dependent for completion upon availability of hydroxyl
groups within the reach of the second functionality, and
that penetration of reagents into cellulosic substrates is
dependent upon hydrogen-bonding and water-associating
capacities of reagent molecules as well as molecular size.
It is clearly evident that much chemistry remains to be
clarified in connection with the chemical modification of
cellulose and that with this clarification substantial
improvements in reactions and end products may be expected.

REFERENCES

1. Rowland, S. P., in Encycl. Polym. Sci. and Technol. (N. M. Bikalis, Ed.) Supplement No. 1, p. 146. John Wiley and Sons, Inc., New York, 1976.
2. Rowland, S. P., in "Textile and Paper Chemistry and Technology" (J. C. Arthur, Jr., Ed.) ACS Symposium Series No. 49, American Chemical Society, Washington, D. C., p. 20, 1977.
3. Rowland, S. P., in "Applied Fibre Science" (F. Happey, Ed.) Vol. 1, Academic Press, Inc., London, Chapter 20, 1977 (in press).
4. Valko, E. I., in "Chemical Aftertreatment of Textiles" (H. Mark, N. S. Wooding, and S. M. Atlas, Eds.) Wiley-Interscience Div., John Wiley and Sons, Inc., New York, p. 5, 1971.
5. Bose, J. L., Roberts, E. J., and Rowland, S. P., J. Appl. Polym. Sci. 15, 2999 (1971).
6. Stone, J. S., Treiber, E., and Abrahamson, B., Tappi 52, No. 1, 108 (1969).
7. Rowland, S. P., and Bertoniere, N. R., Text. Res. J. 46, 770 (1976).
8. Cirino, V. O., Bullock, A. L., and Rowland, S. P., J. Polym. Sci. A-1, 7, 1225, (1969).
9. Wade, C. P., Roberts, E. J., and Rowland, S. P., J. Polym. Sci. B, 6, 673 (1968).
10. Ramnäs, O., and Samuelson, O., Sv. Papperstidn. 71, 82 (1968).
11. Rowland, S. P., Roberts, E. J., and Bose, J. L., J. Polym. Sci. A-1, 9, 1431 (1971).
12. Ramnäs, O., and Samuelson, O., Sv. Papperstidn. 15, 569 (1963).
13. Roberts, E. J., Wade, C. P., and Rowland, S. P. Carbohyd. Res. 17, 393 (1971).
14. Rowland, S. P., Roberts, E. J., and French, A. D., J. Polym. Sci.: Polym. Chem. Ed. 12, 445 (1974).
15. Rowland, S. P., Roberts, E. J., and Wade, C. P., Text. Res. J. 39, 530 (1969).
16. Rowland, S. P., and Roberts, E. J., J. Polym. Sci.: Polym. Chem. Ed. 12, 2099 (1974).
17. Roberts, E. J., and Rowland, S. P., Carbohyd. Res. 5, 1 (1967).
18. Nieduszynski, I., and Preston, R. D., Nature (London) 225, 273 (1970).
19. Frey-Wyssling, A., Biochim. Biophys. Acta. 18, 166 (1955).
20. Meyer, K. H., and Misch, L., Helv. Chim. Acta 20, 232 (1937).
21. Hermans, P. H., DeBooys, J., and Mann, C., Kolloid Z. 102, 169 (1943).

22. Liang, C. Y., and Marchessault, R. H., J. Polym. Sci. 37, 385 (1959).
23. Jones, D. W., J. Polym. Sci. 42, 173 (1960).
24. Gardner, K. H., and Blackwell, J., Biopolymers 13, 1975 (1974).
25. Sarko, A. and Muggli, R., Macromolecules 7, 486 (1974).
26. French, A. D., Carbohyd. Res. in press.
27. Kolpak, F. J., and Blackwell, J., Macromolecules 9, 273 (1976).
28. Kolpak, F. J., Weih, M., and Blackwell, J., Polymer, in press.
29. Stepanovic, A. J., and Sarko, A., Macromolecules 9, 851 (1976).
30. Jeffries, R., Jones, D. M., Roberts, J. G., Selby, K., Simmens, S. C., and Warwicker, J. O., Cellul. Chem. Technol. 3, 255 (1969).
31. Rowland, S. P., and Roberts, E. J., J. Polym. Sci. A-1, 10, 2447 (1972).
32. Rowland, S. P., Cirino, V. O., and Bullock, A. L., J. Polym. Sci. A-1, 9, 1677 (1971).
33. Roberts, E. J., Bose, J. L., and Rowland, S. P., Text. Res. J. 42, 321 (1972).

PART 4 Cellulose Modification by Grafting Techniques

NEW METHODS FOR GRAFT COPOLYMERIZATION ONTO CELLULOSE AND STARCH

B. RÅNBY

Department of Polymer Technology
The Royal Institute of Technology
Stockholm, Sweden

Two new methods for initiation of graft copolymerization onto cellulose and starch have been invented and studied.

1. Peracetic acid, formed in situ from acetic acid and hydrogen peroxide in acid aqueous solution, is found to initiate graft copolymerization of vinyl monomers, e.g., methyl methacrylate and 4-vinyl pyridine, onto wood and cellulose fibers, polyamide film and polyester film in aqueous media at 60°C. The grafting reaction is not specific and considerable amounts of homopolymer are formed. Peracetic acid is also a bleaching agent for wood pulps and has the advantage as an initiator for grafting that it can be applied also to pulp fibers containing lignin.

2. Mn^{3+} ions complexed with pyrophosphate groups and formed by oxidizing Mn^{2+} ions in aqueous solution with permanganate, are found to be very specific and efficient initiators for graft copolymerization of vinyl monomers, e.g., acrylonitrile and methyl methacrylate, onto starch and purified cellulose substrates. The reaction is probably an electron transfer from the substrate to the Mn^{3+} ions (Mn^{2+} ions are formed) and glycol or aldehyde groups may be involved. The grafting occurs at room temperature (30°C) at high rates. It is very specific and gives 98% grafting efficiency (only 2% homopolymer), but it cannot be applied to cellulose substrates containing lignin.

The two new grafting methods offer commercial possibilities for industrial processes of modifying wood, cellulose, and starch to more useful products.

INTRODUCTION

Graft copolymerization of vinyl monomers onto cellulose and starch has been reviewed in the general context of graft and block copolymers by Battaerd and Tregear (1967). A considerable number of initiation methods are available and are generally based on ionizing radiation, ultraviolet and visible light, mechanochemical reactions, thermochemical and other chemical reactions. A few chemical initiation methods have been shown to be well suited to cellulose, starch and related substrates. A great deal of attention has been given the Ce^{4+} electron transfer reaction discovered by Mino and Kaizermann (1958). This grafting technique is operable in aqueous media at moderate temperatures and is highly specific for grafting to polysaccharides, i.e., very small amounts of homopolymer are formed. Another chemical initiation method of considerable interest is the ferrous ion (Fe^{2+}) - hydrogen-peroxide initiation method developed by Ogiwara and Kubota (1968). However, this redox reaction produces Fe^{3+} ions which can cause discoloration of the resulting product if not removed at the end of the graft polymerization. Further, the reaction is not specific for grafting and initiates some homopolymerization as well. Several investigators have reported initiation methods for grafting to cellulose which involve cellulose xanthates. For example, Chaudhuri and Hermans (1961) combined the xanthates with benzoyl peroxide for polymer initiation and Faessinger and Conté (1967) utilized ferrated cellulose xanthates and hydrogen peroxide. A xanthate method for continuous grafting to rayon or cellophane on line in the technical viscose process has been described by Krassig (1971). Recently, two new initiation methods for grafting to cellulose and starch have been invented in our laboratories. One method uses aqueous peracetic acid as initiator for grafting and is described by Hatakeyama and Rånby (1975). The other method is based on Mn^{3+}-complex ions in aqueous solution as electron transfer agents to a substrate containing 1,2-glycol structures as described by Mehrotra and Rånby (1977). This paper presents recent results based on these two new methods for modification of cellulose and starch.

The grafting reactions and resulting grafted samples are characterized according to the parameters defined in Table 1. The two initiation methods are described separately as follows.

PERACETIC ACID AS GRAFTING INITIATOR

Peracetic acid is formed in an aqueous solution of acetic acid and hydrogen peroxide with dilute mineral acid as catalyst:

$$CH_3\text{-COOH} + H_2O_2 \xrightarrow[\text{aq.}]{H^+} CH_3\text{-CO-OOH} + H_2O \qquad (1)$$

TABLE 1

Polymer Grafting Parameters

$$\text{Total conversion (\%)} = \frac{\text{Total weight of polymer formed}}{\text{Weight of monomer charged}} \times 100$$

$$\text{Grafting ratio (\%)} = \frac{\text{Weight of polymer in grafts}}{\text{Weight of substrate}} \times 100$$

$$\text{Grafting efficiency (\%)} = \frac{\text{Weight of polymer in grafts}}{\text{Total weight of polymer formed}} \times 100$$

$$\text{Add-on (\%)} = \frac{\text{Weight of polymer in grafts}}{\text{Total weight of grafted polymer (substrate + grafts)}} \times 100$$

Frequency of grafts = Average number of anhydroglucose units (AGU) per grafted chain.

Peracetic acid is known to have bleaching effects on wood pulps by oxidizing both polysaccharides and lignin. The reaction probably proceeds through formation of free radicals.

In a typical grafting experiment, the wood pulp sample (about 0.5 g) was placed in a reaction vessel together with the monomer (about 5 g) and a dilute solution (100 ml) containing a predetermined amount of acetic acid (80, 8 or 0.8 mmol/l) and trace amounts of sulfuric acid. Sodium hydroxide was added to attain the desired pH. A measured amount of hydrogen peroxide solution (giving a resulting concentration of 100, 10 or 1 mmol/l) was then added and oxygen purged by bubbling purified nitrogen through the solution for about 1 hour. The polymerization was allowed to proceed for specified lengths of time at a temperature of 60°C.

The resulting grafted pulp sample was washed with water and an organic solvent to remove the homopolymer (acetone for methyl methacrylate (MMA) and methanol for 4-vinyl pyridine (VP)) and then transferred to a tared Soxhlet extraction thimble and extracted for 48 hours with the appropriate solvent to remove remaining homopolymer. The thimble was dried and weighed and the weight increase recorded as grafted polymer (chemically bonded).

Two wood pulp samples were used in these experiments: dissolving pulp (DP) from spruce (sulfite process, bleached) and groundwood pulp (GP) from spruce. The pulps were reacted as received wet or purified by extraction (GP extra) with alcohol-benzene solution (1:2) for 18 hours using a Soxhlet apparatus. Fig. 1 shows the percent total conversion of methyl methacrylate (MMA) as a function of time for the three pulp samples. The polymerization reaction was carried out at

60°C with the pH at 6.0 and an initiator concentration of 8 mmol/l HAc and 10 mmol/l H_2O_2. Figs. 2 and 3 depict the grafting ratio and grafting efficiency for this reaction, respectively.

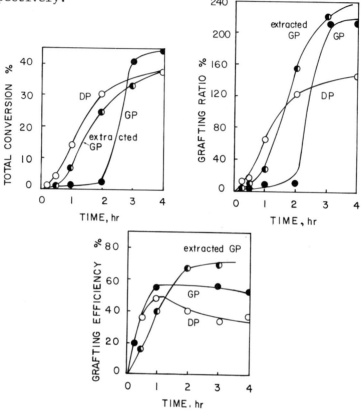

Figs. 1, 2, and 3. Grafting of methyl methacrylate (MMA) onto dissolving pulp (DP), groundwood pulp (GP), and extracted GP. Initiator peracetic acid from HAc (8 mmol/l) and H_2O_2 (10 mmol/l). Polymerization temperature 60°C and pH 6.0. Total conversion vs. time (Fig. 1), grafting ratio vs. time (Fig. 2), and grafting efficiency vs. time (Fig. 3) (Ref. 7).

The above results for polymerization of methyl methacrylate onto the various pulps indicate rather extensive grafting of the MMA to GP and extracted GP. Unexpectedly, the grafting reaction with the DP sample resulted in considerably more homopolymers than the GP pulps. As shown in Fig. 4, the grafting efficiency with the GP pulp is pH dependent and increases with initiator concentration. Detailed data for GP pulp grafting are given in Table 2.

TABLE 2

Graft copolymerization. Effects of initiator concentration and pH of reacting solution for MMA onto groundwood pulp (GP) at 60°C for 4 hours CP samples about 0.47 g and MMA amount 4.68 g (5.00 ml)

Expt. No.	Initiator conc. mmol/l		pH of solu.	Total conv. (%)	Graft. ratio (%)	Graft. eff. (%)
	HAc	H_2O_2				
1	80	100	1.1	9.8	3.5	3.6
2	80	100	2.7	42.6	273.6	64.5
3	80	100	6.0	13.5	90.4	67.8
4	8	10	1.2	19.7	22.7	11.5
5	8	10	3.1	19.8	62.6	31.8
6	8	10	6.2	44.9	206.1	46.2
7	0.8	1	1.0	10.6	6.0	5.7
8	0.8	1	2.8	21.7	18.9	8.7
9	0.8	1	5.5	7.7	33.3	43.2

TABLE 3

Grafting of methyl methacrylate (MMA) onto rayon fibers using HAc (8.4 mmol/l) and H_2O_2 (4.2 mmol/l)[a]

Substrate g	Monomer g	Time h	Total conv. (%)	Graft. ratio (%)	Graft. eff. (%)
0.536	4.45	1.0	7.4	56.4	91.8
0.508	"	2.0	39.4	229	66.3
0.534	" (5.00 ml)	3.0	39.8	240	72.3
0.510	"	4.0	42.7	325	81.9

[a]Temp. 60°C, reaction time varied.

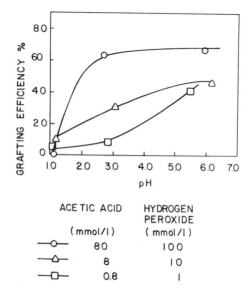

Fig. 4. Grafting efficiency vs. pH of reacting solution for MMA onto groundwood pulp at 60°C for 4 hours (Ref. 7).

The grafting parameters have also been found to vary with the monomer type. Figs. 5, 6, and 7 show the results for grafting the GP and DP pulps with 4-vinyl pyridine (VP) using essentially the same conditions as described for MMA grafting (except pH 5.9). The plots of grafting efficiency, total conversion and grafting ratio versus time indicate differences between DP and GP as substrates, similar to what was found with MMA monomer. However, the patterns of the grafting parameter curves exhibit differences for the two monomers.

Rayon fibers have also been grafted utilizing the peracetic acid initiator. The results in Table 3 indicate that these regenerated fibers are more accessible than wood pulps to grafting of MAA utilizing this initiator system. High grafting efficiencies resulted from the rayon-MMA copolymerization.

A number of other substrates have been shown to be suitable for grafting via this technique. Cellophane has been found to react like rayon as a substrate for grafting (Bhattacharyya and Rånby, unpublished results), while other synthetic material showed different behavior. Thin films of 6,6-nylon and poly(ethylene terephthalate) grafted with MMA and VP using the peracetic acid initiator, exhibited lower

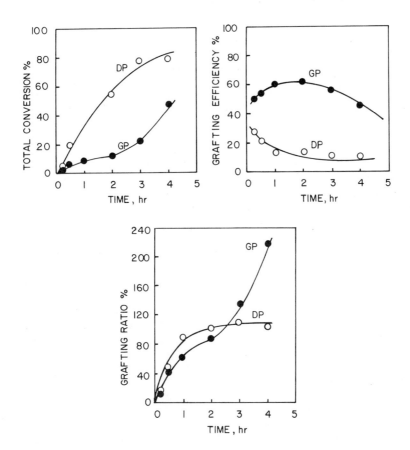

Figs. 5, 6, and 7. Grafting of 4-vinyl pyridine onto dissolving pulp (DP) and groundwood pulp (GP), initiated by peracetic acid from HAc (8 mmol/l) and H_2O_2 (10 mmol/l) at 60°C and pH 5.9. Total conversion vs. time (Fig. 5), grafting ratio vs. time (Fig. 6), and grafting efficiency vs. time (Fig. 7) (Ref. 7).

grafting efficiency compared to wood pulps and rayon (Table 4) (Bhattacharyya and Rånby, unpublished results).

Peracetic acid as an initiator for grafting of wood pulps has the advantage that it is a bleaching agent as well and that it can be applied not only to purified cellulose but also to wood pulps rich in lignin. Its shortcomings as initiator are that homopolymer is formed as a competing reaction with grafting, i.e., the grafting efficiency is far from 100%.

TABLE 4

Grafting of methylmethacrylate (MMA) and 4-vinylpyridine (VP) onto 6,6 nylon film and poly(ethylene-terepththalate) film using HAc (8.4 mmol/l) and H_2O_2 (4.2 mmol/l) at 60°C for 4 h

Substrate (weight g)	Monomer (weight g)	pH	Total conv. (%)	Graft. ratio (%)	Graft. eff. (%)
Nylon (0.504)	MMA (4.74)	6	52.8	82.7	15.8
PETP (0.497)	MMA (4.74)	8.2	62.1	98.6	15.8
Nylon (0.568)	VP (5.00)	4	12.3	22.6	22.8
PETP (0.528)	VP (5.00)	4	12.3	22.0	37.6

Mn^{3+} COMPLEX AS INITIATOR FOR GRAFTING

It has been reported from preliminary experiments (Singh, Thampy, Chipalkatty, 1965) that Mn^{3+} in aqueous solution can be used as initiator for grafting to polysaccharides with reactions similar to those of Ce^{4+} ions (Singh, Thampy, Chipalkatty, 1965). Mn^{3+} ions are stable only in strongly acid solutions (Waters, 1958). To prevent their precipitation and disproportionation at moderate acidities, we found that complex formation with pyrophosphate was quite efficient without interfering with the electron transfer reactions (Mehrotra, Rånby, 1977). Mn^{3+} ions in aqueous solution are prepared by dissolving an Mn^{2+} salt, e.g., $MnSO_4 \cdot H_2O$, and titration potentiometrically with a $KMnO_4$ solution (see below). With a pyrophosphate salt present ($Na_4P_2O_7 \cdot 10H_2O$), a stable red to violet solution of $Mn(H_2P_2O_7)_3^{3-}$ was formed and used as an initiator after adjusting the pH to a desired value. The color is pH-dependent.

Previous attempts to initiate grafting of vinyl monomers to amylose using sulfate complexes of Mn^{3+} ions gave only low grafting efficiencies and degraded chains (Patra, Ghosh, Patnaik, Thampy, 1968). Mn^{3+} pyrophosphate complex ions were immediately successful as initiating agent. The preparation of Mn^{3+} complex ion solutions is illustrated in Table 5 and the first grafting experiments with starch substrates are listed in Table 6. The native potato starch was in granular form and the hypochlorite bleached starch (Farinex A 90) and quarternary amine starch derivative (Posamyl E) were powders, which were all suspended in aqueous systems when grafted.

PREPARATION OF Mn^{3+} INITIATOR

$$4\ Mn^{2+} + Mn^{7+} \xrightarrow{H^+\ aq.} 5\ Mn^{3+}$$

$$(MnSO_4 \cdot H_2O)\quad (KMnO_4) \qquad\qquad + (Na_4P_2O_7 \cdot 10H_2O)$$

$$\downarrow aq.\ acid$$

$$\left[Mn\ (H_2P_2O_7)_3\right]^{3-}$$

$$Mn^{3+}\ complex$$

Stable complex ion in aq. solution:
pH 6 - Dull red
pH 1-2 - Bright red-violet (acid. w. H_2SO_4)

TABLE 5

Quantities of $MnSO_4 \cdot H_2O$ and $KMnO_4$ for the preparation of the manganic pyrophosphate initiator[a]

$MnSO_4 \cdot H_2O$ per 100 ml distilled water, g	0.1217	0.2434	0.4057	0.8113	1.623	2.0285	2.434
$KMnO_4$ per 100 ml distilled water, g	0.0285	0.0569	0.095	0.1897	0.379	0.4745	0.569
Resulting concentration of Mn^{3+} in reaction vessel, mmol/l	0.15	0.3	0.5	1.0	2.0	2.5	3.0

[a] $[Na_4P_2O_7] = 3 \times [Mn^{3+}]$ in each Mn^{3+} solution.

TABLE

Graft copolymerization of vinyl monomers onto starch and starch derivatives initiated by manganic pyrophosphate[a]

Expt. No.	Substrate	Monomer	Product yield, g	Conversion of monomer to polymer, g/batch and (%)	Homopolymer (% of total synthetic polymer formed)	Graft. eff., %	Add-on[b] %	Mol. wt. of grafted chains, M_n	Frequency of grafts AGU/chain
1	Native potato starch	Acrylonitrile	13.6	6.10 (75.3)	4.9	95.1	44.7	86,000	650
2	Native potato starch	Acrylonitrile	13.75	6.25 (77.2)	4.4	95.6	45.3	87,900	655
3	Native potato starch	Acrylonitrile	13.73	6.20 (76.5)	4.2	95.8	45.0	83,500	640
4	Farinex A90 (Hypochl. ox.)	Acrylonitrile	13.0	5.5 (67.9)	5.9	94.1	40.9	112,200	997
5	Posamyl E (Quart. amine)	Acrylonitrile	13.15	5.65 (69.7)	6.8	93.2	41.6	190,500	1,648
6	Native potato starch	Methyl methacrylate	12.4	4.90 (52.1)	13.3	86.7	36.2	2.5×10^6	27,250
7	Native potato starch	Acrylamide	9.31	3.09 (30.9)	41.4	58.6	19.4	--	--

[a] Starch substrate: 7.50 g (dry basis); monomer: acrylonitrile (10 ml = 8.1 g); methyl methacrylate (10 ml = 9.4 g); acrylamide = 10 g (Mn^{3+}) = 1.08×10^{-3} m/l; ($Na_4P_2O_7$) = 21.6×10^{-3} m/l; (H_2SO_4) = 68.61 mmol/l; reaction time = 75 min; temperature = 30°C.

Farinex A90 and Posamyl E are commercial names of hypochlorite oxidized and quarternary cation derivatives of native potato starch, respectively, manufactured by AB Stadex, Sweden.

[b] Definition by G. F. Fanta, ref. (1).

TABLE 7

Reproducibility of grafting reactions[a]

Expt. No.	Product yield, g	Conversion of monomer to polymer, g/batch and (%)	Grafting efficiency, %	Add-on %	Average molecular weights of grafted chains, $\overline{M}_n$	Frequency of grafts, AGU/chain
1	37.7	17.7 (88.5)	98.1	46.5	156,000	1,100
2	37.5	17.5 (87.5)	97.7	46.1	141,000	1,020

[a] Starch substrate: 20 g potato starch (dry basis); $(H_2SO_4) = 80 \times 10^{-3}$M; acrylonitrile (25 ml = 20.5 g); $(Na_4P_2O_7) = 10 \times 10^{-3}$M. Reaction time = 3 hours; temperature = 30°C; $(Mn^{3+}) = 2.0 \times 10^{-3}$M.

Acrylonitrile monomer gave very high grafting efficiencies (93-95%) and rather low molecular weights, while methyl methacrylate gave somewhat lower (86%) grafting efficiency, but much higher molecular weight. Acrylamide gave large amounts of homopolymer (41% total synthetic polymer formed).

The grafting experiments with Mn^{3+} ions are quite reproducible and close to 100% efficient. The 2% homopolymer formed is insignificant (Table 7). We have also observed that no polymer is formed when Mn^{3+} initiator and monomer are charged without starch substrate added, which verifies the specificity of the grafting reaction.

The mechanism of the initiation reaction is probably a glycol cleavage of the kind proposed by Drummond and Waters (1953) and should follow Scheme I given in formula (3). A probable termination reaction is in formula (4). There is a possibility that the initiation reaction involves enolized aldehyde groups according to Scheme II formula (5), e.g., for oxidized starch (cf. Table 6). It seems likely that acrylamide also can take part in an initiation reaction with Mn^{3+} ions, since this would lead to homopolymer formation according to formula (6), (7), and (8) in Scheme III. This postulation is consistent with the experimental results obtained (cf. Table 6).

The physical state of the substrate has a strong influence on the grafting process. Pretreatment of the starch in water at temperature > 60° causes irreversible swelling. In a subsequent grafting reaction with acrylonitrile, the molecular weight ($\bar{M}_n$) and the grafting frequency (AGU/chain) increases sharply at the swelling temperature (Fig. 8), probably due to decreased termination rate in the starch "gel" formed. Increased Mn^{3+} concentration above 0.5 mmol/l has only an insignificant effect on molecular weight and frequency of grafts (Fig. 9). The grafting properties of methyl methacrylate (MMA) are quite different from those of AN. Both molecular weight and grafting frequency are much higher than for AN (Fig. 10) by a factor of 10-15. Conversion of MMA to grafts and add-on (Fig. 11) are both high at Mn^{3+} concentrations above 0.5 mmol/l. Again, this may be due to low termination rate of growing MMA chains.

The grafting initiation reaction with Mn^{3+}-complex ions is a remarkably reproducible and specific reaction (Table 8). With low Mn^{3+} concentration (1 mmol/l) at room temperature for 3 hours, it gives 75-84% conversion of AN to polymer of which more than 97% is grafted.

The Mn^{3+}-complex ion initiator has also been applied successfully to cellulose fibers as substrates (Sallmen, 1977). The experiments were carried out with three monomers (AN, MMA, and n-butyl acrylate) and three substances:

SCHEME I. Glycol cleavage mechanism with Mn^{3+} ions leading to initiation of graft copolymerization.

Initiation:

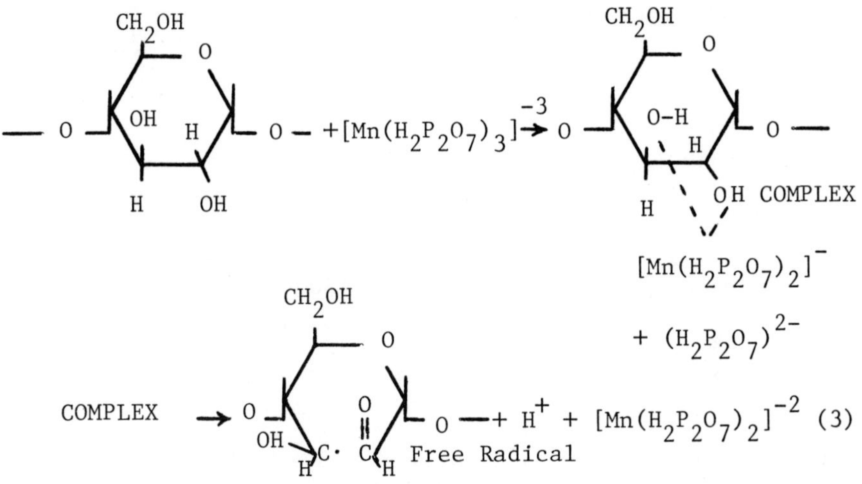

[Structure: glucopyranose unit with CH_2OH, OH, H substituents] $+ [Mn(H_2P_2O_7)_3]^{-3} \rightarrow$ [glucopyranose unit with O-H···OH COMPLEX]

$$[Mn(H_2P_2O_7)_2]^{-}$$

COMPLEX $\rightarrow$ [ring-opened structure with C·C Free Radical] $+ H^+ + [Mn(H_2P_2O_7)_2]^{-2}$ $+ (H_2P_2O_7)^{2-}$ (3)

Propogation: Free radical + Monomer $\longrightarrow$ Graft Copolymerization

Termination: $(M)_x M\cdot + Mn^{3+} \longrightarrow Mn^{2+} + H^+ +$ Polymer (4)

SCHEME II. Enolization of oxidized starch and subsequent oxidation with Mn^{3+} ions leading to initiation of graft copolymerization.

$$R_1R_2CH - \overset{O}{\underset{\|}{C}} - H + H^+ \rightleftharpoons R_1R_2CHCH\overset{OH}{\underset{+}{|}} \rightleftharpoons R_1R_2C = CH\overset{OH}{\underset{|}{|}} + H^+$$

$$R_1R_2C = CH\overset{OH}{\underset{|}{|}} + Mn^{3+} \longrightarrow Mn^{2+} + H^+ + (R_1R_2C = \overset{H}{\underset{\cdot}{C}} - O)$$

Enol

$$\downarrow H$$

$$R_1R_2C - \overset{H}{\underset{|}{C}} = O \quad (5)$$

Free Radical + Monomer   Graft Copolymerization

SCHEME III. Proposed mechanism for homopolymerization of acrylamide by Mn^{3+} initiation.

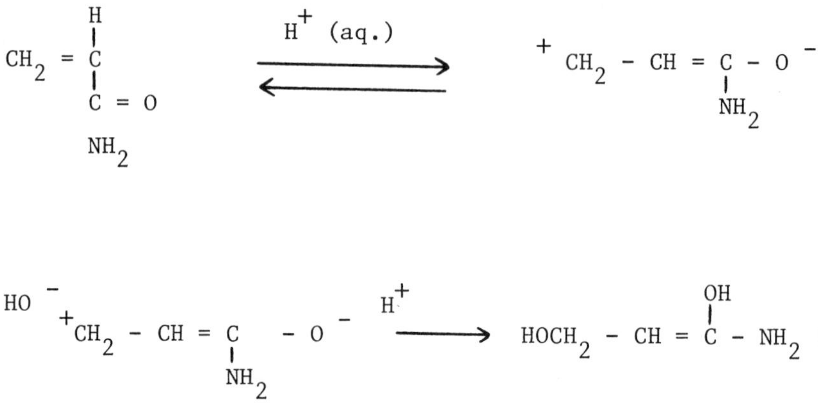

$$CH_2 = \underset{\underset{NH_2}{|}}{\overset{\overset{H}{|}}{C}} \quad \underset{\longleftarrow}{\overset{H^+ (aq.)}{\longrightarrow}} \quad {}^+CH_2 - CH = \underset{\underset{NH_2}{|}}{C} - O^-$$

$$HO^- + {}^+CH_2 - CH = \underset{\underset{NH_2}{|}}{C} - O^- \quad \overset{H^+}{\longrightarrow} \quad HOCH_2 - CH = \underset{\underset{NH_2}{|}}{\overset{\overset{OH}{|}}{C}} - NH_2 \quad (6)$$

Enol of β-hydroxypropionamide

$$Enol + Mn^{3+} \longrightarrow Mn^{2+} + H^+ + \text{Free Radical} \quad (7)$$

$$\text{Free Radical} + \text{Monomer} \longrightarrow \text{Homopolymerization} \quad (8)$$

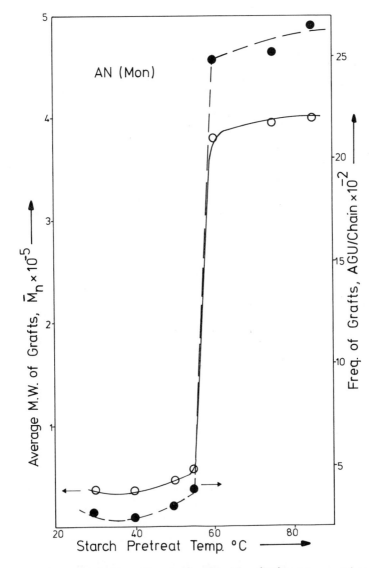

Fig. 8. Grafting of acrylonitrile (AN) onto native potato starch after pretreatment in water at the temperatures indicated. Initiator is Mn^{3+} complex (3 mmol/l), reaction temperature 30°C and reaction time 3 hours. Average molecular weight $\bar{M}_n$ and frequency of grafts (AGU/chain) vs. pretreatment temperature.

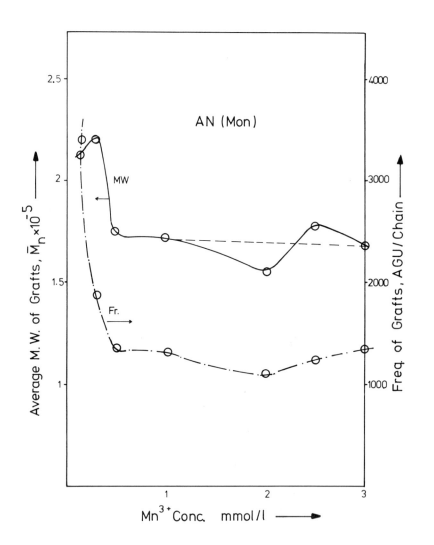

Fig. 9. Grafting of acrylonitrile (AN) onto native potato starch. Initiator is Mn^{3+} complex, reaction temperature 30°C, and reaction time 3 hours. Average molecular weight of grafts ($\bar{M}_n$) and frequency of grafts (AGU/chain) vs. Mn^{3+} concentration.

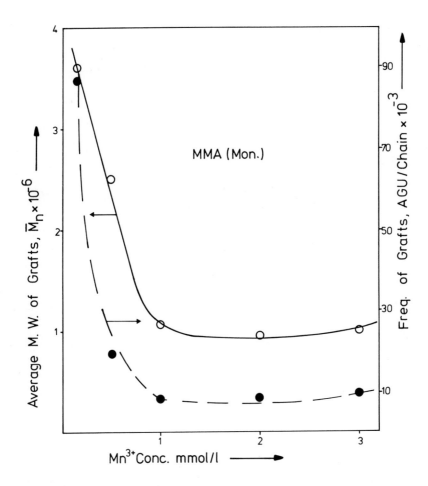

Fig. 10. Grafting of methyl methacrylate (MMA) onto native potato starch. The following conditions were utilized: initiator: Mn^{3+} complex, reaction temperature - 30°C, reaction time - 3 hours, starch substrate - 20 g, MMA - 25 ml. Average molecular weight of grafts ($\bar{M}_n$) and frequency of grafts (AGU/chain) vs. Mn^{3+} concentration.

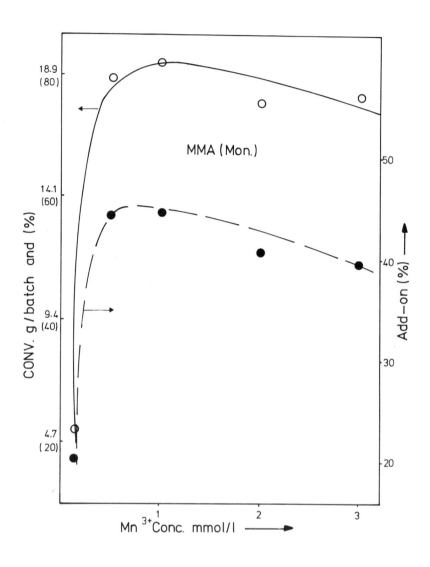

Fig. 11. Grafting of methyl methacrylate (MMA) onto native potato starch. Conditions identical to those described for Fig. 10. Conversion of MMA (g and %) and add-on (%) vs. Mn^{3+} concentration.

TABLE 8

Reproducibility of grafting reactions[a]

Expt. No.	Starch pretreatment temperature, °C[b]	Product yield, g	Conversion of monomer to polymer, g/batch and (%)	Add-on %	Average molecular weight of grafts, $\bar{M}_n$	Graft. frequency AGU/chain
1	30	19.3	9.3(77.5)	47.6	41,400	280
2	30	19.1	9.1(75.8)	47.1	33,000	230
3	60	19.8	9.8(81.7)	48.7	370,000	2,410
4	60	19.9	9.9(82.5)	48.4	391,000	2,570
5	85	20.0	10.0(83.3)	49.2	404,000	2,580
6	85	20.1	10.1(84.2)	49.2	430,000	2,740

[a]Starch substrate: 10 g potato starch (dry basis); $[H_2SO_4] = 85.8 \times 10^{-3}$ M; $[Na_4P_2O_7] = 10 \times 10^{-3}$ M; reaction time = 3 hours; acrylonitrile (15 ml = 12 g); reaction temperature = 30°C; $[Mn^{3+}] = 1.0 \times 10^{-3}$ M.
[b]Starch was treated at this temperature for 1 hour before the graft copolymerization reaction.

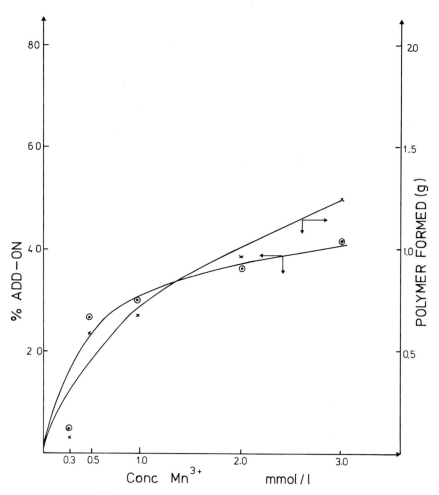

Fig. 12. Grafting of acrylonitrile (AN) onto cotton linters (CL). Initiator is Mn^{3+} complex, reaction temperature 30°C and reaction time 2 hours. CL charged 1.52 g and AN charged 10.0 ml. Add-on (%) and total conversion of AN to polymer (g) vs. Mn^{3+} concentration (Ref. 14).

cotton linters, dissolving wood pulp and bleached paper pulp, the latter two both from spruce wood. The results, summarized in Table 9, show that high grafting efficiencies are obtained with AN and MMA (80-93%) for the three pulps, while n-BA gives a much lower GE value.

The cotton linters (CL) - acrylonitrile combination exhibited the best grafting results for the systems studied. The reaction rates of grafting onto CL at various pH-values are similar to those of native granular starch (Fig. 12).

TABLE 9

Grafting of acrylonitrile (AN), n-butyl acrylate (BA), and methyl methacrylate (MMA) onto cellulose fibers, using Mn^{3+} pyrophosphate complex as initiator, at 30°C for 2 hours

Substrate (weight, g)	Monomer (ml)	Mn^{3+} initiator conc. mmol/l	Add-on %	Graft eff., %	Graft. freq. AGU/chain	$\bar{M}_n$ of grafts
Cotton linters						
(1.519)	AN (5.0)	3	24.5	93.6	952	50,000
(1.519)	AN (10.0)	3	41.5	89.7	1,033	119,000
(1.519)	AN (15.0)	3	49.3	90.5	832	131,000
Diss. pulp from wood						
(4.80)	AN (25.0)	2	62.4	87.6	--	--
(3.16)	AN (25.0)	2	70.1	82.7	--	--
(1.501)	u-BA (10.0)	2	73.0	58.7	--	--
Bleached wood pulp						
(1.498)	MMA (10.0)	2	58.9	79.9	--	--

This system results in add-on of more than 70% (Fig. 13) and a grafting frequency approaching 1,000 AGU units/chain for Mn^{3+} concentrations of 2-3 mmol/l (Fig. 14). The molecular weights of the grafts are in the range 100-120,000.

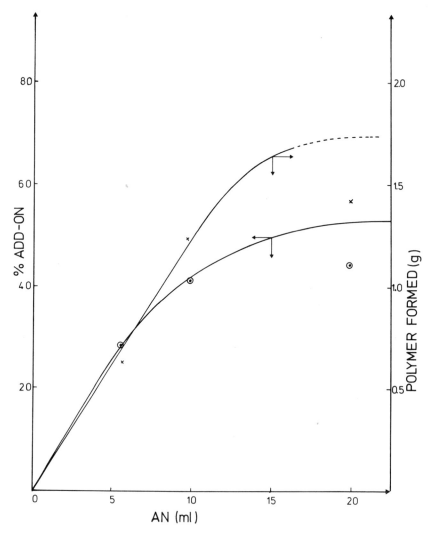

Fig. 13. Grafting of acrylonitrile (AN) onto cotton linters (CL). The following conditions were utilized: initiator - Mn^{3+} complex, reaction temperature - 30°C, reaction time - 2 hours, CL - 1.52 g. Add-on (%) and total conversion of AN to polymer (g) vs. amount of AN charged (ml) (Ref. 14).

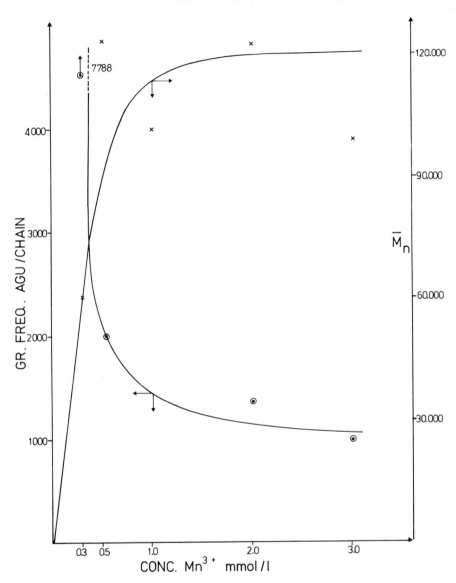

Fig. 14. Grafting of acrylonitrile (AN) onto cotton linters (CL). Conditions identical to those given for Fig. 13. Grafting frequency (AGU/chain) and average molecular weight of grafts ($\bar{M}_n$) vs. Mn^{3+} concentration (Ref. 14).

The grafting reactions with cellulose samples were completed in 2 hours at 30°C. However, pure cellulose substrates of low lignin content were necessary for the reaction to take place. Groundwood pulps gave no grafting reaction at all and no homopolymer was formed.

IV. CONCLUSIONS

Two new initiators for graft copolymerization onto cellulose and starch by radical mechanism have been developed:
1. <u>Peracetic acid</u>, prepared <u>in situ</u> from acetic acid and hydrogen peroxide in acid aqueous solution initiates graft copolymerization to cellulosic fibers of monomers like methyl methacrylate and 4-vinyl pyridine. The method can be applied both to purified wood pulps and paper pulps of high lignin content. The initiation reaction, however, is not specific and considerable amounts of homopolymer are formed.
2. <u>Mn^{3+} complex ions</u> initiate graft copolymerization in aqueous suspensions of starch, purified cellulose, and other polysaccharides. The reaction is a free radical electron transfer mechanism and it is inhibited by the presence of lignin. Initiation of Mn^{3+} ions, complexed with pyrophosphate ions, is a very specific reaction which gives high grafting efficiencies (> 95%) and only very small amounts of homopolymer (a few percent) with monomers like acrylonitrile and methyl methacrylate.

REFERENCES

Battaerd, H.A.J., and Tregear, G. W. 1967. In Graft Copolymers, Polymer Reviews 16 (H. F. Mark and E. H. Immergut, eds), Interscience, New York.
Fanta, G. F. 1973. In Block and Graft Copolymerization 1 (R. J. Ceresa, ed.), Wiley-Interscience, New York.
Mino, G., and Kaizermann, S. 1958. <u>J. Poly. Sci.</u> 31, 242.
Ogiwara, Y., and Kubota, H. 1968. <u>J. Appl. Poly-. Sci.</u> 12, 2575.
Chaudhuri, D.K.R., and Hermans, J. J. 1961. <u>J. Poly. Sci.</u> 51, 573.
Faessinger, R. W., and Conté, J. S. 1967. U.S. Patent 3,330,787.
Krässig, H. 1971. <u>Svensk Pappetidn.</u> 74, 417.
Hatakeyama, H., and Rånby, B. 1975. <u>Cellulose Chem. Tech.</u> 9, 583.
Mehrotr, R., and Rånby, B. 1977. <u>J. Appl. Poly. Sci.</u> 21, 1647.
Bhattacharyya, N., and Rånby, B, unpublished results.

Singh, H., Thampy, R. T., and Chipalkatty, V. B. 1965.
  J. Poly. Sci. A3, 4289.
Waters, W. A. 1958. Quart. Rev. (London) 12, 296.
Patra, S. K., Ghosh, S., Patnaik, B. K., and Thampy, R. T.
  1968. Chem. Soc. (London), Spec. Publ. 233.
Drummond, A. Y., and Waters, W. A. 1953. J. Chem. Soc. 3119.
Sallmén, B. 1977. Graft Copolymerization onto Cellulose.
  Senior Problem in this dept., June.

MODIFICATIONS TO CELLULOSE USING UV GRAFTING PROCEDURES

N. P. DAVIS AND J. L. GARNETT

School of Chemistry, University of New South Wales
Kensington, Australia

A detailed investigation has been performed of the copolymerization of styrene to cellulose initiated by UV. A variety of different experimental conditions have been used including both solvent and solvent-free systems. The simultaneous irradiation procedure has been utilized exclusively for these studies. The variables influencing this grafting reaction and examined in this work include effect of solvent structure, role of sensitizer, effect of oxygen and homopolymer formation. The inclusion of novel additives to improve the photosensitized grafting process is compared with the corresponding technique using gamma radiation initiation. These UV copolymerization studies are shown to be of fundamental value in the study of bond formation occurring during rapid cure UV processing of polymers on cellulose.

INTRODUCTION

Extensive work has previously been published for modifying the properties of cellulose by gamma ray induced grafting of monomers (1-5). The technological development of this process has been restricted for the following essential reasons. The basic capital cost associated with the installation of the large cobalt-60 and related facilities is high. For optimum grafting efficiency, the presence of solvent is usually necessary, thus requiring the installation of expensive solvent recovery plants for recycling. Rates of copolymerization, even with large gamma ray sources, are relatively slow and competing homopolymerization is always a potential difficulty since it is wasteful in the efficient utilization of monomer and, for many applications, must be removed from the product in an additional step at the end of the reaction.

These problems with gamma ray processing have led to the development of rapid cure, solvent free systems initiated by electron beam or ultraviolet light radiation sources. Techniques involving UV are particularly attractive since the UV sources are relatively cheap, flexible and easy to instal. Many of the UV processes currently in use for modifying the properties of cellulose involve a simple polymerization coating technique. The interesting possibility associated with such coating processes is the degree to which surface copolymerization occurs between polymer and cellulose. Preliminary studies indicate that under certain rapid cure, UV polymerization conditions, some sigma bonding between polymer and cellulose does occur. These results indicate that more information is required on the nature of the bonding which occurs in such UV processes. In particular, it would be valuable to know the experimental conditions required to obtain optimum photosensitized grafting yields when cellulose is used as trunk polymer.

It is the purpose of the present paper to report a fundamental study of the mechanism of the photosensitized grafting of styrene to cellulose. Isolated studies of this copolymerization reaction have previously been published, however no detailed report similar to the gamma ray work (3) of the photosensitized grafting process is available. Oster and Yang (6) have reviewed the types of photosensitizers used in photopolymerization but no photosensitized grafting to cellulose was discussed in this article. In previous experiments (7,8), dyes were used to sensitize the UV grafting of monomers such as styrene to cellulose. Ferric ion has also been shown to sensitize the same copolymerization reaction (9) especially in the presence of hydrophilic solvents such as methanol, acetone and dioxan. However, except for a preliminary publication (10) no systematic study of the use of a particular homologous series of solvents such as the alcohols for this grafting process has been reported. In the current work, a detailed investigation of the role of both wetting and non-wetting solvents in the photosensitized copolymerization is described. In addition novel additives which accelerate the grafting reaction will be reported. Mechanistic studies of the role of oxygen, radical scavengers and type of sensitizer have also been investigated. The conditions under which neat styrene with no solvent present may be UV grafted to cellulose have also been studied.

Finally a brief comparison of the present photosensitized results with the corresponding extensive data (3) from the analogous gamma ray induced grafting process has been carried out in order to determine whether similar intermediates are involved in the reaction mechanism.

EXPERIMENTAL

A.   Materials

Styrene and methyl methacrylate monomers were donated by Monsanto Chemicals (Aust.) Ltd. and Rohm and Haas Australia Pty. Ltd. respectively. All remaining monomers were purchased from Polysciences Inc. Inhibitor was removed prior to use by passing the monomer down a column of activated alumina. This process was previously shown to be just as satisfactory as vacuum distillation for purification of monomers for grafting. All solvents in the copolymerization solutions were AR grade and used without further treatment.

B.   Grafting Process

The UV source used in the grafting work was a Phillips 90-W high pressure mercury vapor lamp fitted with a quartz envelope. The samples of cellulose to be treated were cut from Whatman No.41 filter paper (5x4 cm) and were extracted in hot appropriate solvents for 72 hrs. prior to grafting. These strips were weighed, before and after grafting, at a constant temperature of $21°C$ and 65% RH. For the copolymerization experiments, the cellulose strips were completely immersed in a solution (25ml) of monomer in solvent. For most runs the appropriate photosensitizer was present in the grafting solution. Lightly stoppered pyrex tubes were used for simplicity in the present work. Although evacuated quartz tubes give higher grafting efficiencies, reasonable rates of copolymerization could still be obtained in the pyrex vessels. To ensure uniformity of exposure to UV, all tubes were held in a rotating rack and were irradiated at distances of 12 or 24cm from the source. Actinometry was performed with the uranyl nitrate-oxalic acid system, although some calibrations were also carried out with potassium ferrioxalate. At the completion of the reaction, the cellulose strips were quickly removed from the tubes to avoid post-irradiation effects, washed in the appropriate solvents and then extracted in these same solvents in a Soxhlet for at least 72 hours to remove homopolymer. The percent graft was then considered to be the percentage increase in weight of the cellulose strip. In many previous experiments, this procedure was shown to be satisfactory for the complete extraction of homopolymer and residual monomer for grafting by the analogous radiation technique (3). It was also satisfactory in the current ultraviolet radiation work.

C. γ-Grafting Procedure

The experimental procedure for grafting in the presence of gamma rays was similar to the UV technique already described, however smaller tubes containing smaller volumes (5ml) of solution were used. The actual irradiations were carried out in either the spent fuel element source of the Australian Atomic Energy Commission's HIFAR reactor (Sydney) or in a 500 Curie cobalt-60 pond facility at the University.

RESULTS AND DISCUSSION

A. Grafting in Wetting Solvents

Two general classes of wetting solvents have been used in the present work, an arbitrary classification into alcohols and solvents other than alcohols being made for convenience of treatment. Because of the large amount of work carried out with the alcohols, they will be considered first.

1. Alcohols

The results (Tables 1-3) for the photosensitized copolymerization of styrene in the alcohols to cellulose demonstrate the dramatic effect of chain length of these solvents on grafting efficiency. Thus, there is a progressive decrease in grafting yield with increasing chain length and degree of branching of the alcohol, copolymerization being most efficient in methanol and effectively cutting out with the butanols. Thus the size of the alcohol molecule is important in these processes. A small molecule, such as methanol, is not only capable of swelling the trunk polymer but also has the additional advantage of being miscible in all proportions with the styrene monomer, thus simplifying access and diffusion to grafting sites in the cellulose trunk polymer. By contrast, butanol is a relatively poor swelling solvent for cellulose and the grafting is correspondingly low.

The other significant feature of the present alcohol work, is that the grafting yield goes through a maximum under certain solvent conditions, 80% styrene for n-propanol and 90% for ethanol and methanol. This Trommsdorff or gel effect (11) which functions as the viscosity of the medium increases or precipitation of the polymer occurs also depends on the total

TABLE 1

Photosensitized Grafting of Styrene to Cellulose Using Alcohols as Solvents[a]

Monomer Conctn. % v/v	20	40	60	80	90
Alcohol			% Graft		
Methanol	13	28	34	53	64
Ethanol	9	17	22	50	70
n-Propanol	5	12	18	30	20
isopropanol	<5	7	5	14	8
n-Butanol	<5	<5	<5	<5	5
isobutanol	<5	<5	<5	<5	<5
t-Butanol	<5	<5	<5	<5	<5
n-Octanol	<5	<5	<5	<5	<5

[a] *Solutions contained 1% w/v of uranyl nitrate and irradiated for 24 hr. at 24 cm. from 90-W high-pressure ultraviolet lamp.*

TABLE 2

Photosensitized Grafting of Styrene to Cellulose Using Methanol-Isobutanol and Methanol-Octanol in a Fixed Ratio (1:1) as Mixed Solvents[a]

Monomer Concn. %	20	40	60	80	90
Alcohol			% Graft		
n-Octanol	<5	<5	<5	<5	<5
isobutanol	<5	<5	<5	<5	<5
Methanol	13	28	34	53	64
Methanol/isobutanol (1:1)	9	26	45	67	
Methanol/n-Octanol (1:1)	35	49	60	78	

[a] *Solutions contained 1% w/v of uranyl nitrate and irradiated for 24 hr. at 24 cm. from a 90-W high-pressure ultraviolet lamp.*

TABLE 3

Photosensitized Grafting of Styrene to Cellulose by Using Methanol-Octanol in Different Ratios as Mixed Solvents[a]

% Octanol in Solvent	0	10	25	33	40	50	55	60	66	75	80	85	90	95	100
Monomer Concentration %						% Graft									
80	53	57	61	-	68	76	80	82	90	78	32	16	5	4	5
60	34	38	48	-	-	54	60	75	89	51	-	-	-	-	4
40	28	-	30	33	29	48	41	50	50	32	-	-	-	-	3

[a] Experimental conditions as in Table 2.

ultraviolet radiation dose delivered (12). The observation of a gel effect is important especially if copolymers are to be prepared utilizing the maximum efficiency of the UV. This accelerated polymerization process is particularly related to the change in viscosity of the medium. At higher viscosities, bimolecular chain termination of the radical chains is hindered, whereas initiation, chain propagation and radical transfer processes are not affected to the same degree with increasing viscosity because the molecules involved are smaller and more mobile. The Trommsdorff effect is thus responsible for a rapid consumption of monomer and the production of high molecular weights. The gel effect is further accentuated in the present grafting work because the mobility of the polymer chains is restricted by the cellulose structure. When grafting commences, the cellulose fibre swells (13) and thus assists the diffusion of monomer to the growing chains and active sites on the cellulose. Swelling predominantly occurs when the gel effect has begun to operate inside the fibres but not in the external solvent system.

An interesting feature of the alcohol results is the increase in grafting which is observed when methanol, an active solvent, is mixed (1:1) with a poor solvent such as isobutanol or n-octanol (Table 2). In the methanol-isobutanol mixture, an increase in grafting yield when compared with the methanol system is only observed above 50% monomer concentration, whereas in methanol-n-octanol, copolymerization is enhanced at all monomer concentrations studied. With the methanol-n-octanol solvent system, a peak in grafting is found at 66% n-octanol in methanol (Table 3).

A plausible explanation for the mixed alcohol effect is that the longer chain alcohol is a better solvent than methanol for styrene homopolymer. Thus in a mixed methanol-n-octanol solvent, homopolymer would remain in solution, the turbidity effect which inhibits grafting would be effectively eliminated and enhanced grafting observed. Turbidity effects are most important in photosensitized grafting, since precipitation of homopolymer from the grafting solution can lead to turbidity and termination of the copolymerization reaction. The reason for an optimum in enhancement occurring at a certain octanol concentration is presumably due to a compromise in the role of methanol, since sufficient methanol must still remain in the solvent mixture to allow efficient swelling and permit grafting.

However, an examination of the effect of chain length of alcohol on the degree of grafting enhancement in mixed alcohols shows that isobutanol is less effective than n-octanol (Table 3). Thus, turbidity effects and solubilization of homopolymer alone do not seem to completely account for the magnitude of the enhancement effect. A further plausible explanation is that octanol and/or isobutanol complex preferentially with the monomer during the actual grafting process. Complexes between alkanes and aromatics are known (14), spectroscopic evidence shows that alkanes approach to within 4 Å of the aromatic in mixtures of the two, such as in solutions of hexane and benzene. Linear alkanes are particularly effective and virtually complex across the benzene ring in such mixtures (14). Thus, in the present grafting system, it is suggested that a contribution to the enhancement in copolymerization is due to complex formation between the alkyl portion of the octanol and the aromatic ring of the monomer. The resulting complex diffuses into the cellulose which has already been preswollen by methanol. The polar hydroxyl group on the octanol would assist diffusion of this alcohol into the swollen cellulose, while the complexing properties of the alkyl portion of this alcohol would assist diffusion of monomer into cellulose. On this model, octanol, alone as solvent, would not swell the cellulose sufficiently to yield appreciable grafting, consistent with observations. These data suggest that monomer-solvent complexes may be mechanistically important in these grafting reactions. This material may thus provide further evidence for the concept that charge-transfer complexes play a significant role as intermediates in polymerization (15,16). This aspect will be discussed further in Section I concerning the mechanism of the reaction.

2. Wetting Solvents other than Alcohols

Five representative wetting solvents other than the alcohols namely dimethyl sulfoxide, dimethyl formamide, dioxan, acetic acid and acetone have been investigated in the present photosensitized grafting work. Of these five solvents, dimethyl sulfoxide is easily the most efficient with uranyl nitrate as sensitizer (Table 4) and appears to be even marginally better than the best of the alcohols. These results reflect the well known sensitizing properties of dimethyl sulfoxide in the ultraviolet. A Trommsdorff effect is also observed at 90% monomer concentration in this solvent, similar to the gel effect found with the alcohols. Of the remaining solvents, acetic acid and then dioxan are the most useful, however with these two solvents the gel effect is now observed at 40-60% monomer concentration.

TABLE 4

UV Grafting of Styrene in Dimethyl Sulfoxide, Dimethyl Formamide, Dioxan, Acetic Acid and Acetone.[a]

% Styrene	DMSO	DMF	Dioxan		Acetic Acid	Acetone
10	-	-	-	-[b]	5.1	3.2
20	6.2	5.2	3.6	3.1	3.5	3.1
30	-	-	-	-	7.9	2.2
40	13.8	9.5	11.7	3.2	26.9	1.8
50	-	-	-	-	28.3	1.4
60	37.5	15.6	20.3	1.8	15.8	1.5
70	66.4	11.3	-	-	10.1	1.3
80	87.6	8.7	7.9	4.2	2.9	1.4
90	99.1	8.2	4.9	5.7	1.9	2.6

[a] Solutions contained 1% w/v uranyl nitrate and irradiated for 24 hr. at 24cm.

[b] No sensitizer for runs in this column.

The interesting development in grafting occurs when solvents, such as acetic acid, are added to methanol. There is a marked increase in copolymerization efficiency at all additions of acetic acid studied (Table 5). Even with 10% acetic acid in methanol, there is a large increase in copolymerization reaching 121% at 80% styrene in solvent mixture. At 25% acetic acid in methanol, the graft at 90% monomer concentration is over twofold higher than copolymerization observed in methanol alone. The enhancement in grafting for this solvent mixture also builds up to the 90% monomer concentration where a Trommsdorff effect is found.

This acetic acid enhancement is not confined to the alcohol solvents, since acetone, dimethyl formamide and dimethyl sulfoxide also exhibit the same property (Table 6). Formic acid is almost as efficient as acetic acid when dimethyl sulfoxide is used as solvent. In the presence of these two organic acid additives, dimethyl sulfoxide remains the best solvent even being superior to the most efficient of the alcohols. Of the following five organic acids studied, namely formic, acetic, propanoic, butanoic and oxalic acids, the last compound is the most effective for enhancement in

TABLE 5

Grafting of Styrene in Methanol/Acetic Acid Mixtures[a]

% Styrene	Acetic	10% Acetic	25% Acetic	50% Acetic	75% Acetic
10	5.1	5.9	–	25.6	–
20	3.5	11.3	–	42.5	–
30	7.9	15.7	–	48.7	–
40	26.9	29.9	–	64.2	–
50	28.3	53.7	54.1	77.8	–
60	15.8	59.1	66.0	105.8	–
70	10.1	83.2	81.2	55.5	107.5
80	2.9	120.6	113.8	–	96.0
90	1.9	–	134.9	115.5	–

[a] *Solutions contained 1% w/v uranyl nitrate and irradiated for 24 hrs. at 24cm.*

grafting of styrene in methanol (Table 7). The fact that oxalic acid, itself, is a sensitizer for the grafting reaction, particularly in the 40-80%

TABLE 6

Grafting of Styrene in Acetone, Dimethyl Formamide and Dimethyl Sulfoxide in the Presence of Formic and Acetic Acids[a]

% Styrene	Acetone	DMF	DMSO	DMSO[b]
9	2.8	1.6	3.2	–
19	4.7	3.6	8.6	6.1
29	12.9	–	18.6	23.4
39	18.6	13.8	37.0	35.7
49	28.4	16.2	63.7	50.6
59	30.2	33.8	85.1	74.9
69	12.3	37.6	–	97.4
79	12.4	29.4	147.8	154.7
89	8.5	38.7	–	–

[a] Solutions contained 1% w/v uranyl nitrate and irradiated for 24 hrs. at 24cm; 1% formic acid added to mixtures except b where 1% acetic acid added.

range of styrene in methanol concentrations (Table 8), suggests that the predominant mechanism of the enhancement observed in Table 7 for this acid and related homologous organic acids is synergistic involving both uranyl nitrate and oxalic acid sensitization. The process thus appears to be significantly different to that observed in Tables 2 and 3 where the addition of relatively long chain, inactive alcohols (octanol) to a short chain, active alcohol (methanol) significantly increased the styrene copolymerization yield.

B. Grafting in Non-Wetting Solvents

The importance of polarity and wetting power of solvents in the current copolymerization studies is demonstrated by the results obtained when the photosensized process is carried out in benzene and hexane as solvents (Table 9). This is the first report of the effect of non-wetting solvents in this photosensitized grafting reaction. Because uranyl nitrate is not soluble in either benzene or hexane, benzoin ethyl ether was used as sensitizer. An anologous experiment in methanol using this sensitizer was necessary for comparative purposes. Benzoin ethyl ether sensitizes appreciable styrene grafting, especially in the 40-60% monomer in methanol concentration region, however at 90% monomer in methanol, the copolymerization yield is only marginally better than the blank without

TABLE 7

Effect of Organic Acids (1% w/v) on UV Grafting of Styrene in Methanol[a]

% Styrene	Formic	Acetic	Propanoic	Butanoic	Oxalic
9	10.7	5.3	5.7	–	8.3
19	6.8	10.0	19.2	16.1	18.4
29	13.0	14.7	19.2	13.2	19.9
39	21.3	7.5	20.8	22.5	16.6
49	22.2	19.4	28.8	25.0	20.5
59	34.2	30.6	36.5	33.1	34.6
69	58.6	47.8	55.2	47.8	43.1
79	79.4	60.1	72.8	66.3	66.0
89	123.4	104.5	112.6	–	131.7

[a] *Solutions contained 1% w/v uranyl nitrate and irradiated for 24 hrs. at 24cm.*

TABLE 8

Effect of Acetic and Oxalic Acids (1% w/v) as Sensitizers in UV Styrene Grafting in Methanol[a]

	% Graft	
% Styrene	Acetic	Oxalic
20	1	–
30	1	–
40	2	35
50	2	–
60	6	30
70	7	–
80	5	36
90	17	17

[a] *Solutions irradiated for 24 hrs. at 24cm.*

sensitizer, the data for this non-sensitized grafting being available later in this manuscript. By contrast with the methanol results, virtually no grafting of styrene in either benzene or hexane is observed at <u>all</u> concentrations studied. The low copolymerization data observed with these latter two

TABLE 9

UV Grafting of Styrene in Benzene and Hexane[a]

	% Graft		
% Styrene	Methanol	Benzene	Hexane
20	7	1	1
40	33	1	1
60	34	1	1
80	28	1	1
90	31	2	-

[a] Solutions contained 1% benzoin ethyl ether and irradiated at 24cm. for 24 hrs.

solvents may be attributed to their poor cellulose swelling characteristics, grafting when it does occur, being limited essentially to the surface of the trunk polymer.

C. Effect of Scavengers and Oxygen on Grafting

Conventional radical scavengers significantly reduce the grafting yields with styrene in methanol, the order of effectiveness being acridine > thiourea > hydroquinone (Table 10). The magnitude of the depression in copolymerization yields caused by these efficient scavengers is an indication of the contribution of free radical processes to the photosensitized grafting reaction. The fact that even at relatively high scavenger concentrations, the grafting is not completely suppressed is interesting. It presumably reflects either the extent to which radicals from the photosensitizer compete directly with monomer and/or trunk polymer radicals for the scavenger or the extent to which different processes such as energy transfer reactions are involved in the grafting mechanism.

The effect of oxygen on the grafting yield appears only to be of significance at the 20% styrene in methanol solution where oxygen is a distinct inhibitor (Table 11). These results are important in a preparative sense, since for maximum efficiency of utilization of the ultraviolet light in synthesizing chemically useful quantities of copolymer, there is no necessity to work in an oxygen free atmosphere since at the

TABLE 10

UV Grafting of Styrene in Methanol Using Acridine Thiourea and Hydroquinone as Scavengers[a]

Additive	% (w/v)	% Graft at Monomer Concn. (% u/w)	
		60	80
Acridine	0.1	5	4
	1.0	4	5
Thiourea	1.0	10	8
Hydroquinone	0.1	18	29
	1.0	15	16
Nil	–	34	53

[a] Solutions contained 1% w/v uranyl nitrate and irradiated for 24 hrs. at 24cm.

TABLE 11

UV Grafting of Styrene in Methanol with Air Excluded

% Styrene	% Graft	
	Air[a]	Vacuum[b]
20	13	30
40	28	19
60	34	33
80	53	42
90	64	66

[a] Solutions contained 1% w/v uranyl nitrate in stoppered tubes and irradiated for 24 hrs. at 24cm.

[b] Same as a except tubes evacuated to 10^{-2} torr.

monomer concentration where copolymerization is highest (90%), yields in both air and evacuated tubes are almost the same. The results suggest that at all styrene concentrations studied (except 20%), oxygen does not detrimentally scavenge the active radicals responsible for grafting. Oxygen could however react with the primary radicals from the initiator to give peroxy species which are still capable of abstracting protons from the trunk polymer to create copolymerization sites. The retarding effect of oxygen in the 20% solution may reflect the degree to which the oxygen concentration in the various monomer solutions in equilibrium with air influences the grafting. The Bunsen absorption coefficient at $20°C$ for oxygen in methanol is 0.175 atm^{-1} (17). As the concentration of styrene in methanol increases, this absorption coefficient should decrease. Thus of the monomer solutions studied, only in the 20% solution would the oxygen concentration be high enough to reduce grafting.

D. Effect of Sensitizer in Grafting

The results in Tables 12 and 13 show that grafting of styrene in methanol can be accomplished to cellulose in the presence of UV without the necessity of using photosensitizers. In Table 12 (run 1), virtually no graft was observed after UV irradiating 30% styrene in methanol solutions without sensitizer for three hours. If the time of irradiation on this monomer solution is increased to 24 hours, 6.0% copolymerization is observed. With the 80% styrene in methanol solution, grafting of 24.2% is found in unsensitized solution after 24 hours irradiation.

However, the data also show that if the appropriate sensitizer is added to these solutions there is a very large increase in % graft for each of the monomer solutions studied. Uranyl nitrate is the best of the photosensitizers used for copolymerization of styrene in methanol (Table 12). If a non-alcoholic or non-aqueous grafting medium is used, then the best sensitizers appear to be biacetyl and benzoin ethyl ether. The anthraquinone dye similar to the compound used by other workers (7) is not as satisfactory for sensitization as the previously mentioned materials. Manganese carbonyl, although useful in simple photopolymerization work, (6) is not suitable for sensitizing grafting under the experimental conditions used in Table 12. Copolymerization in quartz, as expected, is more efficient than grafting in pyrex (Table 12), however, in the latter vessels, the level of graft is still satisfactory

TABLE 12

Effect of Sensitizer in UV Grafting of Styrene in Methanol to Cellulose^a

Run	Sensitizer	Distance from UV lamp (cm)	Graft
1	Nil	24	0.0
2	Nil	24	0.0^c
3	Nil	24	0.0^c
4	Uranyl nitrate	24	1.6^c
5	Uranyl nitrate	24	15.2
6	Uranyl nitrate	24	3.4
7	Uranyl nitrate	12	10.9
8	ADS^b	24	1.8
9	ADS^b	12	3.7
10	Biacetyl	12	0.0^c
11	Biacetyl	24	3.8
12	Biacetyl	12	11.0
13	Benzoin ethyl ether	24	0.0^c
14	Benzoin ethyl ether	24	6.7
15	Benzoin ethyl ether	24	1.4
16	Benzoin ethyl ether	12	5.6
17	Manganese carbonyl	24	0.9^c
18	Manganese carbonyl	24	1.5

^{a}*Solutions of styrene (30% v/v) in methanol contained in pyrex tubes with cellulose, irradiated for 3 hrs. with 90W high pressure ultraviolet lamp except runs 4, 5, 13 and 14 where quartz tubes used and runs 17 and 18 where styrene (25% v/v) solutions were utilized. Sensitizer concentration (1% w/v).*

^{b}*ADS = anthraquinone -2, 6- disulfonic acid, disodium salt.*

^{c}*Ultraviolet light excluded.*

for general use.

E. Homopolymer Formation during Grafting

One of the essential problems unique to photosensitized grafting is the formation of homopolymer and its subsequent effect on the grafting process. If the homopolymer formed is

TABLE 13

Homopolymer Formation During UV Grafting of Styrene in Methanol to Cellulose[a]

% Styrene	% Sensitizer[b]	% Graft	% Homopolymer
30	0	6.0	0.3
30	2×10^{-6}	7.6	-
30	2×10^{-5}	14.8	-
30	2×10^{-3}	17.2	-
30	1.0	21.0	-
80	0	24.2	6.0
80	2×10^{-6}	29.5	4.5
80	2×10^{-5}	42.3	4.8
80	2×10^{-4}	48.5	6.5
80	1.0	53.0	-
80	2×10^{-4} [b]	28.3	-
80	2×10^{-3} [b]	31.7	23.9

[a] *Irradiation 24cm. from 90W lamp for 24 hrs.*

[b] *Sensitizer 1% (w/v) of uranyl nitrate except last two runs where benzoin ethyl ether (1% w/v) used.*

insoluble in the monomer solution, turbidity occurs and the grafting becomes erratic or even terminates. The formation of large amounts of homopolymer, even though soluble at low concentrations in the grafting solution, can also lead to turbidity. The yield of homopolymer which is always a competing reaction to grafting is thus extremely important for the photosensitized process.

Homopolymer formation is relatively low in the unsensitized grafting of styrene in methanol (Table 13) and increases slowly with increase in percentage graft and addition of sensitizer to a particular level of sensitizer. Above that level of sensitizer increase in graft is small with a corresponding large yield of homopolymer. Thus in Tables 13 and 14 for 80% styrene in methanol, 48.5% graft with 6.5% homopolymer is achieved using 2×10^{-4}% uranyl nitrate. With 1.0% of this sensitizer the graft is increased only to 53%, but the homopolymer yield has reached at least 39.3% (Table 14). Thus at a certain level of sensitizer the grafting can only be marginally increased for a corresponding large increase in homopolymer yield.

At a fixed sensitizer concentration (1% uranyl nitrate in Table 14), both graft and homopolymer yields increase steadily from the 10% styrene in methanol solution reaching a maximum for the 90% monomer solution. Thus for high grafting yields under the present photosensitized conditions, significant amounts of homopolymer require to be tolerated.

TABLE 14

Kinetics of Homopolymer Formation During UV Grafting of Styrene in Methanol to Cellulose^a

% Styrene	12 hr. Irradiation		18.5 hr. Irradiation	
	% Graft	% Homopolymer	% Graft	% Homopolymer
10	9.8	6.1	8.0	15.8
20	14.2	8.9	18.4	10.0
30	16.4	7.2	17.7	13.0
40	18.8	–	20.0	17.7
50	19.4	15.0	25.6	22.0
60	32.7	18.9	33.9	24.4
70	27.4	18.8	36.8	26.1
80	33.1	–	46.2	39.3
90	53.6	24.7	56.3	33.5

^{a}*Uranyl nitrate (1% w/v) as sensitizer at 24cm. from 90W lamp.*

This is important since the overall copolymerization reaction when homopolymer forms is not only wasteful in monomer consumption, but turbidity may also occur during the grafting process. More importantly, the homopolymer must be removed from the copolymer at the completion of the reaction and at high graft this is not always facile.

F. Grafting of Styrene without Solvent

The question of whether neat styrene can be directly grafted to cellulose under photosensitized conditions is important since recent commercial UV coating processes which have been installed use the principle of solvent free rapid cure sensitized polymerization onto substrates such as cellulose. Fundamentally it is of significance to know whether a carbon-carbon bond between polymer and substrate is capable of being formed under these rapid cure solvent free conditions.

The data in Table 15 show that styrene without solvent can be readily grafted to cellulose using either benzoin ethyl or biacetyl as sensitizer, the former being more efficient. In order to extend this work to the rapid cure process, it is essential to develop special styrene

TABLE 15

UV Grafting of Neat Styrene to Cellulose[a]

Time (hr)	Photosensitizer	% Graft	% Homopolymer
3	Benzoin ethyl ether	1.7	10.3
3	Biacetyl	1.3	-
24	Benzoin ethyl ether	37.9	-
24	Biacetyl	9.8	-

[a]*Solutions contained 1% w/v sensitizer and irradiated for 24 hr. at 24cm.from 90W lamp.*

prepolymers for this purpose. Because the degree of copolymerization under rapid cure conditions may be small, it is also essential to label these styrene prepolymers with tritium as tracer in order to detect this low level of graft if it does occur. Because of the results in Table 15, the present authors are now proceeding with the above tritium experiments in order to discover whether any grafting occurs during rapid cure UV coating and will report the data elsewhere in the future.

G. γ-Ray Induced Grafting

For comparison purposes with the present photosensitized grafting work relevant representative data from the corresponding gamma ray induced grafting are summarised in Tables 16 and 17. Some of the material in Table 16 has previously been published in detail (3), however these data are required here for discussion purposes.

In the radiation induced grafting (Table 16) of styrene to cellulose, solvents which wet the paper are more suitable for copolymerization than non-wetting solvents. Thus the low molecular weight alcohols such as methanol are extremely

TABLE 16

γ-Ray Induced Grafting of Styrene in Representative Solvents to Cellulose

Solvent	Dose (Mrad)	Dose Rate (rad/hr)	Graft (%) at Styrene Conctn. (%)			
			20	40	60	80
MeOH	0.2	8.3×10^3	114	210	103	102
n-BuOH	0.2	8.3×10^3	1	4	4	1
DMF	1.0	6.7×10^5	0	5	23	34
Acetone	1.0	6.7×10^5	0	32	29	38
Dioxan	1.0	6.7×10^5	3	9	21	42
Hexane	1.0	7.7×10^4	0	0	0	0
Benzene	1.0	1.5×10^5	-	0	-	-

[a] All irradiations carried out in stoppered vessels.

efficient. If the molecular weight of the alcohol is increased, there is a decrease in graft such that at n-butanol copolymerization is very low. The other wetting solvents such as dimethyl formamide, acetone and dioxan are also suitable for grafting but their efficiencies are significantly lower than methanol since much higher radiation doses are required with the former solvents to achieve graft comparable to that with methanol. By contrast, copolymerization in the non-wetting solvents such as benzene and hexane is extremely small even at relatively high radiation doses (1.0 Megarads). Under the appropriate dose rate and total dose conditions, Trommsdorff effects are observed in these ionizing radiation processes, however, with methanol the position of the gel peak occurs at 40% monomer concentration. If a long chain, non-active alcohol (octanol) is mixed with a short chain active alcohol (methanol), there is an enhancement in the original methanol graft.

TABLE 17

γ-Ray-Induced Grafting of Styrene to Cellulose
Using Methanol and Methanol-Octanol in a Fixed
Ratio (1:1) as Solvents[a]

	Graft (%) in	
Styrene (% v/v)	Methanol	Methanol/Octanol (1:1)
10	3.2	3.4
20	7.7	11.1
30	16.5	23.0
40	21.7	30.1
50	25.4	33.2
60	28.2	33.4
80	34.0	34.0

[a] Solutions of styrene in solvent irradiated at 40,000 rad/hr to 0.2 Mrad total dose in stoppered vessels.

There are thus many common properties between photosensitized and ionizing radiation grafting systems hence a direct comparison of the two techniques is justified and will be treated in Section I.

H. Mechanism of Photosensitized Grafting

The essential difference between sensitized and unsensitized UV grafting reactions is that in the latter process, radical sites for grafting in the trunk polymer are formed as a result of direct absorption of UV by the cellulose. Monomer and solvent are also capable of absorbing appropriate wave lengths of UV and thus also may participate primarily in the unsensitized grafting mechanism via excited states and energy transfer processes. The mechanism of the unsensitized copolymerization is thus similar to that already proposed for grafting in the ionizing radiation system (18) and will be further discussed in relevant Section I. In the photosensitized reaction, further pathways are available for grafting in addition to those already mentioned for the unsensitized technique. A convenient method for discussing the mechanism of the photosensitized grafting to cellulose is to treat the mode of action of inorganic and organic sensitizers separately.

1. Inorganic Systems - Uranyl Salts

Uranyl nitrate has been the most successful of the inorganic sensitizers used in the current studies, thus the present discussion will be restricted to uranyl salts. Fundamental studies of the photochemistry of the uranyl ion and its role as a photosensitizer have been reviewed (19). For copolymerization work, two predominant reactions are important: (a) intermolecular hydrogen atom abstraction and (b) energy transfer. In the styrene-methanol-cellulose grafting system, intermolecular hydrogen abstraction can be utilized in several ways to promote copolymerization. Thus radicals can be formed in solvent methanol [eqs. (1) and (2)].

$$UO_2^{2+} + h\nu \rightarrow (UO_2^{2+})^* \qquad (1)$$

$$(UO_2^{2+})^* + CH_3OH \rightarrow CH_3O\cdot + H^+ + UO_2^+ \qquad (2)$$

With liquid methanol, the methoxy radical ($CH_3O\cdot$) is the principal species formed whereas with aqueous methanol $\cdot CH_2OH$ predominates (19). With all other alcohols, ESR studies show that $R\dot{C}HOH$ is the essential species formed even in liquid alcohol (19). These solvent radicals can then abstract hydrogen atoms from the trunk cellulose polymer (CeH) to yield grafting sites [eq.(3)].

$$CH_3O\cdot + CeH \rightarrow CH_3OH + Ce\cdot \qquad (3)$$

In an analogous manner sensitizer can diffuse into the alcohol preswollen cellulose and either directly abstract hydrogen atoms [eqs. (1) and (4)] or alternatively rupture bonds (20) as shown in eq. (5) to form additional grafting sites.

$$(UO_2^{2+})^* + CeH \rightarrow UO_2^+ + H^+ + Ce\cdot \qquad (4)$$

In terms of this concept the importance of small solvents
such as methanol which not only wet and swell the cellulose,
but can also readily diffuse into the trunk polymer is obvious.
The rapid cut-off in grafting with the alcohols at n-
butanol (Table 1) may well be explained by the size of the
molecule and difficulty with subsequent diffusion. The significance of the mixed alcohol work (methanol/octanol) reported in Tables 2 and 3 may also be partly explained in an
analogous manner. Inclusion of air in the grafting system may
lead to reactions in addition to eqs. (1) to (5). Thus in the
presence of air, these processes may be further modified by
peroxy-radical formation. The presence of such species may
explain the observation of a small Trommsdorff peak only in
evacuated tubes for the grafting of styrene in methanol at
low monomer concentrations (20%).

Finally, the data in Table 10 suggest that energy transfer processes may contribute to the mechanism of the graft.
Thus, even at relatively high acridine concentrations all
grafting is not suppressed which suggests that even considering the concentration of photosensitizer used, acridine does
not scavenge all free radical precursors to grafting or, more
likely, the residual grafting after acridine scavenging is due
to independent reactions such as energy transfer. Energy transfer processes have previously been proposed for other photopolymerization reactions (21). Thus it has been suggested
that energy transfer occurs from $(UO_2^{2+})^*$ to monomers such as
acrylamide to give a triplet state which is responsible for
the UO_2^{2+} photosensitized polymerization of acrylamide.
Obviously in analogous grafting reactions, such as the present
cellulose system, more work is required to clarify unequivoc-
ally the possible mechanistic role of energy transfer.

2. Organic Photosensitizers

Similar mechanisms to the above can be proposed for ultraviolet grafting initiated with organic photosensitizers. Thus,
both radical and energy-transfer processes are possible. Generally, radicals can be formed by homolytic cleavage [eq.(6)]
where AB is the sensitizer.

$$AB \xrightarrow{h\nu} AB^* \longrightarrow A\cdot + B\cdot \qquad (6)$$

Hydrogen abstraction by these radicals from trunk polymer can
then occur to give grafting sites [similar to eq. (3)]. Alternatively, direct hydrogen abstraction from trunk polymer is

possible [eq. (7)].

$$AB \xrightarrow{h\nu} AB^* \xrightarrow{CeH} ABH + Ce\cdot \qquad (7)$$

The essential difference in operation of the various organic photosensitizers is predominantly the relative emphasis of processes depicted by eqs. (6) and (7), and also the nature of the radicals formed in homolytic cleavage [eq. (6)]. Thus, in the two most successful organic photosensitizers used in the present work (namely, benzoin ethyl ether and biacetyl), the types of radicals formed are shown in eqs. (8) and (9).

$$C_6H_5-\overset{O}{\underset{\|}{C}}-\underset{H}{\overset{OEt}{\underset{|}{C}}}-C_6H_5 \xrightarrow{h\nu} C_6H_5\overset{O}{\underset{\|}{C}}\cdot + C_6H_5\underset{H}{\overset{\cdot}{C}}-OEt \qquad (8)$$
$$\qquad\qquad\qquad\qquad\qquad\qquad (I) \qquad\qquad (II)$$

$$CH_3-\overset{O}{\underset{\|}{C}}-\overset{O}{\underset{\|}{C}}-CH_3 \xrightarrow{h\nu} 2\ CH_3-\overset{O}{\underset{\|}{C}}\cdot \qquad (9)$$

Stability of the resulting radical and steric factors then predominantly determine the relative efficiencies of the two sensitizers. However there are difficulties in making comparisons of initiator efficiencies, unless extensive care is taken to standardise the conditions with respect to light absorption characteristics, solution viscosity, temperature etc. (22).

The benzoin group of sensitizers, of which benzoin ethyl ether is one of the most important, are among those aromatic carbonyl compounds which fragment when exposed to UV light. They have thus been used as sensitizers in UV rapid cure polymerization and much work on the mechanism of the fragmentation has been carried out. Thus in the benzoin photo-initiated polymerization of styrene, Melville and coworkers (23) obtained kinetic evidence for the photochemical production of radicals from benzoin. The reaction was assumed to proceed by a Norrish Type I cleavage (eq. (8) using the ether derivative). This conclusion is supported by recent CIDNP data (24) where it was shown that the radical pair formed on the irradiation of benzoin was identical to that formed on irradiation of benzaldehyde. By using diamagnetic radical scavengers III and IV [eqs. (10) and (11)] as spin traps

$$C_6H_5\ CH = N-C_4H_9 + R^\bullet \rightarrow C_6H_5\ CHR - \overset{\overset{O^\bullet}{|}}{N}-C_4H_9 \quad (10)$$
(III)

$$C_4H_9\ N = O + R^\bullet \rightarrow C_4H_9\ RN - O^\bullet \quad (11)$$
(IV)

for radicals produced from the photolysis of benzoin methyl ether, Ledwith and coworkers (25) obtained further evidence for the Type I cleavage. From benzoin methyl ether, $C_6H_5\dot{C}O$ and $C_6H_5\dot{C}HOCH_3$ were trapped and characterized by ESR.

Of particular importance in the present grafting work, were the results of similar experiments to the above for reactions carried out in benzene and methanol (26). In methanol, fragmentation of photo-excited benzoin methyl ether occurs much more rapidly than hydrogen abstraction. With methanol, such abstraction reactions, even though not the predominant photolysis pathway, would still yield methanol radicals capable of further reaction with the trunk cellulose to give more grafting sites.

The final problem relevant to the current work in the Norrish Type I cleavage of benzoin ethyl ether is the relative efficiency of each of the resulting radicals [eq. (8)] for initiating photopolymerization and photosensitized grafting. With respect to species (I) in eq. (8), it appears that the unpaired electron on the benzoyl radical is not delocalized into the aromatic ring (27), thus benzoyl radicals should be very reactive and efficient initiators. The reactivity of the more stable species (II) is less certain, however indirect evidence (26) suggests that this radical can also initiate monomer polymerization and also presumably grafting.

I. Comparison of Ultraviolet with γ-Ray Grafting

There are a number of common mechanistic aspects associated with the current ultraviolet radiation grafting work and copolymerization initiated by gamma rays, especially if the simultaneous irradiation technique is used. When the alcohols are utilized as solvents, the effect of the alcohol is very similar in both systems, the lower alcohols being the most effective with graft effectively cutting out at n-butanol. Preliminary work (Table 17) also shows that there is a somewhat similar mixed solvent effect in the γ-system analogous to the ultraviolet process. Further work, i.e., dose and dose-rate studies, will be needed to discover whether the magnitude of the mixed solvent effect in the γ-system is as large as that observed with ultraviolet radiation. The effect of wetting and non-wetting solvents is also analogous for both radiation systems. Homopolymer formation appears to be more of a problem with the photosensitized process, since significant levels of sensitizer are required to obtain finite grafts in a reasonable irradiation time. As the concentration of sensitizer is increased, homopolymer formation can increase markedly (Tables 13 and 14) leading to problems which have already been discussed and are unique to the photosensitized system.

One of the most important differences between the present ultraviolet and γ-systems seems to be the appearance of a Trommsdorff effect. With γ-radiation, this peak is usually observed at approximately 30% concentration of monomer in methanol, whereas in the analogous current ultraviolet system, the peak is observed at approximately 80-90% monomer concentration. The reason for this difference in position of peaks may be associated with the nature of the formation or activation of radicals in the trunk polymer. With ionizing radiation, complete penetration of the solution and cellulose is readily achieved during grafting. Thus radicals are directly formed in the trunk polymer and are immediately available for termination and therefore grafting when a monomer diffuses to the site. The charge-transfer mechanism (3) already proposed for such grafting would be applicable. By contrast, with the ultraviolet system, radical sites on the trunk polymer are predominantly formed only after sensitizer and/or solvent radicals have diffused into the polymer and abstracted hydrogen atoms. Because of this diffusional limitation only in the ultraviolet system, chain length of grafted polymer will be increased before termination. Since the higher molecular weight chain to be grafted will be more soluble in a solvent containing high percentages of monomer, the Trommsdorff peak in ultraviolet grafting of styrene in methanol is shifted to the 80-90% monomer concentration region. However, the actual mechanism of the ultraviolet grafting appears to possess many similar general facets to the ionizing radiation system and most probably involves analogous charge-transfer type intermediates of the type already proposed for γ-radiation induced grafting to cellulose (3,18). The salient feature of this mechanism is that relatively long lived radicals formed by ionizing irradiation are available for bonding in the trunk polymer. Then charge-transfer adsorption of monomer or growing polymer to the trunk polymer (eq.12) facilitates subsequent copolymerization.

$$2\dot{P} + \text{Ph-CH=CH}_2 \longrightarrow \text{Ph-CH=CH}_2 \text{ (with } \dot{P}, \dot{P}\text{)} \quad (12)$$

Using styrene and irradiated cellulose as representative model, the complex in eq. (12) shows the delocalized Π-bonding between styrene and free valencies of irradiated cellulose. From this intermediate charge-transfer complex, a number of specific grafting mechanisms are possible. For radical sites that are easily accessible at the surface, the Π-complex may react further by a Π-σ conversion with either the ring (eqs. 13 and 14) or side-chain (eq. 15) to give σ-bonded species.

$$\text{Ph-CH=CH}_2 + \dot{P} \longrightarrow \text{Ph-CH=CH}_2 \text{ (with } \dot{P}\text{)} \quad (13)$$

$$\text{Ph-CH=CH}_2 \text{ (with } \dot{P}\text{)} \longrightarrow [\ldots] \longrightarrow \ldots \quad (14)$$

$$\text{Ph-CH=CH}_2 \text{ (with } \dot{P}\text{)} \longrightarrow \text{Ph-CH-}\dot{C}H_2 \text{ (with } \dot{P}\text{)} \xrightarrow[\text{R=M or R=H}]{+R} \text{Ph-C(H)-CH}_2R \text{ (with } \dot{P}\text{)} \quad (15)$$

Grafting if R=M

Where the mobility of the polystyrene chains and radicals is impeded, Π-σ conversion processes are restricted; however, grafting could still occur through charge-transfer bonding of the type shown in eq. (16). Such bonding would keep homopolymer locked between the chains as "graft" and could explain the app ent anomalies observed when homopolymer is extracted from copolymerized celluloses.

$$-CH-CH_2-CH-CH_2-CH-CH_2-CH-CH_2-$$
$$||||$$
$$\bigcirc\bigcirc\bigcirc\bigcirc$$
$$||||$$
$$\dot{P}\dot{P}\dot{P}\dot{P}$$

(16)

The extension of this mechanism to photosensitized grafting is obvious since radical sites can also be formed in the trunk polymer cellulose by direct photolysis (8) or by abstraction reactions involving solvent radicals or radicals from the photosensitizer. The subsequent sequence of events is then analogous to the radiation-induced grafting system.

The possibility that energy transfer processes contribute to the mechanism especially in the photosensitized reaction should also be considered, however current evidence suggests that any such contribution from these competing processes will be minimal.

ACKNOWLEDGEMENTS

The authors thank the Australian Institute of Nuclear Science and Engineering and the Australian Atomic Energy Commission for the irradiations. They are also grateful to the Australian Research Grants Committee for continued support. One of them (N.P.D.) wishes to thank Sidney Cooke Chemicals Pty. Ltd. for the award of a Fellowship.

REFERENCES

1. Krassig, H.A., and Stannett, V., Adv.Polym.Sci. 4, 111 (1963).

2. Moore, P.W., Rev.Pure Appl.Chem. 20, 139 (1970).

3. Dilli, S., Garnett, J.L., Martin, E.C., and Phuoc, D.H., J.Polym.Sci. C37, 57 (1972).

4. Arthur, J.C., and Blouin, F.A., Proceedings of the International Symposium on Radiation-Induced Polymerization Graft Polymerization, Battelle Memorial Institute, 1962, T1D7643, p.319.

5. Guthrie, J.T., and Haq, Z., Polymer 15, 133 (1974).

6. Oster, G., and Yang, N.L., Chem.Rev. 68, 125 (1968).

7. Geacintov, N., Stannett, V., Abrahamson, A.W., and Hermans, J.J., J.Appl.Polym.Sci. 3, 54 (1960).

8. Arthur, J.C., Polym.Prep. 16 (1), 419 (1975).

9. Kubota, H., Murata, T., and Ogiwara, T., J.Polym.Sci. 11, 485 (1973).

10. Davis, N.P., Garnett, J.L., and Urquhart, R., J.Polym.Sci.Polym.Letters Ed. 14, 537 (1976).

11. Trommsdorff, E., Kohle, H., and Lagally, P., Makromol.Chem. 1, 169 (1948).

12. Davis, N.P., Garnett, J.L., and Urquhart, R., J.Polym.Sci.Symp.No.55, 287 (1976).

13. Dlugosz, J., Polymer 427 (1965).

14. Lamotte, M., Joussot-Julien, J., Mantione, M.J., and Claverie, P., Chem.Phys.Letters 27, 515 (1974).

15. Plesch, P.H., Progr.High Polym. 12, 139 (1968).

16. Gaylord, N.G., Deshpande, A.B., Dixit, S.S., Maiti, S., and Patnaik, B.K., J.Polym.Sci.Polym.Chem.Ed. 13, 467 (1975).

17. "International Critical Tables, Volume III", 1st ed., McGraw-Hill, New York, 1928, p.262.

18. Dilli, S., and Garnett, J.L., J.Polym.Sci. A-1, 2323 (1966).

19. Burrows, H.O., and Kemp, T.J., Chem.Rev. 3, 139 (1974).

20. Greatorex, D., Hill, R.J., Kemp, T.J., and Stone, T.J., JCS Faraday I 68, 2059 (1972).

21. Venkataras, K., and Santappa, M., J.Polym.Sci. A-1, 8, 1785, 3429 (1970).

22. Pappas, S.P., and Chattopadhyay, A.K., J.Polym.Sci.Polym.Letters Ed. 13, 483 (1975).

23. Chimmayanandam, R.B., and Melville, H.W., Trans.Faraday Soc. 50, 73 (1954).

24. Closs, G.L., and Paulson, D.R., J.Am.Chem.Soc. 92, 7229 (1970).

25. Ledwith, A., Russell, P.J., and Sutcliffe, L.H., J.Chem.Soc.Perkin II 1925 (1972).

26. Ledwith, A., J.Oil Col.Chem.Assoc. 59, 157 (1976).

27. Solly, R.K., and Benson, S.W., J.Am.Chem.Soc. 93, 1592 (1971).

WOOD PULP GRAFTING WITH DIFFERENT MONOMERS BY THE XANTHATE METHOD

V. HORNOF

Department of Chemical Engineering
University of Ottawa, Ontario, Canada

and

C. DANEAULT, B. V. KOKTA, AND J. L. VALADE

Department of Engineering
University of Quebec at Trois-Rivières
Quebec, Canada

Kraft semibleached softwood pulp was copolymerized with different monomers using the xanthate grafting process. Results with styrene, acrylonitrile, methyl methacrylate, ethyl acrylate, butyl acrylate, vinyl acetate, methacrylic acid and acrylamide showed marked differences in both total conversion to polymer and grafting efficiency obtained. While most monomers did not require an inert atmosphere, acrylamide and methacrylic acid gave better results in the absence of oxygen. No copolymers were obtained when vinyl acetate was used. Similarly, additions of vinyl acetate to styrene resulted in decreased polymer yields. The stability in respect to hydrolysis of xanthate graft copolymers was compared with the stability of similar products prepared by using ceric ammonium nitrate as initiator. Almost no difference in the extent of hydrolysis was detected, both with methyl methacrylate and ethyl acrylate as monomers. This observation would indicate that the xanthate grafts are not bound to the backbone material via the easily hydrolyzable xanthate links. In general, better copolymerization parameters were obtained when using the xanthate process. The ceric process was also found to bring about more oxidative degradation of the pulp, resulting in poorer mechnical properties.

INTRODUCTION

Cellulose xanthates, discovered in 1891 by Cross, Bevan and Beadle (1) are best known as intermediates in the manufacture of viscose fibers. The possibility to use them for the preparation of graft copolymers has been recognized only recently by Faessinger and Conte (2). Their process is based on the formation of a redox couple between the xanthates and a peroxide. Free radicals are formed on the macromolecular chain of the substrate, which can subsequently give rise to free radical polymerisation, provided a suitable monomer is introduced into the system. A graft copolymer is formed, whose composition and properties depend on the type of monomer and on the conditions of the reaction.

The applicability of the xanthate process is not limited to pure cellulose. Various other materials such as wood pulp, cotton and starch can also be used as substrates. However, it is in the textile field that this process has appeared to show most promise (3, 4).

Several groups of researchers have investigated the xanthate grafting process in some detail. Notably, Dimov and Pavlov (5) studied the effect of reaction conditions on the grafting of acrylonitrile on partly hydrolyzed viscose. Later studies by Hornof, Kokta and Valade, (6-8) carried out using partially thiocarbonated wood pulp ($\gamma \simeq$ 5-6), showed that apart from being dependent on reaction conditions such as pH, temperature and initiator concentration, the grafting reaction was strongly influenced by the character of pulp used. The parameters most strongly affected included graft yield as well as the homopolymer-copolymer ratio.

The present paper investigates the possibility to copolymerize pulp with different monomers and compares the xanthate grafting process with that employing ceric ions as initiators.

EXPERIMENTAL

A. Copolymerization

The procedure employed for xanthate grafting has been described previously (9,5). Unless specified otherwise, the experimental conditions used were as follows: pulp, 4.5 g (oven-dry weight); monomer, 9 g; surfactant (Tween-40), 0.9 g; hydrogen peroxide, 1.5 g (diluted into 25 ml before adding); mercerization, 45 min. in 150 ml 0.75 N NaOH (room temperature); xanthation, 2 hr. (room temperature); ion exchange, 2 min. in 150 ml 0.004% solution of $(NH_4)_2Fe(SO_4)_2$; initial pH, 5 ± 0.1; reaction temperature, 25°C; reaction time 2 hrs.

The copolymerizations initiated by ceric ions were carried out as described by Mansour and Schurz (10). The experimental conditions were as follows: pulp, 4.5 g (oven-dry weight); monomer, 9 g; ceric ammonium nitrate, 1.37 g; reaction time, 1 hr.

The copolymerizations were terminated by additions of hydroquinone. In the case of the xanthate method, the excess of hydrogen peroxide was destroyed by adding 1% solution of potassium meta-bisulfite.

B. Extraction

The quantity of homopolymers in the reaction products was determined by a Soxhlet extraction of 2 - 3 g samples of the products with suitable solvents. The calculations of grafting parameters were made taking into account pulp losses during mercerization (about 3.5%). Only polymer loading was determined in the case of water-soluble polymers (polyacrylamide and poly(methacrylic acid)).

The grafting parameters are defined as follows:

Total conversion, % = (D-B)/C x 100,
Polymer loading, % = (A-B)/B x 100,
Grafting efficiency, % = (A-B)/(D-B) x 100,

where A is weight of products after copolymerization and extraction, B is weight of pulp, corrected for solubility in NaOH solution, C is weight of monomer charged and D is weight of unextracted products.

C. Molecular Weights

Grafted polymer was isolated by first removing the cellulose according to a method described by Nakamura et al. (11). This was followed by dissolving the residue in acetone. Homopolymer was isolated by shaking dry reaction products with acetone during three hours and by collecting the clear solution of the polymer. Molecular weights were determined by viscosimetry.

D. Materials

Kraft semibleached pulp (Consolidated - Bathurst Co., Division Waygamack) was used as the grafting substrate.

Monomers, with the exception of acrylamide, were purified by distillation on a column. The center cut was collected and stored in the refrigerator. All other reagents used in this work were used as supplied by the manufacturers.

RESULTS AND DISCUSSION

A. Dependence of Grafting on Reaction Conditions

The results published in a previous article (6) have shown the effect of reaction conditions on the grafting of one particular monomer, viz: acrylonitrile. It became evident from further work that the results could not be generalized to systems involving other monomers. For example, copolymerizations with acrylonitrile gave the same results regardless of whether the experiments were carried out under an inert atmosphere or not. This was not true with other monomers.

TABLE I

Grafting with Various Monomers in the Presence of Air[1]

Sample	Monomer	Polymer Loading %	Grafting Efficiency %	Total Conversion %
26	Styrene	116.9	72.3	81.0
32	Methyl methacrylate	121.0	60.5	98.1
33	Ethyl acrylate	127.7	76.9	83.1
35	Butyl acrylate	99.4	70.0	71.0
16	Acrylamide	11.9	-[2]	-[2]
30	Vinyl acetate	0	-	0
37	Methacrylic acid	0	-[2]	-[2]

[1] Time of polymerization = 2 hours

[2] Total conversion and grafting efficiency not determined.

TABLE II

Grafting with Various Monomers under Inert Atmosphere[1]

Sample	Monomer	Polymer Loading %	Grafting Efficiency %	Total Conversion %
27	Styrene	119.2	70.5	84.5
29	Methyl methacrylate	119.2	64.6	92.3
34	Ethyl acrylate	121.7	78.5	78.1
36	Butyl acrylate	100.7	68.5	74.0
24	Acrylamide	44.2	-[2]	-[2]
31	Vinyl acetate	0	-	0
38	Methacrylic acid	133.0	-[2]	-[2]

[1] Time of polymerization = 2 hours.
[2] Total conversion and grafting efficiency not determined.

Table I shows the grafting results obtained in the presence of air, while those obtained under the inert atmosphere of nitrogen are compiled in Table II. Comparing the two sets of data, it is evident that styrene, methyl methacrylate and the two acrylate esters show very little difference. In the case of methyl methacrylate, total conversion seems to be even higher in the presence of oxygen, although lower grafting efficiency reduces the corresponding polymer loading. In the case of ethyl acrylate, both total conversion and polymer loading are slightly higher in the presence of air.

The three remaining monomers behave quite differently. Acrylamide shows a very low conversion in the presence of air, which increases four-fold when the reaction is carried out under nitrogen. Methacrylic acid, on the other hand, gives no polymer loading in the presence of air, but a very high one under an inert atmosphere. Finally, vinyl acetate was found to be non-reactive under the both sets of conditions.

Brickman (3) has recently stated that one of the advantages of the xanthate grafting process is that it does not require a special atmosphere. The present results indicate that, at least for certain monomers, this is not always true.

TABLE III

Molecular Weights of Grafted Polymer and of Homopolymer[1]

Monomer	Type of polymer	Intrinsic viscosity	Molecular weight[2]
Styrene	Homopolymer	0.371	69,000
	Graft	0.388	74,000
Methyl methacrylate	Homopolymer	0.397	201,000
	Graft	0.442	235,000
Ethyl acrylate	Homopolymer	0.450	103,000
	Graft	0.565	146,000
Butyl acrylate	Homopolymer	0.669	229,000
	Graft	0.682	236,000

[1] Products prepared by the xanthate process.

[2] Viscosity average.

Table III lists the intrinsic viscosities and the corresponding molecular weights of several pairs grafted polymer - homopolymer. The molecular weights found are in general higher than in the case of acrylonitrile as monomer (6). There is only a small difference between the molecular weights of homopolymers and those of corresponding grafted polymers, which once again appears to be in a disagreement with the previous work (6). One must not however overestimate the accuracy of these results. With poly (alkyl acrylates) in particular, some hydrolysis of the polymer probably takes place during the treatment with 72% sulfuric acid, resulting in a serious experimental error.

A variation of pH as well as of the peroxide concentration has been shown previously (6) to exert a strong influence on copolymerization yield. The effect of these two parameters on the grafting of acrylamide is shown in Table IV. Polymer loading is observed to increase significantly both with increasing pH and with rising H_2O_2 concentration. Also with acrylamide as the monomer, an attempt was made to improve grafting by additions of organic solvents (isopropyl alcohol and tert-butanol). The results, however, were almost identical with those obtained in the absence of solvents.

TABLE IV

Grafting with Acrylamide
in the Presence of Air[1]

Sample	pH	H_2O_2 (g)	Polymer Loading %
16	5	1.5	11.9
9	6	1.5	35.2
14	7	1.5	43.4
10	5	3.0	21.8
11	5	4.5	48.7

[1] Time of polymerization = 2 hours.

B. **Hydrolytic Stability of Copolymers**

According to the reaction mechanism proposed by Dimov and Pavlov (5), the first product of the reaction between cellulose xanthate and hydrogen peroxide is a dithiocarbonate (xanthate) radical:

$$\text{Cell} \begin{array}{c} \text{OH} \\ \text{O-C} \end{array} \begin{array}{c} \text{S} \\ \text{S:}^{\ominus} \end{array} + \text{HO:OH} \rightarrow \text{Cell} \begin{array}{c} \text{OH} \\ \text{O-C} \end{array} \begin{array}{c} \text{S} \\ \text{S.} \end{array} + \cdot\text{OH} + \text{OH}^{\ominus} \quad (1)$$

In the next step, the dithiocarbonate radical decomposes into a hydroxy radical and free carbon disulfide:

$$\text{Cell}\begin{array}{c}\text{OH}\\\text{O}\end{array}\!\!\!\!\!\!\!\!\!\begin{array}{c}\text{S}\\\diagup\\\text{C}\\\diagdown\\\text{S}\end{array} \longrightarrow \text{Cell}\begin{array}{c}\text{OH}\\\text{O}\cdot\end{array} + CS_2 \qquad (2)$$

The copolymerization is then initiated by the hydroxy radicals in the presence of a suitable monomer:

$$\text{Cell}\begin{array}{c}\text{OH}\\\text{O}\cdot\end{array} + \text{Monomer} \longrightarrow \text{Cell}\begin{array}{c}\text{OH}\\\text{O - Polymer}\end{array} \qquad (3)$$

The graft would thus be attached to the substrate by a relatively strong ether-type linkage.

Under favourable conditions however, the dithiocarbonate radicals could be stable enough to initiate copolymerization:

$$\text{Cell}\begin{array}{c}\text{OH}\\\text{O-C}\diagup^{\!\!\text{S}}\!\!\diagdown_{\text{S}\cdot}\end{array} + \text{Monomer} \longrightarrow \text{Cell}\begin{array}{c}\text{OH}\\\text{O-C}\diagup^{\!\!\text{S}}\!\!\diagdown_{\text{S - Polymer}}\end{array} \qquad (4)$$

The presence of such linkages in the material would have a negative effect on its resistance against hydrolysis.

Studying the alkaline hydrolysis of acrylonitrile - grafted pulp, Lepoutre et al. (12) reported that a considerable loss of polymer occured in the case of copolymers prepared by the xanthate process. They explained this loss by the presence of weak dithiocarbonate linkages between polymer branches and the substrate.

Ceric ion initiation of graft copolymerization does not produce such latent instability. A comparative study of the hydrolysis of xanthate and ceric copolymers would thus show a greater weight loss for the former, if reaction (4) actually takes place. For this purpose, a series of graft copolymers was prepared using ceric ammonium nitrate as initiator under an inert atmosphere. The results are shown in Table V.

TABLE V

Grafting with Various Monomers Using the Ceric Ion Process

Sample	Monomer	Polymer Loading %	Grafting Efficiency %	Total Conversion %
85	Styrene	0	-	0
80	Methyl methacrylate	81.1	68.1	59.6
87	Ethyl acrylate	87.5	51.0	85.9
83	Butyl acrylate	81.2	58.7	69.3
78	Acrylamide	0	-	-
86	Vinyl acetate	0	-	0
82	Methacrylic acid	67.4	-[1]	-[1]

[1] Total conversion and grafting efficiency not determined.

Dithiocarboxylic esters are known to be easily hydrolysed under both acid and basic conditions (13). Of the two, acid hydrolysis appeared to be more selective and was selected for the comparison. Copolymers with ethyl acrylate and with methyl methacrylate were chosen for the hydrolytic study together with their xanthate counterparts. Samples were carefully extracted with acetone, then treated with 0.01 N - H_2SO_4 during 3 hours. After drying, the samples were once more extracted with acetone and their final weight recorded. The results are summarized in Table VI and compared with blank experiments, in which pure pulp was subjected to the same treatment.

TABLE VI

Comparison of the Hydrolytic Stability
of Xanthate and Ceric Grafts

Monomer	Grafting Process	Polymer Loading %	Weight before Treatment g	Weight after Treatment g	Weight Loss %
Ethyl acrylate	Xanthate	115.2	1.5818	1.5264	3.5
	Ceric	87.5	1.5882	1.5266	3.9
Methyl methacrylate	Xanthate	119.2	1.1290	1.0971	2.8
	Ceric	81.1	1.1298	1.0975	2.9
Blank experiment	-	-	1.5945	1.5855	0.6
	-	-	1.5931	1.5825	0.7

The results show clearly that while the weight loss of all copolymerized samples is higher than for the blank, its absolute value is relatively low. One observes also that the xanthate copolymers are in fact more stable than the ceric ones. This becomes even more obvious when one considers that the weight loss shown in Table VI refers to the total original weight of a sample. As the copolymers prepared by the xanthate method are characterized by a considerably higher polymer loading, their relative polymer loss is in fact substantially lower than that shown by their ceric-initiated counterparts. It thus appears that the contribution of reaction (4) to the overall production of copolymer is very small if not zero. The reasons for the lower hydrolytic stability of the ceric-ion initiated copolymers is not clear; it is possible that the degradative character of the reaction of ceric ions with cellulose renders the latter less stable towards acid hydrolysis.

C. <u>Grafting with Mixtures of Monomers</u>

It has been shown by several authors (16, 17, 18), that

certain monomer mixtures copolymerize with cellulose more rapidly than each monomer alone. Such effects were also observed by Kokta and Valade (9) using the xanthate grafting process. Several experiments have been carried out in the present work using styrene as the principal monomer. The results are compiled in Table VII. Additions of vinyl acetate (a monomer which is non-reactive when alone) result in a significant reduction in polymer yield. As little as 10% of vinyl acetate in the reaction mixture reduces total conversion and polymer loading by one half.

With the mixture using ethyl acrylate as comonomer, an improvement in total conversion and polymer loading is observed at 10% and 20% of the comonomer, while grafting efficiency remains essentially the same as for styrene alone. Conversely, small additions of methacrylic acid bring about a rise in grafting efficiency resulting in high polymer loading, although total conversion is lower than for pure styrene. The best result is obtained at 20% of methacrylic acid; when this is increased to 50%, a significant decrease in all grafting parameters ensues.

The effects observed with ethyl acrylate and with methacrylic acid are thus qualitatively different. Being the most reactive monomer among those tested (Table I) ethyl acrylate probably accelerates the overall reaction. Methacrylic acid, on the other hand, is a water-soluble monomer and its effect most likely consists in improving the contact between pulp fibres and a hydrophobic monomer such as styrene. This in turn would result in increased grafting efficiency.

D. Mechanical Properties of Grafted Pulp

The mechanical properties of copolymerized pulps were in general much lower than the properties of the original pulp. The main cause of this however, was found to be pulp degradation during the various stages of the copolymerization procedure.

In the xanthate grafting method, only mercerization had a positive effect on mechanical properties. Xanthation, when carried out in the presence of air, reduced pulp strength by about 60%. This could be improved considerably by carrying out the xanthation process under nitrogen. The most degrading effect on pulp properties was traced to the presence of hydrogen peroxide in the reaction mixture. This effect was very strong even at low H_2O_2 concentrations and it affected especially the folding endurance. Similar and perhaps even

TABLE VII

Grafting with Binary Mixtures of Monomers

Sample	Monomers				Polymer Loading %	Grafting Efficiency %	Total Conversion %
	I	Weight g	II	Weight g			
26	Styrene	9.0	–		116.9	72.3	81.0
66	Styrene	4.5	Vinyl acetate	4.5	48.0	62.1	38.6
68	Styrene	9.0	Vinyl acetate	1.0	63.5	73.2	39.1
40	Styrene	8.1	Ethyl acrylate	0.9	124.6	68.0	91.6
41	Styrene	7.2	Ethyl acrylate	1.8	125.3	69.1	91.0
71	Styrene	9.0	Methacrylic acid	1.0	105.1	80.6	65.4
70	Styrene	8.0	Methacrylic acid	2.0	134.2	88.8	80.6
69	Styrene	5.0	Methacrylic acid	5.0	80.9	56.1	72.1

stronger pulp degradation was observed when the ceric ion method was applied.

Most of the copolymerized pulps were not homogeneous and contained small lumps of polymer which could not be completely disintegrated. This resulted in poor handsheet quality and hence in poor reliability of mechanical tests. Relatively the best strength properties were observed in the case of pulps copolymerized with mixtures of methacrylic acid and another monomer (acrylonitrile, butyl acrylate). Addition of such hydrophylic copolymers improved strength in comparison to pulp which was subjected to the whole procedure in the absence of monomer. Further work is being carried out with the aim to lessen the degradative effect of the xanthate process and to find monomer mixtures, which would bring about the best improvement in mechanical properties as well as in grafting efficiency.

ACKNOWLEDGMENT

The authors wish to acknowledge with thanks the participation of Mr. A.R. Plourde in this project. The financial support of the National Research Council of Canada is also gratefully acknowledged.

REFERENCES

1. Cross, Bevan and Beadle, Ber. 26, 1090 (1893).
2. Faessinger, R.W., and Conte, J.S., U.S. Patent 3,359,224 (Dec. 19, 1967).
3. Brickman, W.J., Tappi, 56, 97 (1973).
4. Brickman, W.J., Paper presented at the ACS Symposium on New Developments in Cellulose Technology, 168th ACS National Meeting, Sept. 8-13, 1974, Atlantic City, N.J.
5. Dimov, K., and Pavlov, P., J. Polymer Sci., A-1, 7, 2775 (1969).
6. Hornof, V., Kokta, B.V., and Valade, J.L., J. Appl. Poly. Sci., 19, 545 (1975).
7. Hornof, V., Kokta, B.V., and Valade, J.L., J. Appl. Poly. Sci., 19, 1573 (1975).
8. Hornof, V., Kokta, B.V., and Valade, J.L., J. Appl. Poly. Sci., 20, 1543 (1976).
9. Kokta, B.V., and Valade, J.L., Tappi, 55, 366 (1972).
10. Mansour, O.Y., and Schurz, J., Svensk Paperstidning 76, 258 (1973).

11. Nakamura, Y., Arthur, J.C., Jr., Negishi, M., Doi, K., Kageyama, E., and Kudo, K., J. Appl. Polym. Sci., 14, 929 (1970).
12. Lepoutre, P., Hui, S.H., and Robertson, A.A., J. Appl. Poly. Sci., 17, 3143 (1973).
13. "The Chemistry of Carboxylic Acids and Esters", Ed. by S. Patai, Interscience Publishers, 1969, p. 754.
14. Arthur, J.C., Jr., Baugh, P.J., and Hinojosa, O., J. Appl. Polym. Sci., 10, 1591 (1966).
15. Arthur, J.C., Jr., Macromol, J., Sci. - Chem., A10, 653 (1976).
16. Stanchenko, G.I., Livshits, R.M., Toffe, T.A., and Rogovin, Z.A., Cellulose Chem. Technol., 3, 567 (1969).
17. Geacintov, N., Stannett, V., Abrahamson, E., and Hermans, G., J. Appl. Polym. Sci., 3, 54 (1960).
18. Rapson, W.H., and Kvasnicka, E., Tappi, 46, 662 (1963).

A NOVEL PROCESS FOR THE TREATMENT OF PULP MILL EFFLUENTS USING THE GRAFTED PULP FIBRES

P. LEPOUTRE, S. H. HUI, AND A. A. ROBERTSON

Pulp and Paper Research Institute of Canada
Pointe Claire, Quebec, Canada

Pulp fibres grafted with 2-dimethylaminoethyl methacrylate have been found very effective in removing colour, BOD and COD from the first caustic extraction liquor of bleached kraft mills. pH was the main controlling factor. Below pH 9.5, the coloured materials were absorbed by the fibres. Above pH 9.5, they were released and the fibres were regenerated. Colour, BOD, and COD removal increased with the total amount of amino groups present, with decreasing pH and with increasing temperature, but it was independent of the graft level. Regeneration of the fibres by alkali could be achieved even in the presence of a high concentration of colour in the regenerating solution. This should theoretically permit the concentration of the coloured materials into a volume 1/38 of that of the treated effluent. Preliminary experiments indicated no loss in fibre efficiency after 40 cycles.

INTRODUCTION

In recent years, graft-polymerisation has been recognized as a simple and effective technique for introducing large quantities of desirable functional groups into a host substrate. As an example, the grafting of a polymer containing carboxylic groups onto startch [1] or onto cellulose fibre [2] has been shown to confer on these materials an outstanding affinity for water. An added advantage of grafting over competitive methods is the retention of the fibrous characteristics of cellulose without having to resort to cross-linking.

Another type of functional group which as been found to have very useful properties is the amino group.

This paper deals with a novel process for removing the coloured materials dissolved in pulp-mill effluents, based on cellulosic fibres grafted with a polymer bearing tertiary amino groups.

BACKGROUND

Several years ago, workers at the Centre Technique du Papier in Grenoble, France, discovered that when tertiary amines, dissolved in kerosene, were stirred in an accidified kraft bleachery effluent, the coloured materials went into the kerosene-amine phase, leaving a clear acqueous phase and the amine could be regenerated. The decolourisation potential of tertiary amines had been demonstrated. After a joint study with PPRIC, the project was eventually abandoned primarily because of problems inherent in the liquid-liquid extraction process that it involved.

However, the discovery that amino groups were effective in removing colour prompted Carles[3] to replace the liquid-liquid extraction by a solid-liquid extraction, using cellulosic fibres substituted with tertiary amino groups (diethylaminoethyl cellulose). This was at a time when an investigation of the properties of grafted cellulosic fibres was being carried out at this Institute. It was felt that a grafting technique might be a better way to introduce a large number of amino groups without the adverse effects on the fibrous properties of the cellulose that often occur in substitution reactions.

Grafting is an *in situ* polymerisation of a monomer within the structure of the fibre. The polymer formed is thus an intimate part of the modified fibre and cannot be removed by solvent extraction.

Experiments using cellulosic fibres grafted with a polymer containing tertiary amino groups have confirmed their potential in the clarification of the effluents from bleached-pulp mills and their technological advantage over the liquid-liquid amine process.

PROPERTIES OF THE GRAFTED FIBRES

In most of this work, fibres grafted with poly(2-dimethylaminoethyl methacrylate) were used. The tertiary amino groups of the polymer are in the free amine form at a pH above 9.5. As the pH is lowered, they are converted to the ammonium salt form which ionises, confering on the polymer the properties of a polyelectrolyte. At pH 4.5, conversion to the ammonium salt form is complete.

It is this pH-dependent change from a non-ionic to a cationic character of the polymer that is utilized in this process. At a pH where the amino groups are ionised, the fibres absorb the coloured materials in the effluent. When they are not ionised, they desorb whatever colour was picked up and are regenerated. This is illustrated in Fig. 1. At this point, however, it is not known whether the colour removal process is purely an ion-exchange pheonomenon or whether adsorption also occurs.

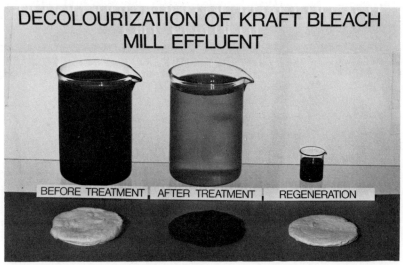

Fig. 1. Illustration of the process. The coloured materials present in the effluent are removed by the grafted fibres. The colour is then desorbed in a concentrated form and the fibres are regenerated.

EXPERIMENTAL

Preparation of the Grafted Fibres

A softwood bleached kraft pulp was grafted according to the method described in reference [4]. Graft level (g polymer/100 g cellulose) was adjusted by varying the concentration of monomer. The monomer used in most of this study was 2-dimethylaminoethyl methacrylate referred to here as (DMM). The anion exchange capacity (AEC) of the fibers, expressed in milliequivalents per gram of dry fibre, was determined by either titration [4] or nitrogen analysis.

Standard Treatment Procedure

A known quantity of grafted fibres was added to 300 ml of raw effluent and the pH was adjusted to 4.5 with 5 N HCl to avoid dilution. After 10 minutes stirring, the fibres were filtered off. The filtrate was examined for colour, BOD, and

COD.

Standard Regeneration Procedure

The dark filtered fibres were regenerated by stirring them in sodium hydroxide at pH 11.5. After filtration, they were washed with distilled water.

Effluents

Effluents were obtained fresh from various kraft pulp mills, mostly in eastern Canada, which use conventional bleaching sequences. They were stored in a cold room under nitrogen. Most of the work was done with the liquor from the first caustic extraction stage of the bleaching sequence. This very dark effluent contains approximately 75% of the colour discharged by the mill, although it represents only 7-10% of the total effluent volume of the kraft mill.

Determination of BOD, COD, Colour and Toxicty

BOD and COD were determined by CPPA standard methods. Colour was measured by relating the absorption by the effluent of monochromatic light at a wavelength of 465 nm to that of platinum cobalt standard colour solutions, and was expressed in chloroplatinate colour units. All measurements were done at pH 7.6

$$\% \text{ Reduction} = \frac{\text{Initial colour units} - \text{Final colour units}}{\text{Initial colour units}} \times 100$$

Fish bioassay tests were performed at PPRIC using speckled trout *(Salvelinus fontinalis)* fingerlings.

FACTORS INFLUENCING TREATMENT EFFICIENCY

pH of the Treatment

As was indicated earlier, pH is a key parameter in this process. Fig. 2 shows the very rapid increase in colour removal that took place in the pH range 9.5 - 8. This pH is the apparent pH that was measured in the effluent containing the slurried fibres at the time of the treatment. A discussion of this apparent pH will be given elsewhere.

Above pH 9.5, no or very little decolourisation took place. Below pH 8, a further improvement in colour removal was obtained at lower fibre concentrations. It was immaterial whether this apparent pH was reached by acidifying separately the fibres and effluent or by acidifying the whole slurry.

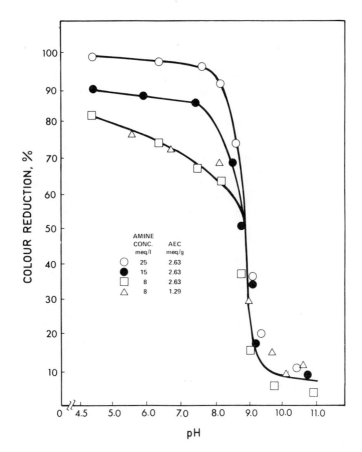

Fig. 2. Influence of the pH on the efficiency of colour removal. Initial colour was 10,800 C.U. The fibre concentration (g/l) is equal to the concentration of amine groups (meq/l) divided by the anion-exchange capacity (AEC, meq/g).

Concentration of Grafted Fibres

Typical curves of colour, BOD and COD removal as a function of grafted fibre concentration are given in Fig. 3. The colour removal increased at a decreasing rate as the fibre concentration was increased.

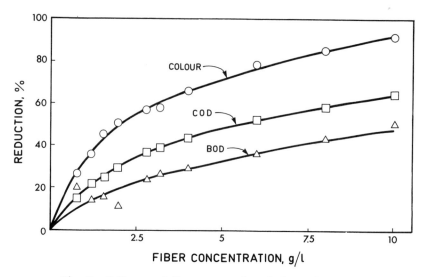

Fig. 3. Influence of the concentration of the grafted fibres on the reduction of colour, BOD and COD. (Initial values: colour 7616 C.U; BOD 550 mg/1; COD 1950 mg/1. Fibre ACE 2.5 meq/g, pH 4.5).

Influence of Anion-Exchange Capacity or Graft Level

The same total number of milliequivalents of amino groups per litre of effluent can be obtained by using more fibres of low graft level, i.e., low anion exchange capacity, or by using fewer fibres of high graft level.

There are advantages, both in the preparation of the fibres, and in the handling of the effluent slurry, to operate with a low concentration of high graft-level fibres. In addition, swelling measured by the water retention value[4] increases with AEC, so that one might have expected that the accessibility, and therefore, the efficiency in decolourisation, would also increase.

To examine this effect, fibres were prepared, covering a range of graft level from 22% to 105%, i.e., a range of AEC from 0.93 to 2.64 meq/g. Fig. 4 shows that the percent colour reduction achieved was a function of the total amount of amino groups available rather than of the AEC or graft level of the fibres.

Temperature and Time

The influence of the temperature on the colour removal efficiency has been studied by performing experiments at 25° and at 60°C (the approximate temperature of the effluent in the mill). The colour removal and the rate of decolourisation improved at the higher temperature as shown in Fig. 5.

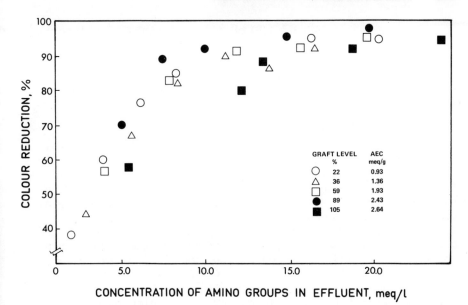

Fig. 4. Influence of the anion-exchange capacity, or grafted level, of the fibres on their colour removal efficiency. (pH 4.5; Initial colour 20,900 C.U.)

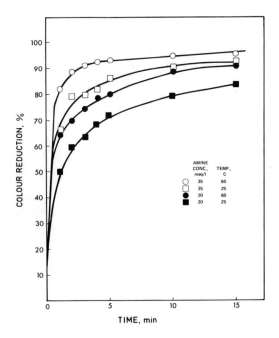

Fig. 5. Influence of time and temperature on the colour reduction (Initial colour: 14,525 C.U.; pH 4.5).

The time to reach equilibrium colour removal is important for process design. Fig. 5 also shows the colour removal after various times of contact, at 25 and 60°C. It seems that under conditions of high dosage and high temperature, a contact time of 2 mins would provide adequate colour removal (90%).

Type of Effluents

During the course of this work, effluents from several different sources were tested. Results on BOD, COD and colour reductions using the same dosage of fibres are shown in Table I. The results are comparable to those of most of the other external decolourisation processes, such as ion exchange, alum coagulation, and adsorption.

It is interesting to note in Table I that, overall, the percent colour removal was fairly independent of the initial colour level. This has been further verified by experiments in which the effluent was diluted with water to change its initial colour level. This indicates that highly coloured effluents should not be diluted with less coloured ones prior to the treatment.

FULL UTILISATION OF THE COLOUR REMOVAL POTENTIAL OF THE FIBRES

The concave shape of the curve of colour removal versus fibre dosage (Fig.3) suggested that, when they were used at high concentrations, the decolourisation potential of the fibres was not fully utilized. This was confirmed in experiments where "used" fibres were used again, without being regenerated, to treat more effluent. The results, shown in Table II, indicate clearly that it is more efficient from the standpoint of colour removal to treat 3 litres of effluent, one litre at a time with the same, say, 30 g fibres per litre without regeneration, than to use these 30 g once to treat the whole 3 litres.

Fig. 6 gives the cumulative number of colour units absorbed per meq of amino groups when several volumes of effluent were successively treated with the same, unregenerated fibres. It is seen that the capacity of the fibres is in the order of 4000 C.U./meq.

These data indicate that a multi-step, countercurrent process, in which the clear effluent meets incoming, freshly regenerated fibres, would be particularly efficient although, perhaps, not economical.

TABLE I
Results of BOD, COD and colour removal from various effluents[1]

Effluent Source	Wood Type	Bleaching Sequence	Initial Colour, C.U.	Colour Reduction, %	Initial BOD, mg/l	BOD Reduction %	Initial COD, mg/l	COD Reduction, %
Mill A	Softwood	CEHDED	20,900	95.0	1105	67.0	4530	78.0
	Hardwood	CEHDED	6,512	88.6	400	48.3	1755	64.3
	Hardwood	CEHDED	7,616	89.8	550	50.0	1950	64.0
Mill B	Softwood		7,366	89.3	636	65.4	1753	66.9
Mill C	Softwood[2]		14,690	91.0	920	57.7	3120	62.0
Mill D	Softwood		11,466	97.8	478	33.9	2305	69.6

1) Fiber dosage: 25 meq/l of effluent. Standard treatment conditions.
2) Fiber dosage: 20 meq/l of effluent. Temperature: 60°C.

TABLE II
Effect of recycling the fibres prior to their regeneration.

Concentration of fibers, meq/l	Number of times fibres are reused	Volume treated each time, l	% Colour Removal			Cumulative colour Removal	Total Fibres Used, meq.	Total Effluent Treated l
			1st cycle	2nd cycle	3rd cycle			
20	3	1	84.6	63.9	52.4	65.2	20	3
6.66	1	3	56	-	-	56	20	3
25	3	1	88.4	69.2	57.5	72.5	25	3
8.33	1	3	61	-	-	61	25	3
35	3	1	98.8	83.0	69.2	83.4	35	3
11.66	1	3	70	-	-	70	35	3

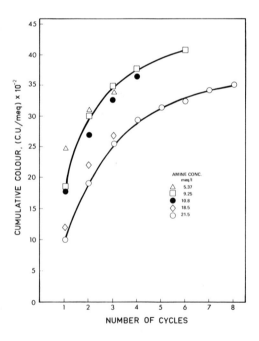

Fig. 6. Accumulation of colour on the fibres when they are reused without being regenerated (pH 4.5).

REGENERATION OF THE FIBRES

Influence of pH

As mentioned earlier the coloured materials picked up by the fibres were desorbed by converting the quaternised amino groups back to the amine form. As in the case of the decolourisation, pH was the countrolling factor.

Fig. 7. shows the amount of colour remaining in the fibres after they have been regenerated by dispersing them in sodium hydroxide at the indicated apparent pH. The similarity between this curve and that of Fig. 2 is striking and suggests that the colour sorption-desorption phenomenon may be reversible.

Regeneration was completed after 1 min and was not affected by temperature in the range 25-60°.

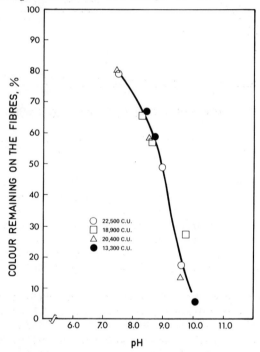

Fig. 7. Influence of pH on the regeneration efficiency. Figures refer to the level of colour present in the "used" fibres.

Influence of the Amount of Colour in the Regenerating Liquor

Obviously, one should not regenerate the fibres with a volume, V, of liquid as great as that of the effluent treated. The volume, v, of the liquor to be disposed of, and which contains the colour removed from the effluent, should be small compared to V, i.e., the ratio V/v, the concentration factor, should be high as possible.

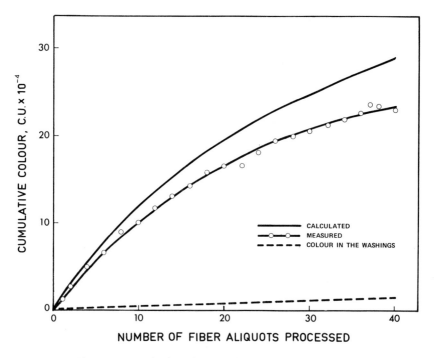

Fig. 8. Accumulation of colour in the regeneration liquor as a function of the number of 1 g aliquots of used fibres processed.

When dealing with fibre slurried, there is a limitation to the ratio of liquid to fibre that can be handled on conventional equipment and this puts a limit on the concentration factor. However, one way to increase this concentration factor is to recycle the regenerating liquor continuously while bleeding off only a fraction of it, in order to keep the colour balance in the whole system at equilibrium.

To do this, however, one must ensure that, even when it contains a high concentration of colour, the regenerating solution is still capable of performing its task. To check this, forty 1 g lots of fibers loaded with coloured materials were regenerated successively with one volume of regenerating liquor. Similarly, each 1 g portion of regenerated fibres was washed successively in the same water.

The build-up of colour in the regenerating liquor and in the wash water was measured and the results are shown in Fig. 8. Comparison with the calculated colour level curve show that, even when the regenerating liquor contained a high level of colour, it retained its regeneration efficiency.

RECYCLABILITY OF THE FIBRES

The economic feasibility of this process hinges upon the possibility of recycling the regenerated fibres a great many times. In fact, in an industrial process, one would like to recycle them indefinitely, only making up for the loss occurring in the filtration steps.

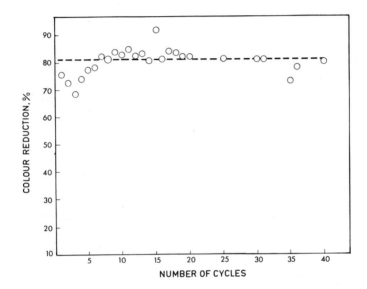

Fig. 9. *Colour removal efficiency of the fibres as a function of the number of times they have been regenerated. (pH of treatment 4.5: AEC 2.44 meq/g; Fibre concentration 3.6 g/l).*

Preliminary tests have been performed in the following fashion: The fibres, after the treatment, were regenerated in an NaOH solution, washed and reused. 40 cycles were performed. (Each time, however, a fresh NaOH solution was used for the regeneration). As seen on Fig. 9, after the 40th cycle the colour removal efficiency was unchanged.

The composition of the fibres did not significantly change during these 40 cycles, as shown in Table III.

It is realized that these conditions do not simulate the industrial process where the regeneration would be carried out in a highly coloured alkaline liquor and where the possibility of slow poisoning of the fibres is quite real.

USE OF MILL WASTE ACID AND CAUSTIC FOR pH ADJUSTMENT

When alkaline white liquor and chlorine dioxide waste acids, readily available at low cost in the mill, were used to adjust the pH in the regeneration and treatment stages, there was little change in the colour removal efficiency, as seen in Table IV.

However, as will be seen later, the dilution this creates, compared to the use of NaOH, lowers the concentration factor.

FISH BIOASSAY STUDIES

Few experimental data are available (Table V). They indicate that this method is only partially successful in detoxifying the caustic extraction liquor. For an effluent with medium initial toxicity, the requirement of the Canadian Federal Standard of 65% 98-hr LC50 could be met. However, for highly toxic effluents, detoxification was not sufficient. These results, nevertheless, are similar to those obtained with other physical-chemical treatment methods.

TABLE III

Anion exchange capacity (AEC) of grafted fibres after recycling.

Number of Cycles	Nitrogen[1] Found, %	Calculated AEC from Nitrogen meq/g	Calculated AEC by Titration Method, meq/g
0	3.39	2.42	2.44
5	3.05	2.18	2.22
10	3.02	2.16	2.43
20	3.05	2.18	N/A
30	3.07	2.19	2.44

1) Nitrogen was determined by the Kjeldahl method.

TABLE IV

Effect of types of acid and alkali on colour reduction.

Alkali Used in Regeneration	Acid Used in Treatment	Colour Removed in the 2nd Cycle, %
NaOH	HCl	94
NaOH	H_2SO_4	91
NaOH	Waste Acid	82
White Liquor	H_2SO_4	92
White Liquor	Waste Acid	90

TABLE V
Results of fish bioassays

Effluent Source	Wood Type	Fibre Dosage meq/l	Initial Toxicity, 96-hr LC50	Toxicity after Treatment 96-hr LC50
Mill A	Softwood	6.6	4.0	28
Mill A	Hardwood	15.0	26	65
Mill A	Hardwood	23.0	8.8	20
Mill C	Softwood	27.6	1.7	37

In a separate series of experiments, model toxic compounds such as cinnamic acid and chorophenols were used. Compounds with carboxyl groups were easily removed from the aqueous solution, but this was not true for phenolic derivatives. Toxic compounds without acidic carboxyl groups, e.g. tetrachloro-O-benzoquinone (which has been positively identified in such effluents [5]) were not removed by this process.

OTHER AMINO MONOMERS

Attempts were made to graft other monomers containing amino groups: (allylamine, 2-and 4-vinylpyridine, and 2-aminoethyl methacrylate). However, under the conditions used, grafting efficiency was very low. No effort was made to optimize it.

A quaternary amine polymer was also grafted, trimethylamino-ethyl methacrylate, obtained by reacting (DMM) with methyl sulfate. However, while quite efficient in decolourisation over a wide range of pH, the grafted fibres could not be regenerated and therefore, would have little economic interest.

CONCLUSION

The potential of cellulosic fibres, grafted with a polymer containing tertiary amino groups, in removing colour and BOD from kraft bleachery effluents has been demonstrated. The solid-liquid extraction process involves only conventional preper-making equipment.

It remains to confirm the long-term recyclability of the fibres, key to an economical process. Work is now underway to answer this question and will be reported at a later date.

REFERENCES

1. Weaver, M. O., Bagley, E. B., Fanta, G. F. and Duane, W. M., US Patent 3,935,099 (1975).
2. Lepoutre, P., Hui, S. H. and Robertson, A. A., J. Appl. Polym. Sci. 17: 3143 (1973).
3. Carles, J. Pulp and Paper Research Institute of Canada. Private Communication (1973).
4. Hui, S. H. and Lepoutre, P., J. Appl. Polymn. Sci. 19: 1771 (1975).
5. Das, B. S. et al, J. Fish. Res. Bd. Canada 26 (1969).

PART 5 General Cellulose Modification Reactions

Modified Cellulosics

ANTIBACTERIAL FIBERS

T. L. VIGO

USDA Textiles and Clothing Laboratory
Knoxville, Tennessee

Representative classes of compounds investigated for their ability to impart antibacterial activity to fibrous products are reviewed. Various chemical and physicochemical approaches for affixing these agents to fibers, and mechanisms by which the treated products are active against micro-organisms, are also discussed. Other topics are: test methods for evaluating antibacterial and antifungal activity of treated fibers; species and genera of bacteria, dermatophytes and other fungi, and viruses evaluated on the modified fibers; specific considerations important for end uses of the textile or paper products, especially durability of antimicrobial activity and safety to the user or wearer of said materials.

INTRODUCTION

The application of antibacterial agents to textile and paper products began in the early 1940's as an outgrowth of related research in protection of fibers from outdoor microbiological degradation and in techniques for disinfecting and sterilizing fibrous materials contaminated with various microorganisms. The initial efforts in this area of research focused on development of a reliable test method to evaluate bacteriostatic and bactericidal properties of fabrics impregnated with various agents (1). Textile and paper products treated with antibacterial and antimicrobial agents have been advocated for many end uses, but current applications are primarily for medical, hygienic, and aesthetic functions.

CLASSES OF CHEMICAL COMPOUNDS

Most of the known antibacterial, antifungal, and broad spectrum antimicrobial agents have been investigated for their ability to impart microbiocidal and microbiostatic activity to textile and paper products. Some formulations and compositions have received considerable attention, whereas others have only been cursorily evaluated. Table I lists major classes of compounds used for this purpose, with a specific compound or composition shown for each class and the end use for which it was described.

A. Metal Salts and Organometallics

1. Mercury and Silver

The salts of mercury and silver are excellent bacteriostats and fungistats even at very low concentrations, and have been extensively tested for incorporation into fibrous materials. Inorganic mercury salts have been affixed to fibers by use of urea-formaldehyde resins (2), graft copolymers (3), and reactive dyes (4). Phenylmercuric acetate (5) was one of the first reagents used to prepare washfast germicidal textiles. In one novel and practical method, textiles were treated with mercury-containing allyl triazines to produce antibacterial and antifungal properties (6).

Silver compounds have been incorporated into fibers by similar techniques. In early investigations insoluble compounds were tested. For example, silver oxide (7) was used for antibacterial gauzes, and silver thiocyanate (8) for washfast antibacterial fabrics. Fibrous products treated with thiourea resins (9) or bearing anionic substituents (10) had good affinity for silver salts, and were therefore effective in preparing fibers with antibacterial activity.

2. Tin and Zinc

Organometallic tin compounds and inorganic zinc compounds have been used as antibacterial agents for textiles and paper. Tributyltin oxide (11) has been the most frequently used tin compound. A mixed organometallic tin and titanium salt was used to produce washfast antibacterial acrylic fibers (12). Although zinc salts have only moderate antibacterial activity, they have been combined with other antiseptic or germicidal agents, such as ethylenediaminetetraacetic acid (13) or hydrogen peroxide (14), to produce washfast antibacterial textiles.

3. Various Metals

A variety of other metal salts and organometallics, usually in combination with other reagents, have been used in the antibacterial treatment of fibers. These include practically all the transition metals not previously mentioned: vanadium, chromium, iron, cobalt, nickel, copper, zirconium, molybdenum, cadmium, rhenium, and gold. Chrome mordants (15) were initially used to protect wool from bacterial damage;

polyureas (16) and aziridinyl phosphine oxides (17) were
effective in binding a variety of metal ions to fibers to
impart antibacterial activity. Current approaches include
incorporation of such exotic materials as formylferrocene (18)
and rhenium salts (19) into fiber spinning baths to produce
antimicrobial products. In addition to transition metals,
organobismuth (20) and organoarsenic compounds (21) have been
used to impart antibacterial activity to textiles and paper.

B. Iodine and Iodophors

Germicidal and bacteriostatic activity have been imparted
to fibers by treatment with iodine and iodophors (solubilizing
agents which slowly release iodine). Surgical fibers for
wound dressing packages having germicidal activity were prepared from a variety of iodine-containing cellulose derivatives (22). Copolymers of poly(vinylpyrrolidone) and poly
(vinyl alcohol) produced fibers having the ability to take up
iodine (23), and thus produce materials having antibacterial
activity.

C. Phenols and Thiophenols

The "phenol coefficient" has long been the basis for
evaluating relative potencies of germicidal and disinfecting
compositions. Bisphenols, particularly those containing
halogen, are usually much more effective microbiostats than
monophenols, and have therefore been used extensively in
rendering textile and paper products bacteriostatic and
fungistatic. Hexachlorophene [(2,2'-methylenebis(3,4,6-
trichlorophenol)], dichlorophene [(2,2'-methylenebis(4-
chlorophenol)], and bithionol [(2,2'-thiobis(4,6-dichloro-
phenol)] are the bisphenols most often used. For example,
the potassium salt of hexachlorophene was affixed to cotton
fabric previously treated with epichlorohydrin, to produce
textiles with bacteriostatic properties (24); in an alternative method, hexachlorophene is bound with a crosslinking
agent and a zirconium salt (25) to produce cotton fabrics that
retain antibacterial activity after 20 launderings. Blends of
untreated cotton or wool fibers with polypropylene fibers
containing hexachlorophene or bithionol were suitable for
antibacterial end uses (26). Recently, phenolic diphenyl
ethers, such as 5-chloro-2-(2,4-dichlorophenoxy)phenol, were
used in dyebaths containing phthalic or terephthalic esters
as carriers to produce antimicrobial synthetic fabrics durable
to laundering (27).

D. Quaternary Ammonium and Other Onium Salts

Quaternary ammonium salts, particularly those having at least one aliphatic substituent containing 12-18 carbon atoms, were initially used for disinfecting hard surfaces and textiles. Those compounds also have been used as bacteriostatic finishing agents for fibers. For durable finishes, binding agents were used, either as anionic groups in the treating bath or as part of a graft copolymer on the fiber; another method polymerized materials containing pendant ammonium groups onto the fiber. Those approaches are necessary because ammonium salts are readily washed out during laundering.

For production of antiseptic bandages, cellulosics were midly oxidized to create carboxyl groups to more readily absorb quaternary ammonium salts onto the fiber (28). Germicidal carpets were produced by binding alkyldimethylbenzylammonium chlorides with formaldehyde (29). Durable antibacterial nylons were prepared by application of benzimidazolium salts having an aliphatic chain of 10-12 carbon atoms (30). These methods of binding are useful, but fixation of the ammonium or other onium linkages to the fiber as a pendant group of a graft or homopolymer was more effective. Poly (vinylpyridines) grafted onto rayon or mercerized cotton fabrics, then subsequently immersed in hexachlorophene, were washfast (31), as were all types of textile fibers treated with an organosilicon polymer containing pendant quaternary ammonium groups (32). Cotton fabrics, when treated with tetrakis(hydroxymethyl)phosphonium chloride and methylolated melamines, were flame retardant and antibacterial even after 50-100 launderings (33); it was not determined whether the antibacterial activity was due to the phosphonium salt, the free formaldehyde liberated from the resins, or a synergistic effect of all reagents.

E. Antibiotics

In addition to their widespread use for combating internal or external infections, antibiotics have been applied to fibrous materials to inhibit or kill a variety of microorganisms. Neomycin was one of the first antibiotics evaluated for this purpose (34). The durability of this substance to laundering was improved by first modifying the fibers with anionic groups, such as polyacrylic acid for nylons (35), or triazines with sulfonic acid groups for cellulosics (36). One recent and innovative approach involved chelation of cellulosics with transition metal salts of titanium^{III}, iron^{III},

TABLE 1
Representative Classes of Chemicals Used to Impart Antibacterial and Antifungal Activity to Fibers

Type of Compound	Example	End Use or Product	Reference
Metal salts	AgCNS	All textiles	(8)
Organometallics	Mercurated allyl s-triazines	Cotton fabrics	(6)
Organometallics	$(\underline{n}\text{-}C_4H_9)_3SnO$	All textiles	(11)
Iodine and iodophors	Iodinated celluloses	Surgical gauzes	(22)
Phenols and thiophenols	Hexachlorophene	Cotton fabrics	(25)
Onium salts	3-(Trimethoxysilyl)-propyl dimethyloctadecyl ammonium chloride	Fibers, glass surfaces and polymers	(32)
Antibiotics	Neomycin sulfate	All textiles	(34)
Heterocyclics with anionic groups	Acrylamido-methylene-8-hydroxyquinoline polymers	Cotton fabrics	(39)
Nitrofurans	B-(5-nitro-2-furyl) acrolein	Synthetic fibers	(44)
Ureas and related compounds	$CH_3(CH_2)_{11}NH\text{-}\overset{NH}{\underset{\|}{C}}\text{-}NH_2 \cdot HCl$	Paper	(52)
Formaldehyde derivatives	Polyformaldehyde	Synthetic fibers	(55)
Amines	Hexetidine	Chemotherapeutic bandages	(58)
Miscellaneous	Chlorophyll	Shoe insoles	(63)
Miscellaneous	Tetrahydrofuran polymer and sulfanilamides	Dressing for burns	(65)

tin^{IV}, $vanadium^{III}$, or $zirconium^{IV}$, to immobilize antibotic substances on the fiber. The use of ampicillin, gentamicin, kanamycin, neomycin, paromycin, polymyxin B, and streptomycin produced antibacterial effects, while amphotericin B and natamycin exhibited antifungal activity (37).

F. Heterocyclics Containing Anionic Groups

Heterocyclic substances such as 8-hydroxyquinoline and 2-mercaptobenzothiazole, have been used extensively to protect fibers from mildew and rot in outdoor applications. These types of substances have also been applied to fibers to confer antibacterial activity. Substituted hydroxyquinolines containing chloro- and iodo- groups have long been used in surgical gauzes (38). Cotton fabrics treated with homo- or copolymers of acrylamidomethylene-8-hydroxyquinoline metal complexes, had antibacterial activity (39). The cadmium salt of rhodanine was applied to cellulosic materials to produce bactericidal properties in the fibers (40). Paulus and Pauli (41) reviewed their research on the chemical modification of cotton with substituted triazines, benzothiazoles, and pyrimidines as antimicrobial finishing agents.

G. Nitrofurans and Other Nitro Compounds

The use of a nitro compound as an antibacterial agent for synthetic and natural fibers, 3-(5-nitro-2-furfurylidene) oxindole, was first described in the patent literature (42). However, about three years later, the discovery by Vol'f and Hillers (43) that polyvinyl alcohol fibers acetalized with B-(5-nitro-2-furyl)acrolein had excellent antibacterial activity, led to the production of such a modified fiber called "Letilan." This fiber is claimed to be excellent for its effectiveness against bacteria, dermatophytic fungi, and yeasts, and also is washfast. Other synthetic fibers, such as poly(acrylonitrile) and poly(vinyl chloride) have also been treated with the above nitro compound to achieve antimicrobial activity (44). Smirnova et al, (45) has described the optimum reaction conditions for producing Letilan, and attributed the activity of the modified poly(vinyl alcohol) fiber to the slow release of an aldehyde (46). Research in this area is continuing, as reflected by a recent investigation into applying 5-nitrofurylacrolein to di- and triacetate cellulosic fibers (47).

H. Ureas, Carbanilides, and Guanidines

Compounds with amido- ($-\underset{\underset{O}{\|}}{C}-NR_2$) or imino- ($-\underset{\underset{NR'}{\|}}{C}-NR$) groups have also been used as antibacterial finishing agents for fibers. Carbanilides, such as 3,4,4'-trichlorocarbanilide, frequently used in soaps and other cosmetic products, were added to spinning baths to prepare synthetic fibers with microbiocidal activity durable to 25 launderings (48). Halogenated salicylanilides were effectively bound to cotton and rayon fabrics by durable press agents (49). Halogenated arylthioureas were also used in antimycotic and antibacterial finishes for textiles (50). Pseudoureas containing 8-18 carbon atoms were effective in preparing antimicrobial paper suitable for bandages, diapers, and related items (51).

Guanidines and biguanide salts have recently attracted interest as effective bacteriostatic reagents, particularly for diapers and paper products. In the first such application, dodecylguanidine hydrochloride was used to prepare bacteriostatic paper (52). More recently, emphasis has been on utilizing salts of the bisbiguanide called "chlorohexidine." Hydrochlorides of this compound have been added to polyamide spinning baths to manufacture bacteriostatic fibers (53), whereas, digluconate salts were applied to cellulosics for production of bacteriostatic diapers (54).

I. Formaldehyde and Formaldehyde-Containing Derivatives

The use of formaldehyde and urea-formaldehyde resins as durable and permanent press agents for textiles is well known. However, these types of reagents have also been used to enhance binding of antibacterial agents to fibers. Many of the investigations using formaldehyde-containing resins as binding agents have not assessed the role of free formaldehyde in antibacterial effects imparted to the fiber. A few studies, however, demonstrated that formaldehyde itself is an effective antibacterial agent for textile and paper products. A synthetic fiber based on incompletely acetylated polyformaldehyde is claimed to exhibit bactericidal properties (55). More conventional approaches have utilized N-methylol amides and sulfonamides as antimicrobial finishing agents for natural and synthetic fibers (56).

J. Amines

A variety of compounds and polymers containing aminogroups have been applied to fibers as antibacterial finishing agents. On nylon and wool fabrics, halogenated aryl diamines and piperazine salts imparted microbicidal activity against bacteria and dermatophytic fungi even after 20 launderings (57). Hexetidine, a substituted hexahydropyrimidine, was used for producing medicinal cotton bandages (58). Antibacterial toilet paper was prepared by impregnation of pulp fibers with polyethyleneimine (59). A formulation containing amines with 12-24 carbon atoms, an epoxide polymer, and a dicarboxylic acid, provided, for nylon carpets, an antimicrobial finish that was resistant to dry cleaning and shampooing (60).

K. Miscellaneous

The chemical composition of agents used to impart durable antibacterial and antifungal activity to fibers is not disclosed in two successful investigations on this subject. Hausam (61) describes the formulation called "R-52" as an effective agent for producing cotton and synthetic hosiery materials resistant to bacteria and dermatophytic fungi even after 100 launderings. Radford (62) discusses antimicrobial agents, in the "Steri-Septic" series, that confer antibacterial and antifungal activity durable for 75 launderings on a variety of fiber and polymer surfaces. These compositions may be applied by exhaustion, padding, or by spraying and incorporation into adhesive or resin bonding systems from cationic, anionic, or nonionic solution.

Other chemical approaches for imparting antibacterial activity to fibrous materials have included attachment of chlorophyll with aluminum or barium salts to cellulose for sanitary products or shoe insoles (63), incorporation of a thiocyanate group in cellulosics (64), and preparation of a polyester fabric containing a bioerodible tetrahydrofuran polymer and 4-homosulfanilamide to prevent bacterial infection from wounds caused by burns (65). Discovery of new antibacterial and antifungal agents probably will lead to further studies to determine the best methods for binding these agents to achieve antimicrobial activity on the treated surfaces.

PRINCIPLES OF ANTIBACTERIAL FINISHING

Many of the reagents and polymers used for imparting antibacterial and antifungal activity to fibers have been applied by the techniques suggested by Gagliardi (66). These include: (i) deposition of insoluble products in the fiber, (ii) formation of metastable bonds, (iii) interaction with thermosetting agents, (iv) formation of coordination compounds, (v) ion-exchange methods, and (vi) production in fibers of an active germicidal species that is continually regenerated by addition of bleaching agents during laundering or by photochemical exposure. The last method, called "the regeneration principle," is favored over the other five techniques for producing permanent antibacterial activity in fibrous products. All of the above methods are based on the controlled-release concept--if an antimicrobial agent is to be effective on a fiber, it must be slowly released from the fiber surface so it will contact bacteria or other microbiota. Such release is usually brought about by moisture present in the fiber, but may also be accomplished by air oxidation or photochemical exposure of the textile or paper product.

Although no chemical or photochemical method of achieving regeneration has yet been developed, the other five methods have been used with varying degrees of success to produce durable antimicrobial finishes. It is perhaps now appropriate to reevaluate some of the chemical compounds previously listed with respect to each of the five techniques. Silver salts, such as silver thiocyanate (8), have been insolubilized on fibers to produce antibacterial effects. An excellent example of an antimicrobial fiber employing the metastable bond technique is Letilan. This modified poly(vinyl alcohol) fiber is produced by acetalization with 5-nitrofurylacrolein, and is biologically active because of bond breakage to release aldehydes (46). Coapplication of melamine-formaldehyde resins with a phosphonium salt effectively produced antibacterial cotton fabrics durable to multiple launderings (33). Coordination compounds of chelating agents, such as the zinc salt of ethylenediaminetetraacetic acid (EDTA), when applied to fibers, provided a textile finish exhibiting antibacterial activity after 50 launderings (13). Chelation of cellulose with transition metal salts, then replacement of the unsubstituted ligands of transition metal ions with electron-donating groups of antibiotic molecules, rendered cellulosic surfaces resistant to attack by bacteria and fungi (37). Application of compounds containing cationic and anionic groups having germicidal activity, such as alkyltrimethylammonium salts of hexachlorophene (67), typifies an ion-exchange method for imparting antibacterial activity to fibers.

Strictly speaking, graft or homopolymerization of monomers onto fibers for the purposes of imparting antimicrobial or antibacterial activity may be classified under the above five techniques. However, because of the great improvement in durability of activity when polymeric coatings or matricies, rather than simple chemical substances, are present in the fiber, it is appropriate to consider separately the techniques based on this approach. Grafting of cellulosics to affix cationic poly(2-methyl-5-vinylpyridine) or anionic poly(acrylic acid) groups to the fibers, followed by subsequent immersion in solutions containing counterions, such as hexachlorophene or Cu^{II} salts, produced an antimicrobial finish with excellent resistance to laundering (31). More recent investigations demonstrated that the formation of a metal-peroxide polymeric coating on cotton fabrics produces an antibacterial finish durable for 20 launderings when zirconium salts are used (68), or for 50 launderings when zinc salts are used to bind hydrogen peroxide to the fibers (14). The durability of antibacterial activity was attributed to the slow release of peroxide when the fabrics were laundered.

Other methods for achieving antibacterial activity by the controlled release of agents from the fiber have recently been reviewed (69). Encapsulation, a physicochemical method devellped for the controlled release of pesticides, has been successfully used to produce antibacterial textile and paper products. Such methodology represents a promising way of achieving long-lasting antibacterial effects, but the present drawbacks are the intricacy of the process and its lack of applicability to all fiber end uses. Examples in which such a technique has been used to prepare antibacterial fibers are: hospital mattresses containing antibacterial agents that retained their activity after several years' use (70), and the preparation of bactericidal and insecticidal paper containing pine oil and toluamide (71).

Recent studies have claimed that effective antimicrobial or antibacterial action can be exhibited by treated fibrous surfaces via the permanent barrier mechanism. This mode of activity is based on affixing an inherently microbiostatic agent to fibers, thus preventing by contact the growth of microorganisms on the coated surface, in contrast to diffusion of the antimicrobial agent from the fiber via a controlled-release mechanism. An organosilicon polymer, containing a pendant quaternary ammonium group, when bound to glass, textile, metal, and other surfaces, did not diffuse or leach out from the modified surface, yet exhibited antimicrobial activity against a variety of bacteria, yeasts, fungi, and algae (32). This represents the first successful demonstration of the permanent barrier concept. In 1975, polymeric "self-sterilizing" surfaces were prepared by utilizing ionomers of poly (acrylic acid) and 8-hydroxyquinoline copolymerized with polyethylene (72). This methodology may be developed further, as suggested by Radford (62).

TEST METHODS

Development of reliable test methods for evaluating antibacterial activity of fibers has received a great deal of attention. A 1976 review (69) describes the advantages and disadvantages of the most commonly used tests. One way of evaluating antibacterial tests for fibers is to assess them in terms of increasing degrees of sophistication and scope. The Warburg respiration test (73), perhaps the simplest method for rapid screening of bacteriostatic activity of fibers, is based on the increase or decrease of oxygen consumption of bacteria in the presence of treated fibers. The method is quantitative, but does not measure the effect of modified fibers on other bacterial functions. Another way of quickly assessing bacteriostatic activity is the agar plate method (74a). This test has been criticized for its limited scope, since it is reliable only for agents that diffuse off the fiber into the agar medium to give a positive zone of inhibition.

The parallel streak test, in which inhibition of bacterial growth is determined by observing growth-free areas around the specimen of fabric placed perpendicular to streaks of inoculum of agar, has been recently accepted as the standard AATCC method for qualitative evaluation of bacteriostatic activity of modified fibers (74b), and yields reproducible results in interlaboratory testing (75). Control fabrics should be tested with treated fabrics in this method, since one drawback of the test is false-positive results obtained with untreated fibers (75).

Inoculation of the surface of agar with bacteria affords direct contact between the fabric surface and the test microorganism and is advocated as a technique for evaluating activity of fibers treated with non-diffusible and chemically bonded antimicrobial agents (76). This (76) and an earlier study (77), visually demonstrated the advantages of direct contact between the surface of the treated fabric and the test microorganisms over the agar plate method, in which such contact is not made. This contact plate method was used to quantitatively evaluate the antibacterial activity of permanent coatings, such as organosilicon polymers applied to a variety of surfaces (32); this type of test method is consistent with the permanent barrier concept for achieving long-lasting antibacterial effects in fibers.

Most of the methods mentioned above, use representative bacteria such as gram-positive Staphylococcus aureus and gram-negative Klebsiella pneumoniae or Escherichia coli, as test organisms for evaluating antibacterial activity of treated fibers. A semi-quantitative method, developed by Majors (78), is useful for evaluating antibacterial activity of fibers against urealytic organisms, such as Brevibacterium ammoniagenes and Proteus mirabilis. This test is based on determination of titratable acid or alkali produced by growth of test organisms in a highly buffered medium containing urea or glucose. The test has not gained wide acceptance because of its limited scope. A more recent AATCC method is an outgrowth of the Majors test, and was modified to quantitatively evaluate activity of fabrics against S. aureus and K. pneumoniae. However, this method (74c) is time-consuming and involves sterilization of fabrics, inoculation with the bacteria, pH adjustments, incubation of fabrics, and counts of bacteria per fabric sample. Although the Quinn test is equally as difficult to conduct (79), it offers the only quantitative technique for evaluating activity of treated fibers towards a variety of bacteria or fungi under either -static or -cidal conditions. After the fabric is sterilized, it is inoculated with the desired test organism, embedded in an agar medium and incubated; then colonies of microorganisms are counted directly on the fabric surface by means of a low-powered microscope (23 X). Lashen (80) claims improvement and simplification of Quinn's test by using less time-consuming methods for fabric sterilization and a different technique for suspending the fabric in the agar medium.

The Quinn test is the only method now available for quantitative evaluation of antifungal activity of treated fibers (79); three other methods for assessing such activity are qualitative, and differ only in their exposure time to the fungi. In the agar plate method (74d), sterilized fabrics are inoculated with fungi such as Chaetomium globosum (a cellulolytic fungi), incubated, and inspected visually for fungal growth on the fiber surface. In a method for rapid screening of antifungal finishes (81), several strips of treated fabrics are attached to a perfusion bed by an adhesive, then inoculated and incubated. Fabrics can be ranked by evaluation of growth of fungi on the fibers as a function of time. The humidity jar test (82) involves longer exposure times (up to 28 days) of fabrics to fungi, and is also based on a visual evaluation of the treated fabrics.

No test methods are available for evaluation of the effectiveness of treated fibers in inhibiting or killing harmful viruses. However, Sidwell and co-workers (83) described conditons under which polio and vaccinia viruses may be transferred onto all types of fibrous surfaces.

FIBER END USES

Applications for fibers with antibacterial or antimicrobial activity are a function of benefit to the wearer or user, rather than of preservation of textiles and paper from fiber-degrading bacteria and fungi. A convenient classification of fiber end use based on this scope has recently been advocated (69) and includes: (a) medical--treated fibers that suppress growth or kill pathogenic and/or parasitic microorganisms problematic in hospitals and public institutions, (b) hygienic--modified fibers useful for preventing skin and related infections, and (c) aesthetic--treated fibers primarily useful in suppressing or killing odor-causing bacteria found in body perspiration. Requirements for producing effective antimicrobial fibers are also dictated by the species most deleterious in a particular textile or paper product, the extent of durability of antimicrobial activity to laundering or leaching required in using the fibrous product, and the absence of toxic effects of the antimicrobial agent to the wearer or user of the modified fibers.

A. Representative Microorganisms

Fibrous materials have been inoculated with a variety of microorganisms, particularly bacteria and fungi, to assess their antimicrobial activity. Representative gram-positive and gram-negative bacteria, dermatophytic and other fungi, viruses, and other microbiota utilized for this purpose, are listed in Tables 2 and 3. Characteristics of each of the species listed are taken from standard references (84) and from textbooks on microorganisms and their relationship to human disease (85, 86).

The gram-positive bacteria Staphylococcus epidermidis and Corynebacterium sp. (diptheroids) were identified by McNeil et al, (87) as the species most frequently causing body odor in clothing that contacts the axillae of the body. This study questioned whether an antibacterial fiber finish would prevent body odor unless the garments were tight-fitting; however, these observations were based on treatment of clothing with non-durable quaternary ammonium compounds on specific types of apparel. Radford (62) observed that nylon socks treated with his "Steri-Septic" agents produced much less odor after a normal day's wear than untreated socks; the odor production in this instance is attributed to the potentially pathogenic Staphylococcus aureus. As a recent article indicates, distribution of bacterial population of the normal skin flora varies from one part of the body to another; although S. epidermidis and coryneform bacteria were most frequently isolated from axillae, nasal passages, the head, arms, and legs, the representative bacterial population of the feet was comprised of a different distribution of microorganisms (88).

Staphylococcus aureus, the species most frequently used for testing antibacterial activity of treated fibers, is considered to be the major cause of cross-infection in hospitals (89) and during inadequate home and commercial laundry practices (90). In addition to its ability to cause cross-infection, this microorganism has also come under scrutiny for

TABLE 2
Representative Bacteria Used to Evaluate Activity of Modified Fibers

Species of Bacteria [a]	Comments	Typical End Use
Staphylococcus Aureus	Pathogenic; causes pyogenic infections	Hygienic; medical
Staphylococcus epidermidis	Parasitic; produces body odor	Aesthetic
Corynebacterium diptheroides	Non-pathogenic; produces body odor	Aesthetic
Brevibacterium ammoniagenes	Urealytic; causes diaper rash	Hygienic
Streptococcus pneumoniae	Causes bacterial pneumonia	Medical
Mycobacterium tuberculosis	Cause of tuberculosis	Medical
Escherichia coli	Causes infections of urogenital tract	Medical; aesthetic

Klebsiella pneumoniae	Causes pneumonia and other infections	Medical
Pseudomonas aeruginosa	Formed in wounds and burns	Medical
Proteus mirabilis	Found on putrefying materials; causes urinary infections	Medical; hygienic
Salmonella typhosa	Cause of typhoid fever	Medical
Shigella dysenteriae	Cause of epidemic bacillary dysentery	Medical

a. *First six species listed are gram-positive; remainder of species listed are gram-negative.*

TABLE 3
Other Representative Microorganisms Used to Evaluate Activity of Modified Fibers

Microorganisms[a]	Comments	Typical End Use
Aspergillus niger	Cellulolytic organism	Aesthetic; fiber preservation
Chaetomium globosum	Cellulolytic organism	Aesthetic; fiber preservation
Candida albicans	Yeast; causes diaper rash	Hygienic
Epidermophyton floccosum	Causes infection of skin and nails	Hygienic
Trichophyton interdigitale	Cause of athlete's foot	Hygienic
Trichophyton rubrum	Cause of Chronic infections of nails and skin	Hygienic

Polio virus[b]	Acute infection which may cause paralysis	Medical
Vaccinia virus[b]	Localized disease induced by vaccination against smallpox	Medical
Trichomonas vaginalis[c]	Protozoa causing vaginal infection	Hygienic; medical
Schistosoma japonicum[d]	Parasitic flatworms	Hygienic; medical

a. *First six species listed are fungi.*
b. *Representative viruses.*
c. *Protozoa.*
d. *Parasitic worms.*

its ability to cause diaper rash (62). The other two bacteria commonly implicated in diaper rash are the gram-positive Brevibacterium ammoniagenes (62, 78) and the gram-negative Proteus mirabilis (62).

Other bacteria listed in Table 2 are those typically isolated from hospital patients with infections, and can present problems of disease transmission by clothing in such environments. Gram-negative Escherichia coli and Pseudomonas aeruginosa are the most troublesome microorganisms in this regard, and the latter is particularly prevelant in wounds caused by burns (65, 89).

Prevention of growth of yeasts, such as Candida albicans, and of dermatophytic fungi, particularly the Trichophyton and Epidermophyton genus, is important in fibrous materials used for hygienic purposes. The yeast Candida albicans, along with the bacterial species previously mentioned, has been implicated in diaper rash (62). An article on aspects of footcare and hygiene attributes over 75% of foot infections to the two species of dermatophytic fungi Trichophyton rubrum and Trichophyton interdigitale (91); the latter causes athlete's foot. For hosiery, commercial antimicrobial treatments that prevent the growth of dermatophytic fungi are effective against representative Epidermophyton and Trichophyton species (32, 61, 92). Antifungal activity against fiber-degraders, such as Aspergillus niger or Chaetomium globosum, rarely falls within the scope of medical or hygienic fiber functions, but could be considered an aesthetic function and extends the useful life of such materials as shoe insoles. Antiviral finishes for fibers have not been developed, but all types of natural and synthetic fibers are effective fomites (inanimate objects which may be contaminated with infectious organisms and serve in their transmission) for representative viruses encountered in hospitals and similar environments (83). Activity against other microorganisms, such as protozoa and parasitic worms, has also been evaluated on modified fibers. A fiber blend of cotton and the antimicrobial fiber Letilan was recommended as a trichomonocidal product effective against the vaginal infection caused by the protozoa Trichomonas vaginalis (93). N,N-diethyllauramide was the most effective and washfast compound that could be applied to fabrics to confer resistance against the parasitic flatworms Schistosoma japonicum and Schistosoma masoni (94).

B. Specific Requirements

Several articles (62, 66, 69) have listed desirable features of an antimicrobial fiber of optimum benefit to its wearer or user. These features are: (a) selective activity towards undesirable microorganisms, (b) acceptable moisture transport properties, (c) compatability with other finishing agents, (d) absence of toxic effects to the wearer or user, (e) durability of activity to laundering, dry cleaning, or leaching, and (f) marketability or utilization of readily available reagents for producing an antimicrobial fiber finish.

Specificity of the antimicrobial agent means that specific microorganisms harmful in the end use of the fibrous material are either killed or inhibited in their growth. For example, antimicrobial socks, to constitute an effective product, should effectively inhibit the growth of the dermatophytic fungi Trichophyton interdigitale and the gram-positive bacteria Staphylococcus aureus. The choice between microbiostatic and microbiocidal activity would also depend on the end use of the fiber. Most fibrous products used in hospital environments require germicidal activity to prevent cross infection, whereas, other fiber functions usually require only microbiostatic activity of the treated fibers.

Moisture transport properties are considered important in releasing the antimicrobial agent from the fiber so that it contacts the microorganism. Obviously, fiber finishes based on the barrier concept would not fit into this consideration. The greater antibacterial activity exhibited by cotton fabrics over cotton/synthetic fiber blends when treated with polymeric zinc peroxides (14) or hexachlorophene (95) could be attributed to the superior moisture transport properties of cotton.

Compatability of the antibacterial or antifungal agent with other fiber finishing agents may be important in producing other desirable fiber properties such as durable press and flame resistance. Multiple property finishes have occasionally been used to advantage in preparing antibacterial textiles. Examples are the use of a phosphonium salt with melamine resins to produce an antibacterial and flame retardant cotton fabric (33) and the addition of a chlorinated diphenyl ether to a dyebath to prepare an antimicrobial synthetic fiber (27).

The absence of toxic or adverse dermatological effects in the finished antimicrobial product is another necessary feature. Many of the proposed antimicrobial or antibacterial fiber finishes have not been commercialized or have been discontinued either because reagents used in producing the fiber finish had potential adverse side effects on the user or wearer, or because the chemical reagents were unacceptable due to cost, inconsistent product performance, or to their harmful effects on the environment. However, studies have demonstrated the absence of toxic side effects in fibrous materials given an antimicrobial finish. Hosiery treated with the commercial "R-52" (61) and the mercurated triazines (96) were claimed to be safe to wearers and to exhibit prolonged

TABLE 4
Requirements for Specific Antimicrobial Fibrous Products

Market End Use	Durability of Activity	Type of Activity	Fiber Function
Carpets and rugs	Cleaning and shampooing	-Static	Hygienic
Outerwear apparel	Laundering or dry cleaning	-Static	Aesthetic
Underwear	Laundering	-Static or -cidal	Aesthetic; hygienic
Paper towels & tissues	Leaching	-Static	Hygienic
Sheets & bedding materials	Laundering or dry cleaning	-Static	Hygienic
Hosiery	Laundering	-Static	Hygienic
Towels & related items	Laundering	-Static	Hygienic
Diapers & sanitary napkins	Leaching or laundering	-Static or -cidal	Hygienic
Disposable hospital items	Leaching	-Cidal	Medical
Non-disposable hospital items	Laundering or dry cleaning	-Cidal	Medical

antimycotic activity against representative Epidermophyton and Trichophyton species after 50 to 100 launderings. Commercially available antimicrobial socks, based on the permanent bonding of an organosilicon polymer to the fibers (32), have been registered with the Environmental Protection Agency and have met standards for safety of the material to the wearer (92). The highly touted antimicrobial fiber Letilan, in addition to possessing broad spectrum antimicrobial activity, has also been recommended for its use in sutures and protheses, since it only causes mild tissue reaction, and is gradually resorbed by the organism (97). Another recent study in the Soviet Union claims that persons exposed continuously for three weeks to viscose rayon containing 3% hexachlorophene as an antimicrobial agent, experienced no allergic skin reactions to the modified fiber (98).

As shown in Table 4, requirements for durability of antimicrobial activity of modified fibers to laundering, dry cleaning, or leaching, varies greatly with the type of textile or paper product and its antimicrobial end use. Carpets and rugs used in the home, and given an antimicrobial finish, would usually only require durability of activity to cleaning or shampooing, while outerwear apparel, underwear, sheets and bedding, towels, hosiery, and non-disposable diapers would require durability to multiple launderings. As evidenced by the few chemical approaches that have produced durable antimicrobial activity that persists after 20 launderings, this type of durability is by far the most difficult to achieve. Disposable items such as toilet paper, sanitary napkins, and paper towels, only require durability of their activity to leaching.

Marketability of an antimicrobial fiber finish depends on availability of low cost chemical reagents used in applying the finish, as well as on the requirement that the product have the five favorable features just discussed. The fibrous product must also meet current federal regulations pertinent to the antimicrobial finishing of textiles, as discussed in a recent review article (69). Such information may be found by consulting current Federal Registers under various Codes of Federal Regulations (CFR). The Environmental Protection Agency (EPA) considers the application of antimicrobial agents to textiles as a pesticide use, and requires registration and efficacy data on the agents used, as well as distinctions between the terms disinfectant, bacteriostat, and sanitizer. If claims are made by the manufacturer for performance of the agents on the fiber alone, then only EPA guidelines and regulations have to be followed. However, if claims are made for the utility of the antimicrobial agents in arresting growth of microorganisms on living tissue, e.g., to promote wound healing, then the FDA (Food and Drug Administration) would require evaluation of the fibrous material as an antiseptic. Additionally, the use of an antimicrobial finish on materials such as children's sleepwear and mattresses would require that the treated textiles conform to current flammability standards set forth by the Consumer Product Safety Commission (CPSC). The impending passage of the Toxic Substances Control Act may further influence the availability of chemical reagents for antimicrobial fiber products. This whole area of research offers many possibilities for discovering new and useful antimicrobial fiber products of benefit to all.

REFERENCES

1. Hirschmann, D.J., and Robinson, H.M., Soap Sanit. Chemicals 17(9), 94 (1941).
2. Schickedanz, G., Brit. Pat. 838,727 (June 22, 1960).
3. Bank, A.S., Shaposhnikova, S.T., and Askarov, M.A., Dokl. Akad. Nauk. Uzb. SSR, 26(4), 27 (1969); Chem. Abstr. 71, 51161b (1969).
4. Virnik, A.D., Akovbyan, E.M., Gal'braikh, L.S., Pestereva, G.D., and Rogovin, Z.A., Izv. Vyssh. Ucheb. Zaved., Tekhnol. Tekst. Prom 1968 (4), 108; Chem. Abstr. 70, 20918d (1969).
5. Sowa, F.J., U.S. Pat. 2,423,121 (July 1, 1947).
6. Shaw, J.T., Gross, F.L., and Madison, R.K., U.S. Pat. 3,130,193 (April 21, 1963).
7. Fessler, F., U.S. Pat. 2,689,809 (Sept. 21, 1954).
8. Hill, W.H., Brit. Pat. 835,927 (May 25, 1960).
9. Gagliardi, D.D., Shippee, F.B., and Jutras, Jr., W.J., U.S. Pat. 3,085,909 (April 16, 1963).
10. Snezhko, D.L., Virnik, A.D., Pestereva, G.D., and Rogovin, Z.A., Izv. Vyssh. Ucheb. Zaved., Tekhnol. Tekst. Prom. 1968 (3), 162; Chem. Abstr. 69, 78373t (1968).
11. Permachem. Corp., Brit. Pat. 838,722 (June 22, 1960).
12. Murayama, Y., Nakamura, M., and Yokokoshi, K., Japan Pat. 69 29, 379 (Nov. 29, 1969); Chem. Abstr. 72, 112726j (1970).
13. Mendelsohn, M., and Horowitz, C., U.S. Pat. 3,079,213 (Feb. 26, 1963).
14. Chem. & Eng. News 54 (37), 35 (Sept. 6, 1976).
15. Nopitsch, M., Melliland Textilber 31, 619 (1950).
16. Rakhimova, M.V., Tulyaganov, M.M., Gafurov, T.G., and Usmanov, K.U., Dokl. Akad. Nauk. Uzb. SSR, 27 (8), 33 (1970); Chem. Abstr. 74, 77279h (1971).
17. Gagliardi, D.D., and Kenney, V.S., U.S. Pat. 3,547,688 (Dec. 15, 1970).

18. Vol'f, L.A., Polishchuk, B.O., Koteskii, V.V., Kirilenko, Y.K., Meos, A.I., Vlasova, I.D., and Vishnyankova, T.P., U.S.S.R. Pat. 339,061 (May 24, 1972); Chem. Abstr. 77, 115959b (1972).
19. Kalontarov, I.Y., Konovalova, G.I., Vakhobov, B., Gutorova, I.D., and Meitus, I.E., Izv. Akad. Nauk. Tadzh. SSR, Otd. Fiz.-Mat. Geol.-Khim. Nauk 1974 (4), 73; Chem. Abstr. 86, 26453a (1977).
20. American Cyanamid Co., Brit. Pat. 1,003,685 (Sept. 8, 1965).
21. North, James, and Sons, Ltd., Jap. Pat. 74,103,000 (Sept. 28, 1974); Chem. Abstr. 84, 107037d (1976).
22. Lopatenok, Al.A., Lopatenok, An.A., Petrzhak, K.K., and Denisenko, A.I., Eksperim. Khirurg. i Anesteziol. 8 (5), 21 (1963); Chem. Abstr. 60, 5731b (1964).
23. Pozdnyakov, V.M., Vol'f, L.A., Efremova, T.B., and Meos, A.I., Fiziol. Opt. Aktiv. Polim. Veshchestva 1971, 155; Chem Abstr. 77, 165869t (1972).
24. Musser, D.D., U.S. Pat. 2,559,986 (July 10, 1951).
25. Lifland, L., and Stanley, L.A., U.S. Pat. 3,594,113 (July 20, 1971).
26. Morrison, W.L., Ger. Pat. 1,942,222 (Feb. 26, 1970).
27. Klein, S.E., and Gagliardi, D.D., U.S. Pat. 3,788,803 (Jan. 29, 1974).
28. Doub, L., U.S. Pat. 2,474,306 (June 28, 1949).
29. Breens, L.F.H., Brit. Pat. 1,154,908 (June 11, 1969).
30. Heinroth, K.A., Schnegg, R., Steinfatt, F., and Wiegand, C., U.S. Pat. 3,255,078 (June 7, 1966).
31. Goryachev, V.M., Rogovin, Z.A., and Shcheglova, G.V., Tekst. Prom. (Moscow) 32(5), 56 (1972); Chem. Abstr. 77, 6319d (1972).
32. Isquith, A.J., Abbott, E.A., and Walters, P.A., Appl. Microbiol. 24(6), 859 (1972).
33. Hoch, P.E., Wagner, G.M., and Vullo, W.J., Textile Res. J. 36(8), 757 (1966).
34. Parker, R.P., and Abbey, A., U.S. Pat. 2,830,011 (April 8, 1958).
35. Roth, P.B., and Hallows, L.B., U.S. Pat. 3,140,277 (July 7, 1964).
36. Snezhko, D.L., Virnik, A.D., Pestereva, G.D., and Rogovin, Z.A., Izv. Vyssh. Ucheb. Zaved. Tekhnol. Tekst. Prom. 1968 (2), 104; Chem. Abstr. 69, 37013g (1968).
37. Kennedy, J.F., Barker, S.A., and Zamir, A., Antimicrob. Agents Chemother. 6(6), 777 (1974).
38. Wojahn, H., Chem. Zentr. 1947, II, 919.
39. Schaefer, P., Huber-Emden, H., Hitz, H.R., and Maeder, A., U.S. Pat. 3,391,114 (July 2, 1968).

40. Stergiu, G.K., Stergiu, A.N., Aikhodzhaev, B.I., and Pogosov, Y.I., Tekstil'n Prom. 25(4), 52 (1965); Chem. Abstr. 63, 8530f (1965).
41. Paulus, W., and Pauli, O., Textilveredlung 1970, 5(4), 247.
42. Imperial Chemical Industries Ltd., Brit. Pat. 836,477 (June 1, 1960).
43. Vol'f, L.A., and Hillers, S., Khim. Volokna 1963 (6), 16; Chem. Abstr. 60, 9405d (1964).
44. Koteskii, V.V., Kharit, Y.A., Vol'f, L.A., Meos, A.I., Hillers, S., and Venters, K., Fr. Pat. 1,590,713 (May 29, 1970).
45. Smirnova, T.S., Mazin, L.S., Rumyaneseva, E.A., Koteskii, V.V., and Vol'f, L.A., Khim. Volokna 1973, 15 (6) 55; Chem. Abstr. 80, 109657x (1974).
46. Gavrilova, L.M., Ivanova, L.S., Makarova, T.P., and Vol'f, L.A., Nov. Khim. Volokna Tekh. Naznacheniya 1973, 164; Chem. Abstr. 80, 74307d (1974).
47. Polishchuk, B.O., Polishchuk, L.B., Vol'f, L.A., and Koteskii, V.V., Izv. Vyssh. Uchebn. Zaved. Khim. Khim. Tekhnol. 1976, 19(9), 1436; Chem. Abstr. 86, 18181v (1977).
48. Smith, J.L., and Harrington, Jr., R.C., U.S. Pat. 3,034,957 (May 15, 1962).
49. Turnbull & Stockdale Ltd., Brit. Pat. 1,024,642 (March 30, 1966).
50. Hueble, A., Swiss Pat. 547,055 (Mar. 29, 1974).
51. Hinz, C.F., U.S. Pat. 3,728,213 (April 17, 1973).
52. Calgon Corp., Fr. Pat. 1,414,418 (Oct. 15, 1965).
53. Foster, F.L., Brit. Pat. 1,276,499 (June 1, 1972).
54. Blaney, T.L., Ger. Pat. 2,408,296 (Aug. 29, 1974).
55. Yudin, A.V., Anokhin, V.V., and Egorov, B.A., U.S.S.R. Pat. 181,237 (April 15, 1966); Chem. Abstr. 65, 9085e (1966).
56. Farbenfabriken Bayer, A.G., Belg. Pat. 671,898 (March 1, 1966).
57. Boehme Fettchemie G.m.b. H., Belg. Pat. 667,922 (Feb. 7, 1966).
58. Satzinger, G., and Marinis, S., Ger. Pat. 2,066,771 (May 25, 1972).
59. Bondarenko, N.Y., Frolov, M.V., Gembitskii, P.A., and Skvortsova, E.K., U.S.S.R. Pat. 338,581 (May 15, 1972); Chem. Abstr. 77, 128410r (1972).
60. Toepfl, R., Goethenberg, C., and Hool, G., Swiss Pat. 581,729 (Nov. 15, 1976).
61. Hausam, W., Melliland Textilber. 43, 277 (1962).
62. Radford, P.J., Am. Dyestuff Reptr. 62, 48 (1973).
63. Schmidt, J., and Meyer, P.F.A., U.S. Pat. 2,841,529 (July 1, 1958).

64. Vigo, T.L., Danna, G.F., and Welch, C.M., Carbohyd. Res. 44, 45 (1975).
65. Schmitt, E.E., U.S. Pat. 3,983,209 (Sept. 28, 1976).
66. Gagliardi, D.D., Am. Dyestuff Reptr. 51, P49 (1962).
67. Katsumi, M., and Sato, T., J. Am. Oil Chem. Soc. 46(7), 348 (1969).
68. Vigo, T.L., Danna, G.F., and Welch, C.M., Textile Chem. & Color. 9(4), 77 (1977).
69. Vigo, T.L., Chem. Tech. 6, 455 (1976).
70. Chem. Week. 43 (Aug. 21, 1974).
71. National Cash Register Co., Brit. Pat. 1,273,322 (May 10, 1972).
72. Ackart, W.B., Camp, R.L., Wheelwright, W.L., and Byck, J.S., J. Biomed. Mater. Res. 9, 55 (1975).
73. Van der Toorn, J., Adema, D.M.M., and Hueck, H.J., Int. Biodeter. Bull. 1(1), 15 (April, 1965).
74. "Technical Manual of Am. Assoc. of Textile Chem. & Colorists", Vol. 52, (a) AATCC Test Method 90-1974, p. 268, (b) AATCC Test Method 147-1976, p. 270, (c) AATCC Test Method 100-1974, p. 272, (d) AATCC Test Method 30-1974, p. 276. AATCC, Research Triangle Park, N. C. 1976.
75. Arnold, Jr., L.B., Textile Chem. & Color. 8(9), 130 (1976).
76. Pandila, M.M., Textile Res. J. 45(10), 701 (1975).
77. Silliker, J.H., Schiltz, L.R., and Jansen, C.E., Am. J. Med. Tech. 27, 227 (1961).
78. Majors, P.A., Am. Dyestuff Reptr. 48, P91 (1959).
79. Quinn, H., Appl. Microbiol. 10, 74 (1962).
80. Lashen, E.S., ibid., 21, 771 (1971).
81. Allsopp, D., Eggins, H.O.W., and David, J., Int. Biodeter. Bull. 6(3), 115 (1970).
82. Fed. Register 40 (123), 26850 (June 25, 1975).
83. Sidwell, R.W., Dixon, G.J., Westbrook, L., and Forziati, F.H., Appl.Microbiol. 19, 950 (1970).
84. "Bergey's Manual of Determinative Bacteriology" (R.E. Buchanan and N.E. Gibbons, Eds.), 8th Edition. Williams & Wilkins Co., Baltimore, 1974.
85. Pelczar, Jr., M.J., and Reid, R.D., "Microbiology." McGraw-Hill, New York, 1972.
86. Meyer, E.A., "Microorganisms and Human Disease." Appleton-Century-Crofts, New York, 1974.
87. McNeil, E., Blanford, J.M., Choper, E.A., Graham, R.T., Hoak, F.C., Oliva, E.C., and Smith, J.C., Am. Dyestuff Reptr. 52, P1010 (1963).
88. Kloos, W.E. and Musselwhite, M.S., Appl. Microbiol. 30(3), 381 (1975).

89. Westwood, J.C.N., Mitchell, M.A., and Legace', S., Appl. Microbiol. 21(4), 693 (1971).
90. Walter, W.G. and Schillinger, J.E., Appl. Microbiol. 29(3), 368 (1975).
91. Chalmers, L., Dragoco Report 11, 215 (1972).
92. Knitting Industry 96(1), 4 (1976).
93. Lyubimova, L.K., Markovich, A.V., and Slonitskaya, N.N., Vestn. Dermatol. Venerol. 40(8), 52 (1966); Chem. Abstr. 66, 11713y (1967).
94. Nolan, M.O., Mann, E.R., and Churchill, H.M., Natl. Inst. Health Bull. No. 189, 180 (1947).
95. Borshchenko, V.V., Savinich, F.K., Gorshkov, V.P., and Rogatovskaya, A.P., (USSR). Kosm. Biol. Aviakosmicheskaya Med. 10(6), 73 (1976); Chem. Abstr. 86, 44659b (1977).
96. Gip, L., and Magnusson, B., Nord. Med. 82(44), 1375 (1969); Chem. Abstr. 72, 65745c (1970).
97. Lebedev, L.V., Kukurz, Y.S., Gushchin, V.A., and Plotkin, L.L., Polim. Med. 2(2), 153 (1972) Chem. Abstr. 77, 156327n (1972).
98. Mazurov, K.V., Popov, V.I., Berezin, A.A., Gig. Tr. Prof. Zabol. (11), 52 (1976); Chem. Abstr. 86, 26646r (1977).

Modified Cellulosics

PREPARATION, STRUCTURE, AND PROPERTIES OF SOME DERIVATIVES OF CELLULOSE AND OF THERMOMECHANICAL PULP

S. S. BHATTACHARJEE AND A. S. PERLIN

Department of Chemistry, McGill University
Montreal, Auebec, Canada

Studies on several different types of cellulose derivatives are described. Acetals have been prepared by the reaction of 3,4-dihydro-2H-pyran (or other vinyl ethers) with cellulose and with partially substituted methyl-, carboxy-, methyl-, and hydroxyethylcellulose. Such acetals appear to be suitable for gel-permeation chromatography; i.e., they give elution profiles similar to those of the conventional nitrate derivatives. Other experiments are concerned with the derivatization of thermomechanical pulps; e.g., carboxymethylation has afforded fibres having good water-retention capacity. Also related to water-absorbent properties are observations on cross-linked carboxymethylcellulose subjected to γ-irradiation or radiation-induced grafting of polysytrene: a marked enhancement in water-retention shown by the products is attributed mainly to reactions involving ordered regions of the cellulosic material. Enzymic degradation with cellulase has been utilized extensively in these studies for locating substituents or for examining sites of physical modification.

REACTION OF VINYL ETHERS WITH CELLULOSE AND CELLULOSE DERIVATIVES

A. Preparation of O-Tetrahydropyranyl and Other Acetals

In carrying out chemical modifications of cellulose, it is sometimes desirable to use a hydroxyl-blocking substituent that is stable towards base but is easily hydrolysed by acid. Acetals prepared from vinyl ethers, such as have been obtained with starch (1-3), fulfill this requirement. Thus, treatment of carboxymethylcellulose (1) with 3,4-dihydro-2H-pyran (2) afforded an O-tetrahydropyranyl (THP) derivative (3) having solubility properties suitable for hydride reduction, thereby permitting a facile conversion of 1 into a structurally-analogous hydroxyethylcellulose (4) (4). In another type of application (5), the formation of an acetal derivative of partially-acetylated cellulose, followed by saponification and methylation, has provided a means for locating the positions of the O-acetyl groups.

$$\left[\text{CELLULOSE}\right]{-\text{OH} \atop -\text{CH}_2\text{COO}^-} \xrightarrow[\text{H}^+]{\underline{2}} \left[\text{CELLULOSE}\right]{-\text{OTHP} \atop -\text{CH}_2\text{COO}^-} \longrightarrow \left[\text{CELLULOSE}\right]{-\text{OH} \atop -\text{CH}_2\text{CH}_2\text{OH}}$$

$\underline{1}$ $\underline{3}$ $\underline{4}$

Little reaction was found to occur when dihydropyran (2) and cotton cellulose were suspended in methylsulfoxide in the presence of p-toluenesulfonic acid, as catalyst: the product formed in 48-72 hr. had a degree of substitution (D.S.) of only ∼0.1. By contrast, disordered cellulose (obtained by deacetylation of cellulose acetate in a non-aqueous medium (6)) being highly accessible, afforded acetals much more readily. Depending on the molar ratio of reactants and the reaction time, THP-derivatives of D.S. up to 2.2 have been prepared (Table 1). Acetals of similarly high D.S. were derived also using methyl vinyl ether (CH_3-O-CH=CH_2)(Table 1). In this instance, however, relatively brief reaction periods were employed to limit the extent of concomittant polymerization of the vinyl ether. Probably because of unfavorable steric factors, the long-chained octadecyl vinyl ether ($CH_3(CH_2)_{16}$-CH_2-O-CH=CH_2) reacted with disordered cellulose to only a very limited degree (Table 1).

A typical preparation was carried out as follows: a suspension of dry disordered cellulose (5 g) in anhydrous methyl sulfoxide (50 ml) contained in a stoppered flask was stirred at 90°C for 3 hr., and then cooled to room temperature. p-Toluenesulfonic acid (0.12 g) was introduced, followed by slow addition of the required amount of the vinyl ether, stirring being continued through the reaction period. Neutralization was then carried out with ammonia or sodium bicarbonate, the mixture was poured into water, and the precipitated product was washed thoroughly with methanol, air dried, and finally dried in vacuo at 60°C. (Alternatively, the product was satisfactorily recovered directly from the reaction mixture without prior neutralization, by precipitation and repeated washing with methanol).

In addition to the preparation of acetal derivatives of carboxymethylcellulose (4) and cellulose acetate (5) using this general procedure, both hydroxyethylcellulose (D.S. 0.87) and methylcellulose (D.S. 0.69) have readily afforded mixed ether-acetals as well.

TABLE 1
Reaction of vinyl ethers with disordered cellulose

	moles/mole	time (hr.)	yield[a]	D.S.[b]
Dihydropyran	1	3	98	0.1
	3	3	98	0.2
	10	1	120	0.4
		3	160	1.1
		24	190	2.0[c]
	12	24	190	2.2[c]
Methyl vinyl ether	2	3	98	0.1
	10	3	120	1.2
Octadecyl vinyl ether	10	24	95	0.1

[a] Grams per 100 g. of cellulose; [b] Estimated from yield of aldehyde (as 2,4-dinitrophenylhydrazone) liberated by acid hydrolysis (3); [c] Intrinsic viscosity (dioxan), 1.2-1.3

B. Location of the acetal groups

The cellulose acetals of D.S. < 1.0 were found (Table 2) to reduce close to one mole of periodate per anhydroglucose residue. This showed, therefore, that there is a high degree of selectivity for reaction of a vinyl ether with the primary hydroxyl group of the anhydroglucose moieties, and that appreciable substitution at secondary positions occurs only on prolonged treatment at a high molar ratio of vinyl ether:cellulose. However, the relative reactivities of the 2- and 3-hydroxyl groups have yet to be determined.

Consistent with this evidence for a rather uniform derivatization at C-6, are results obtained by treating the acetals with a cellulase preparation from Streptomyces sp. QM B 814 (7). That is, the yields of cellobiose released (Table 2), indicated that acetalation was not limited to specific regions of the polymer chains, but was broadly distributed, because the incidence of sequences giving rise to cellobiose

TABLE 2
Periodate oxidation and enzymic digestion of cellulose acetals

Acetal	D.S.	Moles periodate /mole	Relative yield of cellobiose[b]
	0[a]	0.95	1.0
Tetrahydro-pyranyl	0.1	0.90	0.8
	0.4	0.80	0.5
	1.1	0.78	tr.
	2.0	0.10	0
Methoxy-ethyl	0.12	0.90	0.8
	1.2	0.65	tr.
Octadecyloxy-ethyl	0.1	0.90	0.4

[a] Disordered cellulose from which acetals were prepared;
[b] Estimated by g.l.c. analysis of the tetramethylsilyl derivative.

(i.e., two (or more) unsubstituted glucose residues (4,8-10)) diminished at a rate corresponding to the number of acetal groups introduced. For example, the yield of cellobiose from the THP-derivatives was reduced by half at a D.S. level of 0.4, and was negligible at a D.S. of 1.1. It is noteworthy that relatively much less of the disaccharide was released from the octadecyl derivative than from the other acetals of D.S. 0.1, which suggests that the bulkier alkyl moiety hinders attack by the enzyme.

C. Solubility properties

Generally, the materials of medium D.S. (∼1.0) prepared from disordered cellulose are soluble in dioxan, whereas those of D.S. ∼2.0 are soluble as well in dichloromethane or tetrahydrofuran. Similar properties are exhibited by the starch acetals (2,3). However, the degree of solubility of the O-(1-methoxyethyl) derivatives was observed to be slightly lower than that of the THP-derivative at the same D.S. level. Probably, this difference is attributable to easier formation of cross-links by methyl vinyl ether, in view of the facile intramolecular transacetalation found with O-(1-methoxyethyl) derivatives of methyl α-D-glucopyranoside (3).

O-Tetrahydropyranyl derivatives prepared from carboxymethylcellulose of D.S. 0.5 or higher, dissolve readily in dioxan (4) or tetrahydrofuran. However, they became insoluble when stored for a prolonged period or heated at 70-80° in vacuo for 24 hr., whereas the solubility of acetals obtained from disordered cellulose, hydroxyethylcellulose or methylcellulose did not decrease under these conditions. This difference may arise, therefore, through inter-chain esterification between carboxyl and unsubstituted hydroxyl groups in the carboxymethyl acetals.

As already noted, the THP-derivative (freshly prepared) of carboxymethylcellulose has been utilized for hydride reduction of the carboxymethyl substituents in dioxan solution. Another application of the fact that these acetals and mixed ether-acetals are soluble in organic media, is examined in the following section.

D. Gel permeation chromatography of O-tetrahydropyranyl derivatives

Molecular weight distribution of cellulosic materials has commonly been examined (12,13) by gel permeation chromatography of the derived nitrates dissolved in tetrahydrofuran. Since the latter is also an effective solvent for THP-derivatives described above, the gel permeation characteristics of several of the acetals were compared with those of nitrates prepared from the same starting material. As shown in Fig. 1A, the elution profiles for the THP- and nitrate-derivative of a hydroxyethylcellulose were closely similar. Analogous behaviour was observed also for the corresponding pairs of derivatives prepared from methylcellulose and disordered cellulose. Hence, it appears that the THP substituent is a suitable alternative to the nitrate group for molecular weight measurements on modified celluloses (although not on such materials as cotton cellulose, which do not undergo sufficient acetalation (see Sect. IA)).

These findings have been usefully applied in studies on the fine structure of cellulose ethers; i.e., in measuring the molecular weight distribution of fragments formed by the

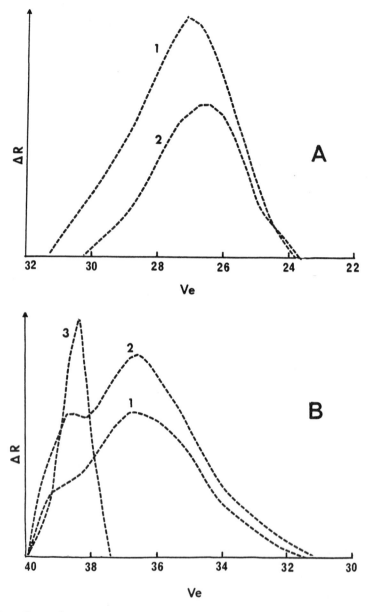

Fig. 1 Gel-permeation chromatography of A) hydroxyethylcellulose acetals (2-THP; 1-nitrate), and B) of acetals (1-THP; 2-nitrate) prepared from enzymic hydrolysate of the hydroxyethylcellulose; peak 3 represents THP-derivative of C_2-C_4 oligosaccharide mixture. ΔR is the refractive index difference (in arbitrary units); V_e is the elution volume (in fraction numbers; 27≃D.P. of 100, 37≃D.P. of 14).

action of cellulase on such ethers. As seen in Fig. 1B, gel permeation chromatography of the enzymic hydrolysate of hydroxyethylcellulose (D.S. 0.87), shows that the majority of the products have a D.P. in the region of 14 residues. For comparison, the nitrate of the same material was examined (Fig. 1B), and shows substantially the same size distribution; this, again, supports the validity of using the THP-derivative. The elution profiles in Fig. 1B also exhibit a shoulder in the lower molecular weight range, which coincides with the chromatographic mobility of the THP-derivative prepared from a mixture of cellobiose, -triose and -tetraose.

By comparing Figs. 1A and 1B, it is apparent that the enzyme treatment of the hydroxyethylcellulose has left little material of a molecular weight overlapping that of the original polymer. Hence, the distribution of hydroxyethyl substituents in the latter must have been such that several sequences of 2,3, or more adjacent unsubstituted residues were to be found in most of the polymer chains (4,8-10). Similar results were obtained with a carboxymethylcellulose (D.S. 0.7). By contrast, the data furnished in the same way by methylcellulose (D.S. 0.7) indicated that a substantial proportion of chains in this preparation were too highly methylated to allow for an appreciable degree of enzymic degradation, and hence that the distribution of the ether substituents in this polymer was less uniform than in the two cellulose ethers.

CHEMICAL MODIFICATION OF THERMOMECHANICAL PULP[1]

Because mechanical pulp is rapidly gaining prominence as a substitute for chemical pulp, a study is being made of its suitability for preparing derivatives analogous to those conventionally derived from purified cellulose. Here, the carboxymethylation of thermomechanical (TM) pulp is examined as an extension of research (14) on carboxymethyl derivatives of cross-linked cellulose.

Sodium carboxymethyl ($-CH_2COO^-Na^+$) substituents induce the cellulose fibre to swell and dissolve in water. By cross-linking carboxymethylcellulose (CMC) an insoluble, highly water absorbant, material is obtained. In TM pulp the cellulose is entwined with lignin (as well as with the hemicellulose) and hence, in principle, such pulp might be

[1]This study has been carried out with the generous collaboration of Dr. D.A.I. Goring.

modified by carboxymethylation so as to give a product resembling cross-linked CMC, but offering a substantial cost advantage.

Treatment of aspen TM pulp with 25% sodium hydroxide and chloroacetic acid in 2-propanol, followed by washing successively with 70% methanol and methanol, gave a range of materials of varying carboxymethyl content. Table 3 describes characteristics of samples having the lowest and highest degrees of substitution. As shown in the Table, these carboxymethyl derivatives exhibited moderately good water- and salt-retention characteristics (WRV and SRV), i.e., about half as effective as cross-linked CMC, although the larger retention values of the more highly substituted samples were accompanied

TABLE 3
Water retention characteristics of carboxymethyl aspen TM pulp

Sample	Yield (%)	Water insoluble (%)	WRV^a (g/g)	SRV^b (g/g)
TM pulp (alkali-treated)	98	95	0.5	0.2
$CM-TM^c$ (6% CM)	95	95	8.0	4.0
$CM-TM^c$ (20% CM)	75	60	14.0	6.0
$CM-TM^c$ spruce-balsam (20% CM)	75	65	10.0	–

[a]Water retention value; [b]1% saline retention value; [c]carboxymethyl TM pulp.

by increased losses due to greater water solubility. Chemical analysis, facilitated by enzymatic digestion with cellulase, indicated that much of the soluble material is carboxymethylated hemicellulose.

In contrast to the behaviour of cross-linked CMC (see Sect. III), gamma-irradiation of carboxymethyl TM pulp did not increase its water retention capacity, although a slight improvement was observed when the pulp was irradiated prior to carboxymethylation.

Thermomechanical pulp from a mixture of spruce and balsam gave carboxymethylated derivatives having generally lower water retention capacities than the aspen product (i.e., by about one-third), possibly because the spruce-balsam mixture has a higher lignin content (24% vs 18%) (Table 3).

Carboxymethyl TM pulp was found to increase the strength of handsheets prepared from untreated TM pulp (Table 4), and also to introduce some bonding strength to rayon fibres.

Another modification examined involves the introduction of long-chain fatty acid groups, such as lauroyl, to TM pulp that has been carboxymethylated to a D.S. level of 0.5. Esterfied materials of this type that have been prepared, were found to absorb 10 gm. of crude oil per gm. from an oil-water mixture.

TABLE 4
Effect of carboxymethylation on hand sheet properties of aspen TM pulp

	TM	CM-TM[a]	TM+CM-TM[b]
Freeness	63	110	55
Breaking L.	4384	4056	5044
Burst Factor	22.7	17.8	24.6
Tear Factor	75	59	61

[a] Carboxymethyl aspen TM pulp; [b] TM pulp admixed with 10% carboxymethyl aspen TM pulp

EFFECTS OF GAMMA-IRRADIATION AND POLYSYTRENE GRAFTING ON WATER RETENTION BY CARBOXYMETHYLCELLULOSE

A. Gamma-irradiation of cross-linked carboxymethylcellulose

The high water-solubility of carboxymethylcellulose can be markedly reduced by the chemical introduction of cross-linkages. This treatment gives rise to products having good water-retention characteristics and, indeed, cross-linked carboxymethylcellulose (CLCMC) finds commercial application as an absorbent. The carboxymethyl content of the material is an important factor in determining the water retention volume (WRV), as seen by comparing the data in Table 5 for samples 1, 5 and 8. Thus, at a constant level of cross-linking (3% of epichlorohydrin), the WRV increases from 11.7 at a D.S. of 0.41 to a value of 32.0 at a D.S. of 0.95. Not surprisingly, also, as the D.S. of the CMC increases, the solubility of the cross-linked product also increases.

In conjunction with a study of the effects of irradiation from a ^{60}Co source on cellulosic material, samples of CLCMC

TABLE 5
Water retention characteristics of cross-linked CMC[a]. Effect of gamma-irradiation (^{60}Co)

Sample	D.S. (CMC)	Irradiation Dosage (MR)	Insoluble (%)	WRV[b] (g/g)
1	0.41	0	80	11.7
2		1.5	75	12.0
3		2.1	75	16.3
4		4.8	70	20.0
5	0.75	0	90	21.0 (11.0)[c]
6		2.1	75	35.0 (15.0)[c]
7		4.8	60	35.0
8	0.95	0	75	32.0
9		1.5	70	32.0
10		2.1	50	35.0
11		4.8	44	35.0

[a]Cross-linked by reaction of CMC with 3% of epichlorohydrin; [b]water retention value; [c]1% saline retention value.

were examined, leading to the observation (14) that γ-rays can induce a substantial increase in water absorption capacity. As shown in Table 5, dosages in the region of 2.1-4.8 megarads promoted enhancements of 50-70% in the WRV values for material of D.S. ∼0.7; with 1% saline solutions, the retention was about one-half those for water, which appears to be a general characteristic of absorbent modified celluloses.

Prolonged irradiation is not necessarily beneficial: hence, an increase from 2.1 to 4.8 megarads dosage for the sample of D.S. 0.75 caused no further increase in WRV, although more of the product was solubilized (Table 5). Effects such as these suggest that the enhanced absorption capacity of CLCMC is due to partial breakdown of the fibres by γ-rays (15), which allows for increased swelling of the cross-linked network.

Related observations concern the effect of irradiation on the water-retention characteristics of non-cross-linked CMC fibres. As an example, a preparation of D.S. 0.7, which formed a thick gel with water, was subjected to a dosage of 2.0 megarads and then heated in an air oven at 190° for 5 min. The material now largely retained its fibrous character in water (15% was water soluble), and exhibited the relatively high WRV value of 27.0. Presumably, the heat treatment (which had little effect without prior irradiation) promoted cross-linking in regions modified by the γ-irradiation, and thereby led to a product having characteristics resembling those of CMC that has been chemically cross-linked, and then exposed to γ-rays.

B. Effects of gamma-irradiation on cellulose[3]

Gamma-irradiation induces several changes in the structure of cellulose itself (15-18). One effect is oxidative, as evidenced by the formation of carbonyl (and carboxyl) groups (Table 6). It is noteworthy that cotton cellulose is oxidized more intensively than is disordered cellulose: e.g., at a dosage level of 7.8 megarads, the carbonyl content of the latter is less than one-third that of cotton. Probably, this difference is attributable (18) to the generation by γ-rays of higher concentration of free radicals in crystalline regions of cellulose, thus producing a lesser effect on the disordered sample.

Accompanying these changes, was the development of resistance towards hydrolysis by cellulase. The enzymic suspectibility of both cotton and disordered cellulose decreased at comparable rates with an increase in the dosage level (Table 6). These decreases are far greater than can be accounted for by the number of anhydroglucose residues modified: e.g., the introduction of only 25 carbonyl groups per 1000 residues of disordered cellulose lowered the enzymic

[2] In collaboration with A. Seltzer.

TABLE 6
Gamma-irradiation (^{60}Co) of cotton and disordered cellulose

	Irradiation Dosage (MR)	Carbonyl[a] (mM/g)	Enzymic[b] Degradation	Mol.[c] Wgt.
Disordered cellulose[d]	0	0	8.0	126
	1.0	-	7.8	99
	2.4	0.08	6.8	-
	4.4	-	5.7	-
	6.0	0.14	5.0	65
	9.0	0.36	3.4	52
	14.0	0.68	1.8	38
Cotton cellulose[d]	0	0	3.1	716
	1.0	-	2.8	468
	4.4	0.58	1.7	-
	7.8	0.84	1.2	-
	9.0	-	1.0	153
	14.0	1.19	0.6	108

[a] Carboxyl content ≈1-2% of carbonyl content; [b] Estimated from % glucose liberated by commercial cellulase under standard conditions; [c] Estimated by g.p.c. analysis of derived nitrate; [d] dry sample.

susceptibility by 50%. Such a marked effect suggests that γ-irradiation promoted the formation of cross-linkages, which could sharply reduce accessibility to the enzyme.

As would be expected (15) irradiation also caused a reduction in molecular weight. Cotton suffered more depolymerization than did the disordered cellulose (Table 6), in parallel with their relative degrees of oxidation.

C. <u>Graft copolymers of carboxymethylcellulose and polystyrene</u>

Cellulosic materials undergo graft copolymerization when

suspended in a mixture of styrene and methanol (1:1) and subjected to irradiation with a ^{60}Co source (17). Differences are observed (Fig. 2) in the rates of grafting for disordered cellulose, CMC (D.S., 0.5) and CLCMC (D.S., 0.7), which are directly related to the proportion of unsubstituted anhydroglucose residues in these materials. Hence, grafting appears to occur most rapidly with disordered cellulose, and least readily with CLCMC. However, a possible factor contributing to these results is the relative effectiveness with which non-grafted polystyrene chains can be extracted (e.g., with

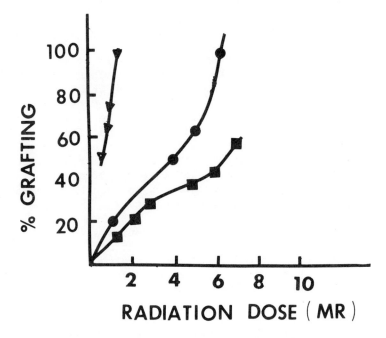

FIG. 2 Graft copolymerization of disordered cellulose (▼), CMC (●) and CLCMC (■) with polystyrene, through gamma-irradiation from a ^{60}Co source.

benzene) from these products; i.e., it is conceivable that when styrene is polymerized in the presence of disordered cellulose, it forms a more firmly associated physical mixture of the two polymers, than with CMC and CLCMC, and not necessarily more graft copolymer.

An examination of one of the CMC products (30% graft level) provided some insight into various aspects of the association between CMC and polystyrene. Gel permeation chromatographic analysis showed the presence in this product of two main components: one (D.P. ~200) corresponded to unreacted CMC, and the other (D.P. ~2500) to grafted material. It appears that, at this graft level, one polystyrene chain was associated with 15 CMC molecules. Noteworthy also was the fact that the elution profile of the unreacted CMC was closely similar to that of the original CMC sample. This showed that the irradiation treatment had caused negligible degradation of the CMC and, hence, that the action of the γ-rays had been confined largely to polymerization of the styrene.

In another experiment on this product, polystyrene-rich material containing < 5% of carbohydrate was isolated by repeated treatment with cellulase, followed by extraction with water. Acid hydrolysis of the insoluble portion, and g.l.c. analysis of the liberated monosaccharides, showed that the carbohydrate component possessed only one-half as many carboxymethyl substituents as the original CMC. This finding indicated, as has already been suggested above, that the generation of polystyrene chains is favored by the presence of unsubstituted anhydroglucose residues. Since the incidence of such residues is higher in crystalline regions of CMC, these data are consistent with evidence (18) that free radicals generated by γ-rays are better stabilized by a crystalline, than by an amorphous, cellulose lattice. No direct proof was obtained, however, that the association between carbohydrate and polystyrene involves covalent bonding.

Striking water retention characteristics were exhibited (14) by the CLCMC samples that had been subjected to the grafting treatment (Table 7). At graft levels up to 15-20% the hydrophobic character of the products increased--as would be expected. However, when the polystyrene content was increased to 25-30%, there occurred a marked increase in the absorption capacity of the fibres, i.e., by a factor of two (see Fig. 1, ref. 14). This enhancement in water retention must originate in a loosening of the cellulosic fibre structure during the grafting process (19), so that greater penetration of water into hydroxyl-rich regions is permitted. The absorbancy then decreased to a graft level in excess of 30%, presumably because the hydrophobic character of the polystyrene component became dominant.

An analogous type of phenomenon was observed when the susceptibility of these products towards cellulolytic attack was measured (Fig. 3). Hence, following an initial decrease in the extent of hydrolysis, susceptibility to the enzyme was enhanced when the polystyrene content reached 25-30%, and then decreased again at higher grafting levels. This suggests, once again, that the grafting process is accompanied by an

TABLE 7
Water retention by cross-linked CMC-polystyrene graft copolymers

Polystyrene graft (%)	Water retention value (g/g)
0	18.0
12	10.5
17	13.0
23	25.0
28	33.0
32	35.0
35	24.0
38	20.0
45	18.0
52	6.0

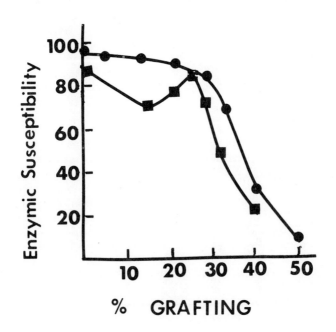

FIG. 3 Changes in susceptibility to hydrolysis by cellulase, induced by graft copolymerization of CMC (●) and CLCMC (■) with polystyrene.

opening up of the fibrous network, so as to expose regions rich in unsubstituted anhydroglucose residues.

ACKNOWLEDGMENTS

The authors express their sincere thanks to the National Research Council of Canada for generous support. They also acknowledge the kind assistance of R.St.J. Manley, A. Seltzer and M. Chang, and gift of TM pulp by W. May and D. Breck.

REFERENCES

1. O. Weaver, C.R. Russell and C.E. Rist. J. Org. Chem., 28, 2838(1963).

2. M.L. Wolfrom, S.S. Bhattacharjee and G.G. Parekh, Die Stärke, 18, 131(1966).

3. M.L. Wolfrom and S.S. Bhattacharjee. Die Stärke, 21, 116 (1969).

4. S.S. Bhattacharjee and A.S. Perlin. J. Polym. Sci., C, 36, 509(1971).

5. H. Björndal, B. Lindberg and K. Rossell. J. Polym. Sci., C, 36, 523(1971).

6. R.St.J. Manley. J. Polym. Sci., A, 1, 1875(1963).

7. E.T. Reese, E. Smakula and A.S. Perlin. Arch. Biochem. Biophys., 85, 171(1959).

8. W. Klop and P. Kooiman. Biochem. Biophys. Acta, 99, 102 (1965).

9. M.G. Wirick. J. Polym. Sci., A-1, 6, 1705(1968).

10. P.J. Garegg and M. Han. Svensk Papperstidn., 21, 695(1969)

11. M.L. Wolfrom, A. Beattie and S.S. Battacharjee. J. Org. Chem., 33, 1067(1968).

12. M. Chang, T.C. Pound and R.St.J. Manley. J. Polym. Sci., A-2, 11, 399(1973).

13. M. Chang, Tappi., 55, 1253(1972).

14. S.S. Bhattacharjee and A.S. Perlin. J. Polym., Sci., Polymer Letters, 13, 113(1975).

15. F.A. Blouin and J.C. Arthur, Jr.. Textile Res. J., 28, 198(1958).

16. F.C. Leavitt. J. Polym. Sci., 51, 349(1961).

17. P.W. Moore. Rev. Pure and Appl. Chem., 20, 139(1970).

18. A.U. Ahmed and W.H. Rapson. J. Polym. Sci., A-1, 9, 2129(1971).

19. Y. Ogiwara, H. Kubota, H, Murayama and A. Sakamoto. Tappi, 53, No. 9, 1685(1970).

THE EFFECT OF LIQUID ANHYDROUS
AMMONIA ON CELLULOSE II.

L. G. ROLDAN

Technical Center, J. P. Stevens & Co., Inc.
Garfield, N.J.

Anhydrous ammonia is a swelling agent capable of transforming, upon evaporation, cellulose I into cellulose III and also cellulose II into cellulose III. However, upon subsequent hydration cell-III reverts to its original cellulose structure. No conversion from cell-I to cell-II or the reverse has been detected. Cell-III appears to be a highly disordered arrangement of cellulosic changes which preserve their original molecular configuration in spite of an increased separation of the chains.

Possible explanations for this behavior are advanced based on two contributing factors: A) The inability of NH_3 to break or penetrate intramolecular H-Bonds of either cellulose I or II, and B) The actual interference of the cellulose Tg on chain internal rotation. A review of the current technological status of the treatment of cellulose with anhydrous ammonia is included.

INTRODUCTION

The effect of liquid anhydrous ammonia (liq. NH_3) on cellulosic materials was recognized from the early 1930's. However, a renewed interest in the last decade toward its technological application in cotton finishing has developed into practical processes of commercial value (1-3).

The basis of the modification of the structure of cotton cellulose by anhydrous ammonia was refined and described by Lewin and Roldan (4) from a structural point of view and by Warwicker (5) via a morphological approach. Cellulose is a polymorphic polymer which in its native state crystallizes in a lattice called Cellulose I (Cell I). This molecular arrangement is altered when strong swelling agents and temperatures are used. Cellulose II (Cell II) describes the lattice present in regenerated or mercerized cellulose, whereas the terms Cellulose III (Cell III) and IV (Cell IV) are used to designate the lattices resulting from treatment of cellulose with liquid ammonia and hot glycerol, respectively.

Transition from Cell I to Cell II is of important commercial value since a stable product with better overall properties than the original cotton is obtained. This effect is due to two independent causes -- the strong swelling effect produced by the NaOH solution used and the stabilization of the morphological changes induced by the swelling, or introduced from external causes during the swelling, by recrystallization. In either Cell I or Cell II liquid ammonia fails to create a permanent lattice change and by re-hydration, the original cellulosic lattice is reformed.

It is the object of this paper to explore and review the effect of liq. NH_3 on Cell II, with emphasis on rayon, and to focus attention onto this structural problem which is not yet fully understood. Technological advantages of liquid ammonia treatment for rayon processing are briefly reviewed to complete the present picture on this subject.

EXPERIMENTAL

The fabric used was a 3.5 oz/yd rayon challis. Samples were immersed for 10 minutes in liquid ammonia and post-treated as follows:
- a) Untreated control of the original fabric
- b) NH_3-treated sample allowed to dry in 0% RH air
- c) Sample b) heated at 200°C for 5 minutes

d) NH_3-treated sample rinsed in water and examined in wet condition
e) Sample d) boiled in water for 15 minutes and examined in wet condition
f) Sample d) heated at $180°C$ for 10 minutes

No tension was applied during these treatments.

X-ray diffraction was the technique chosen for this examination. A General Electric GE-5 diffractometer equipped with a copper tube, nickel filters and a pulse height analyzer was used. The samples were placed in a parafocusing setting with the warp perpendicular to the goniometer diffraction plane. Wet samples were squeezed between filter papers to eliminate the excess water immediately prior to the X-ray analysis. The following parameters were considered during this examination:
a) The nature and relative amount of crystalline phase present
b) The nature of the less ordered regions
c) Water content as observed in the XRD scans
d) Average lateral extension of the crystalline regions

The definition of these variables is as described in a previous publication on a related subject (4).

For this work the average lattice constants, equatorial spacing and indices of the reflections of Cell I, II, III and NH_3-Cell are shown in Table I as reported by Wellard (6) and Hess and Gundermann (7). Approximate relative intensities of the reflections are also included in the table.

RESULTS

The results obtained are summarized in Table II. Reported values are approximate and are used to give a semi-quantitative picture of the variations observed under the assumption of a micellar model. However, as shown throughout the discussion, a paracrystalline model appears more realistic.

The X-ray diffraction (XRD) diagrams of the various samples are reproduced in Fig. 1. The

TABLE 1

Type	d_{002}	d_{101}	$d_{10\bar{1}}$	d_{100}	$a(\text{Å})$	$c(\text{Å})$	β
Cell I	3.90(100)	5.33(20)	5.97(20)	–	8.17	7.85	96°
Cell II	4.07(100)	4.42(90)	7.17(20)	–	7.91	9.15	117°
Cell III$_1$	–	4.33(40)	–	7.59(6)	8.66	–	120°
Cell III$_2$	–	4.28(100)	–	7.45(15)	8.60	–	120°
NH$_3$-Cell I complex	–	–	–	–	12.74	10.75	133.5°
NH$_3$-Cell II complex	–	–	–	–	14.5	14.75	120°

Note: 1) b parameter is equal to 10.3Å for Cell I, II and III's
2) numbers in brackets are relative intensities

TABLE 2

Treatment of Rayon with Liquid Anhydrous Ammonia

	Sample Condition	Phase	Equatorial Crystallinity Index	Swollen Water Index	Apparent Crystallite Size(Å)	
					Cell II	Cell III
a)	Untreated, dry	Cell II	73	--	90	--
b)	NH$_3$-treated, dried	Cell III	65	--	--	17
c)	NH$_3$-treated, dried & heated at 200°C/ 5 min.	Cell III	65	--	--	19
d)	NH$_3$-treated, water exchanged	Cell II & III	90+	50	40	40
e)	NH$_3$-treated, water exchanged, boiled 15 min.	Cell II	90+	40	60	--
f)	NH$_3$-treated, water exchanged, heated at 180°C/10 min.	Cell II & III or Cell II	75 75	-- --	56 28	40 --

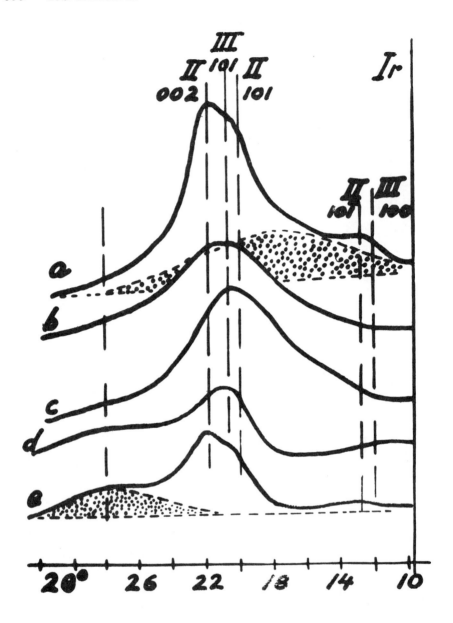

Fig. 1. XRD diagrams taken from liq. NH₃ treated rayon.

original rayon (a) shows a poorly resolved Cell II structure typical of most textile rayons. The equatorial crystallinity index appears high since it refers to the ratio of crystalline to total scattering in the equatorial plane of the fibers rather than the ratio of the integrated values. It is used, however, because the equatorial XRD plane is the most sensitive to the variations induced by the studied treatment.

Apparent crystallite size or the average extension of lateral order is here calculated from the half-intensity breadth of the most prominent crystallographic reflection of the diagram-002 for Cell II. The effect of swelling the sample in liq. NH_3 and its removal by evaporation at room temperature is very pronounced. Fig. 1(b). The extension of lateral order is drastically reduced to the lateral association of a few cellulosic chains though the reduction in crystallinity is small.

The XRD scattering of the 002 and 101 reflections of Cell II have merged into a single Gaussian distribution centered at $2\theta=20.8°$ which is the position accepted for the 101 reflection of Cell III, and accepted in principle as indicative of a phase transition. Heating this sample up to $200°C$ for several minutes did not essentially alter the scattering distribution nor did recrystallization occur. Fig. 1(c).

Exchanging the ammonia immediately after the treatment with water, without an intermediate evaporating step, produced an XRD diagram which, in addition to the presence of water scattering, shows a partial reordering in relation to the dried ammonia-treated samples. Fig. 1(d). Crystallinity of the ordered regions appears exaggerated by the lack of an amorphous scattering where expected (9). If this "sample" is boiled for 15 minutes prior to removal from the water, a clear recrystallization toward the Cell II lattice appears and again the apparent crystallinity is highly increased. Fig. 1(e).

The lack of an amorphous scattering in these water-exchanged samples is indicative of the greater water accessibility of the disordered regions induced by the liq. NH_3 pretreatment. Interatomic vectors in these regions are overrun by the H-bonds from solvation by the water, resulting in a shift of the amorphous scattering toward a new average shorter than 3.95Å and under the strong crystalline reflection. If this shift is accepted, the equatorial crystallinity becomes equal to the crystallinity of the untreated sample, i.e. about 70% for samples d) and e).

From this argument it is not surprising to find that when sample d) is heated and dried well above 100°C a normal amorphous scattering reappears at $2\theta = 18°$ (9). Fig. 2(f). A second feature of this XRD diagram is the perfection of the Gaussian distribution associated with the profile of the main equatorial reflection at $2\theta = 20.6°$ as noted earlier in samples b) and c). Also, the crystallization has advanced enough to show a second reflection with equal intensity and position to that of the $10\bar{1}$ reflection of the original rayon. Sample f), after several wet and dry cycles, fully reverted to the original Cell II structure.

For comparison, the XRD diagram of a sample of cotton treated with liq. NH_3, dried and heated at 105°C is shown in Fig. 2(4). In this case the Cell I structure of the cotton -- shown by the dotted line -- has been converted to Cell III by the liq. NH_3 treatment. As with sample b) of the present study, the extension of lateral order is reduced to a few cellulosic chains and the profile of the main reflection is more in agreement with a skewed Cauchy distribution. This difference in line profile of the 101 reflections of Cell III's obtained from either Cell I or Cell II appears to be a distinctive feature independent of the swelling agent used, i.e. liq. NH_3, ethylamine (10). Attempts to achieve the same profile by subtraction of profile line residues from hypothetical contents of Cell II and Cell I, or by assuming different unit cells than those published for Cell III's, failed.

EFFECT OF LIQUID ANHYDROUS AMMONIA 311

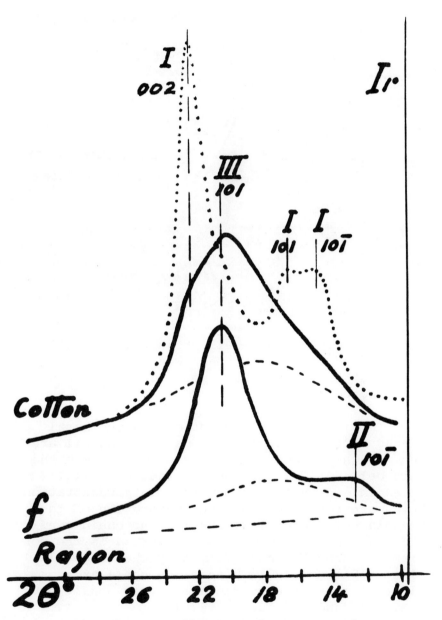

Fig. 2. XRD diagram of cotton and rayon treated with liq. NH_3. The swelling agent was allowed to evaporate and the samples heat-treated.

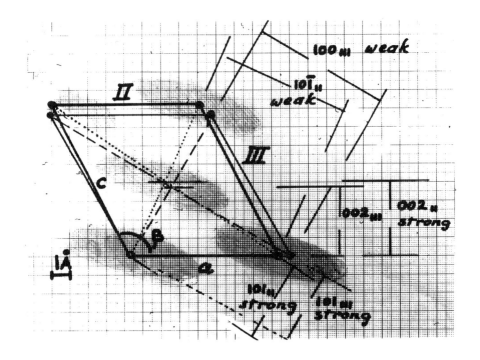

Fig. 3. Unit-cell projection of Cell II and Cell III$_2$. Shadowed areas represent the most likely position of the projected chain.

In Fig. 3., the projections on the basal planes of the unit cells of Cell II and III are shown. The strong similarity shown between the unit cells of Cell II and its induced Cell III may be explained by the theory that interpenetration twins are formed during the evaporation of the ammonia but the Cell II lattice and chain configuration are preserved. This assumption is the most likely since the angle formed by the diagonal plane (10$\bar{1}$) and (001) plane with the <u>a</u> axis or (100) plane -- the twin plane -- are nearly equal, i.e. 115° and 117° respectively.

This twinning should allow up to 3 possible positions in the corners of the projected lattice in which the chain may rest, in agreement with the setting of the surrounding six chains. Fig. 4.

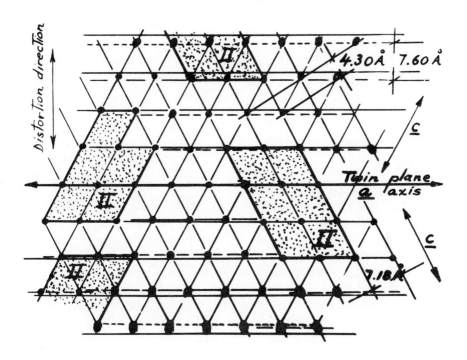

Fig. 4. Proposed twinning mechanism for Cell II.

It is necessary to note that though each cell still maintains the Cell II parameters, the resultant overall lattice -- full of discontinuities, such as voids and faults -- should scatter as an average pseudo-hexagonal arrangement. In Fig. 4., we also see that the 101 reflection becomes averaged to a single 4.30Å spacing and the 10$\bar{1}$ reflection should maintain its position, as experimentally found. The multiplicity of positions in the basal plane in which the chains can be found, in relation to any one as origin, results in a paracrystalline lattice in which fluctuations of inter-chain distances and actual displacement of the lattice points, i.e. lattice distortions of the first and second type, occur (11).

In the present study, comparing the difference between the interplanar spacings of the 002 and 101 reflections of Cell II and the spacing of the actual average reflection in the XRD of sample f), a lattice distortion no greater than 2% can be calculated, in agreement with the displacement produced by twinning. This distortion is of the same order of magnitude as obtained in other polymeric systems from the paracrystalline model (12).

Inter-chain distances of the described model produce a Gaussian distribution centered at the average distance, $\bar{d}$. Its Fourier transform, as obtained by XRD scattering, is also a Gaussian distribution centered at $1/\bar{d}$.

The present experimental data, in spite of being limited to two reflections, supports this explanation. From it, as a corollary, it can be stated that in evaporation of the liq. NH_3 from the treated rayon, no transition to Cell III took place but a defective lattice of Cell II was reformed in which interpenetrating twins of enantiomorphic cells were competing. From this point of view the effect of hydration is understood as a plasticizing effect. Re-swelling and heat helps to increase the lateral extension by reduction of the most disordered regions, though due to the type of disorder involved, it is slower than anticipated.

DISCUSSION OF THE RESULTS

Howsmon and Sisson have suggested that Cell III's are disordered forms of Cell II. In a broad definition of disorder, the present work agrees with their suggestion and credit for it is given to them (8).

The product obtained by the action of liq. NH_3 on rayon is a particular case of Cell II recrystallization different than the product obtained by treating native cellulose with the same swelling agent. The support of this interpretation is the fact that Cell III gives two different IR spectra, depending on its origin -- native or regenerated cellulose. Marrinan and Mann have suggested that they be termed Cell III_1 and Cell III_2 respectively.

Mann and Marrinan (13,14) have made a careful study of this polymorphism by IR spectroscopy and arrived at the conclusion that the molecular configurations of Cell III_2 and Cell II are essentially the same and different from that of Cell I. Cell III_1 differs from both Cell II and Cell I. Intramolecular H-bonds were found to be responsible for maintaining such configurations.

Liq. NH_3 appears unable to alter these intramolecular H-bonds in Cell II, though its interference with intra-molecular H-bonds in Cell I is well documented. A possible reason for this behaviour is that the configuration of the Cell II chain corresponds to a minimum in energy with stronger intra-molecular H-bonds than the other configurations. Thermal vibration amplitudes at a temperature around -33°C, in a material where the assumed Tg is at least 100°C higher, should be restricted. Indirect evidence of this hindrance is found in the mercerization of cotton. A reduction in conversion rate results from a reduction in temperature below 0°C.

It must be concluded that the most obvious effect of liq. NH_3 on Cell II is its plasticizing activity. Being an intrafibrillar swelling agent, it is capable of releasing built-up strains created in the cellulosic lattice during processing and of increasing the accessibility of Cell II to chemicals. From this point of view, it merits some technological exploration, as reviewed in the next section of this paper.

PRESENT STATUS OF LIQUID ANHYDROUS AMMONIA TREATMENT FOR RAYON

Several papers have been published in latter years on Cell I-Cell III-Cell I transitions as a substitution for mercerization. Perhaps the Prograde process of J & P Coats for yarns (1,5), the Sanforset Process of Sanforized Co. (15) for fabrics, or the Tedeco plant for cotton print sheeting (16) are the best known. Several advantages are based on the plasticizing effect rather than a phase transition which is purely intermediate to the final product.

Limiting ourselves to consideration of the effect of liquid ammonia on Cell II, it can be seen that its industrial impact has been of lesser importance, perhaps due to the fact that Cell II-containing fibers are already a man-made product tailored to a particular end application.

The best known patent on this matter was issued to the Sentralinstitute for Industriell Foskning in Norway (16). Viscose rayon fabrics were treated with liquid ammonia/water mixture (90/10 ratio) under tension or pressure. The advantages claimed are:
    a) Elimination of fabric relaxation
    b) Elimination of mussed appearance
    c) Increase in fabric extensibility

Also the fabrics become softer and crease resistant in the wet state. In Table 3 some of the improvements are summarized.

TABLE 3

Effect of Liquid Ammonia on Cellulose II - Viscose Rayon

	Untreated		Treated	
	Warp	Filling	Warp	Filling
Washing shrinkage after 6 launderings (%) (Fabric)	23.3	10.0	4.7	0.0
Extensibility (%) (Staple)	3.0	3.5	20.5	1.5
Non-ironing quality (Monsanto rating)		2		4-3

Another reference to improvements induced in a rayon-containing product comes from the Norwegian Pulp and Paper Research Institute (17). Bohmer and Ormestad studied the effect of liquid ammonia with various contents of water on the properties of rayon-sulfate pulp blend non-woven fabrics. These authors found that a certain amount of water -- up to 25% -- in the ammonia solution used for the treatment produces better effects in the rayon-pulp web than does anhydrous ammonia. Table 4 shows some of their results obtained on a 70% rayon blend web.

TABLE 4

Effect of Ammonia/Water Treatment

	NH_3 evaporated	NH_3 exchanged by water
Bulk, reduction	25%	11%
Tear factor, improvement	2.6X	3.0X
Breaking length, increase	2.7X	2.7X
Stiffness, increase	1.6X	2.0X
Tensile energy absorption, increase	14.0X	9.0X

These results were comparable with those obtained with a more expensive blend of polyvinyl-alcohol fibers and rayon. Therefore they claim a real economic advantage.

Another application, patented by Arthur D. Little and developed as a commercial process by Kane Industries, utilizes the fast absorption of ammonia into cellulosic fibers as a carrier for conventional dyestuffs (18). Rapid dyeing speed and short solvent removal periods, using steam, make the system very attractive in a time of energy and pollution awareness.

To complete the technological picture, the Prograde process (5) of J & P Coats, when applied to viscose rayon yarns, has not produced any drastic improvement in mechanical or physical properties.

SUMMARY

Anhydrous liquid ammonia is an intrafibrillar swelling agent capable of reducing the crystalline order of Cell I and II. However upon subsequent hydration, the original crystalline orders are regained. Although Cell I is capable of being transformed by liquid ammonia into a new crystalline phase -- Cell III_1 -- no evidence for the same transition has been found for Cell II. The form originally called Cell III_2 appears to be an interpenetration twinning of short-range order lattices of Cell II. Possible explanations for this behaviour are suggested based on two contributing factors: a) the inability of NH_3 to alter the intramolecular H-bonds of Cell II, and b) the interference of the cellulose Tg on chain internal rotation at the actual temperature of application.
However, in spite of a defective understanding of the mechanism of ammonia absorption, the ability of liquid anhydrous ammonia, and concentrated ammonia-water solutions, to diffuse and transport other chemicals into the most disordered regions of Cell II has induced the exploration of its technological applications in the processing of rayon and rayon-containing products.

ACKNOWLEDGMENT

The author expresses his thanks to Ms. C. Stevens for reviewing the manuscript and to Dr. F. X. Werber and J.P. Stevens & Co., Inc. for permission to publish this work.

REFERENCES

1. J. & P. Coats, British Patent 1136417 (1967).
2. Sentralinstitute for Industriell Foskning, Norwegian Patent 152995 (1964).
3. Pentoney, R. C., I. & E. C. Prod. Res. & Devel. 5, 105 (1966).
4. Lewin, M. and Roldan, L. G., J. Polym. Sci., Part C 36, 213 (1971).
5. Warwicker, J. O., Cell. Chem. Technol. 6, 85 (1972).
6. Wellard, H. J., J. Polym. Sci. 13, 471 (1954).
7. Hess, K. and Gundermann, Ber. B 70, 1788 (1937).
8. Howsmon, J. A. and Sisson, W. A., in "Cellulose and Cellulose Derivatives," 2nd ed, (E. Ott, Ed.), p. 241. Interscience, New York, 1954.
9. Heritage, K. J., Mann, J. and Roldan, L. G., J. Polym. Sci., Part A-1, 671 (1963).
10. Arthur, J., Hinojosa, O. and Tripp, V., J. Appl. Polym. Sci. 13, 1497 (1969).
11. Hosemann, R. and Bagchi, "Direct Analysis of Matter by Diffraction," p. 239. North-Holland, Amsterdam, 1962.
12. Buchanan, D. R. and Miller, R. L., J. Appl. Phys. 37, 4003 (1966).
13. Marrinan, H. J. and Mann, J., J. Polym. Sci. 21, 301 (1956).
14. Mann, J. and Marrinan, H. J., J. Polym. Sci. 32, 357 (1958).
15. Anonymous, Text. Manuf. 77(4), 36 (1977).
16. Sentralinstitute for Industriell Foskning, British Patent 1084612 (1967).
17. Bohmer, E. and Ormestad, B., "Proceedings of Helsinki's Symposium on Man-made Polymers in Papermaking," p. 273 (1972).
18. Kane, S. M., Amer. Dyestuff Rep. 5, 27 (1973).

EFFECT OF CHANGES IN SUPRAMOLECULAR STRUCTURE ON THE
THERMAL PROPERTIES AND PYROLYSIS OF CELLULOSE

K. E. CABRADILLA AND S. H. ZERONIAN

Division of Textiles and Clothing
University of California, Davis, California

Cotton containing the crystalline forms of cellulose II, III, or IV was prepared from purified starting fiber in the cellulose I form. Microcrystalline cellulose in each crystal form was prepared from these samples by acid hydrolysis. Pyrolysis of the microcrystalline samples showed that the levoglucosan yield of samples in the cellulose II crystal form is less than that of samples in the cellulose I, III, or IV forms. From comparisons of the levoglucosan yields of samples pyrolyzed in the fibrous state and in the microcrystalline state, it appears that on pyrolysis, the amorphous material present in fibrous samples inhibits tar-producing reactions, although such reactions occur primarily in the crystalline areas. It is shown that heat treatments, which increase the crystallinity of fibrous cellulose, will increase the levoglucosan yield on pyrolysis. Comparisons are made of the thermal stability, as indicated by differential thermal analysis and thermogravimetry, of cellulose in the four crystal forms. It appears that the thermal stability of cellulose in the cellulose III crystal form is less than that of cellulose in the cellulose I, II, or IV forms when samples in similar states are compared.

INTRODUCTION

There are several reasons for studying cellulose pyrolysis. For example, such studies are necessary to obtain knowledge needed for the development of flame retardants for cellulosic materials. Again, with the current energy crisis, increasing attention is being paid to novel sources of fuel. One alternative is oil obtained from cellulosics by hydrogenation and pyrolysis processes. To determine fully the potential of cellulose as a source of fuel, the mechanism of its pyrolysis has to be understood including the effect that the fine structure of the cellulose may have on the process.

There is evidence that the supramolecular structure of cotton can influence its pyrolysis. Golova et al. (1) reported that the packing density of cellulose altered the amount of levoglucosan formed during vacuum pyrolysis. Cotton had a levoglucosan content of 60-63%; cotton treated with 10% sodium hydroxide, 36-37%; cuprammonium rayon, 14-15%; and viscose rayon, 4-4.5%. Lewin and coworkers (2,3) have studied the effects of such parameters as crystallinity, orientation and degree of polymerization on pyrolysis of cellulose. They showed that cotton and ramie had significantly higher yields of levoglucosan (22-35%) in air pyrolysis than rayon (13-17%). The emphasis of the cited studies is on the accessibility or crystallinity of the cellulose as a prime factor influencing its pyrolysis. The effect of the crystal form has not been considered.

In a systematic examination of the effect of supramolecular structure on the pyrolysis of cellulose, we have prepared all our samples from a single type, namely purified cotton. We have previously reported on the effect of changes in crystallinity on the thermal properties and pyrolysis of cotton cellulose (4). Unmodified cotton, with its high degree of crystallinity, underwent endothermic decomposition and 28.5% levoglucosan was formed during vacuum pyrolysis at 350°C. However cotton that possessed no crystallinity, as shown by x-ray analysis, underwent exothermic decomposition and little levoglucosan (1.1%) was formed during its pyrolysis. It is generally accepted (5) that when cellulose is heated to high temperatures, it undergoes two principal types of degradation reactions:

1. levoglucosan- or tar-forming reactions, which gather momentum at temperatures above 250°C and result in rapid volatilization.

2. dehydration and char-forming reactions, which are initiated at lower temperatures and continue at elevated temperatures.

Products of these primary reactions, if not removed from the heated environment, can further react and decompose to provide a series of secondary compounds of low molecular weight. Crystallinity affects the type of reaction that occurs. In crystalline regions levoglucosan formation is favored over dehydration and char formation. In amorphous regions, dehydration and char formation are the favored reactions.

In this study we report on the effect of the crystalline form of cotton cellulose on its thermal properties and pyrolysis.

EXPERIMENTAL

Materials

Partially purified cotton fiber, Deltapine-Smoothleaf variety of American Upland cotton, obtained from the Southern Regional Research Center, USDA, was used. Cupriethylenediamine hydroxide solution was obtained from Ecusta Paper Division, Olin-Matheson Chemical Corp., Pisgah Forest, North Carolina. Other chemicals were reagent grade.

Methods of Treatment

The cotton fiber was further purified by refluxing for 8 hr in 1% sodium hydroxide solution (40 g per 2 liter solution) in a nitrogen atmosphere, washed with distilled water until free from alkali, and then dried. All subsequent treatments were done on the purified cotton.

Preparation of cellulose II (mercerization). Cotton cellulose was mercerized in 5 N sodium hydroxide solution (10 g per 500 ml solution) in a stoppered ground glass bottle at 0°C for 1 hr (6). The sample was washed with distilled water, then steeped in 10% acetic acid solution for 15 min, and washed again in distilled water until it was neutral. The product was air dried.

Annealing cellulose II. Cellulose II was annealed in a manner similar to that of Atalla and Nagel (7). Cotton (cellulose I form) was mercerized in 17% sodium hydroxide solution (10 g per 500 ml solution) in a resin reaction kettle under nitrogen for 1 hr at 0°C. The solution was then diluted to 15% and the temperature, increased to 80°C. At 50 min intervals, the solution was further diluted to 12.5%, 7.5%, 5% and 2.5% while the temperature was maintained at 80°C. The sample was then washed with distilled water at 80°C, steeped in 10% acetic acid solution for 15 min at 80°C, and washed again with distilled water at 80°C until it was neutral. The sample was immersed in glycerol in a resin reaction kettle at 100°C under nitrogen and the temperature was increased to

150°C. The sample was heated in glycerol under nitrogen at 150°C for 14 days. After 14 days, the glycerol was cooled to 100°C and the sample washed with boiling distilled water until it was free of glycerol. The sample was air dried.

Preparation of cellulose III. Cotton (cellulose I form) was treated for 2 hr with liquid ammonia (10 g per 500 ml) in a resin reaction kettle immersed in a dry ice-acetone bath. The sample was then transferred to an Erlenmeyer flask, which had been previously cooled in dry ice. To protect the sample from carbon dioxide and water, the Erlenmeyer flask was stoppered with a loose cotton plug. The ammonia slowly evaporated off the sample over a 48 hr period. The sample was kept over phosphorus pentoxide to prevent conversion back to cellulose I.

Preparation of cellulose IV. Cotton (cellulose I form) was mercerized in 5 N sodium hydroxide solution (10 g per 500 ml solution) in a stoppered ground glass bottle at 0°C for 1 hr. The solution was filtered off. The sample was washed in glycerol and allowed to soak in glycerol for 1/2 hour; this was repeated. Next the sample, immersed in glycerol, was heated at 270°C under nitrogen for 1/2 hr. The sample was then allowed to cool to room temperature. Finally, the sample was washed several times with isopropanol and with water. The sample was air dried. The conversion from cellulose II to cellulose IV was found by calculating the ratio between the intensities of the combined (101) and (10$\bar{1}$) reflections of cellulose IV to the (101) reflection of cellulose II present in the sample and determining the percent conversion from the data of Hermans and Weidinger (8).

Heat treatment of cellulose I. Cotton (cellulose I form) was soaked in glycerol for 1 hr and then was heated in glycerol under nitrogen at 270°C for 1/2 hr. The sample was then allowed to cool to room temperature. The sample was washed with isopropanol, followed by distilled water, and then air dried.

Microcrystalline samples. Samples of cotton in the cellulose I, II, III and IV crystal form were hydrolyzed in 2.5 N hydrochloric acid solution (10 g per 500 ml solution) at the boil for 15 min. The procedure has been previously described (9).

Levoglucosan (1,6-anhydro-β-D-glucopyranose). The synthesis of levoglucosan has been previously described (4).

Characterization of Product

Descriptions have been given previously of the procedures used to determine the following: moisture regains at 59% R.H. and 21°C, degree of polymerization (DP), copper number, and carboxyl content (4). The procedure used to obtain x-ray diffractograms of cotton samples has also been described previously (4).

Differential thermal analysis. Differential thermal analysis (DTA) was run on samples (4 mg) in a nitrogen atmosphere at a heating rate of 10°C per min on a Deltatherm III differential thermal analyzer. Alumina was used as the reference material. Two measurements were calculated from the DTA curves: temperature at the onset of the major endotherm (T_{onset}) and temperature at the peak of this endotherm (T_{peak}). The procedure has been previously described (4).

Thermogravimetry. Thermogravimetry (TG) was run on the samples (4 mg) in a nitrogen atmosphere at a heating rate of 5°C per min on a Perkin-Elmer TGS-1 thermobalance. Temperature at which the weight loss reaction began (T), percent residue for the reaction found at the cessation of rapid loss in sample weight, and percent char at 500°C were calculated from the TG curves. The procedure used has been previously described (4).

Levoglucosan analysis. Gas chromatography (GC) was used to determine the amount of levoglucosan in the tar of pyrolyzed samples. The methods used for vacuum pyrolysis at 350°C and for GC analysis of the silylated tar have been previously described (4). The levoglucosan content was calculated by:

$$\text{levoglucosan content} = \frac{\text{weight of levoglucosan in tar}}{\text{dry weight of cellulose}} \times 100$$

Ash content. Ash contents were determined by AATCC Test Method 78-1974 (10).

RESULTS AND DISCUSSION

Cotton with its crystallites in the cellulose I form (cotton cellulose I) was used to prepare cotton containing crystallites in the cellulose II, III, and IV forms (cotton cellulose II, cotton cellulose III and cotton cellulose IV, respectively). X-ray diffractometer tracings showed that there was essentially complete lattice transformation from cellulose I to cellulose II and from cellulose I to cellulose III (Fig. 1). In the case of cotton cellulose IV, which was

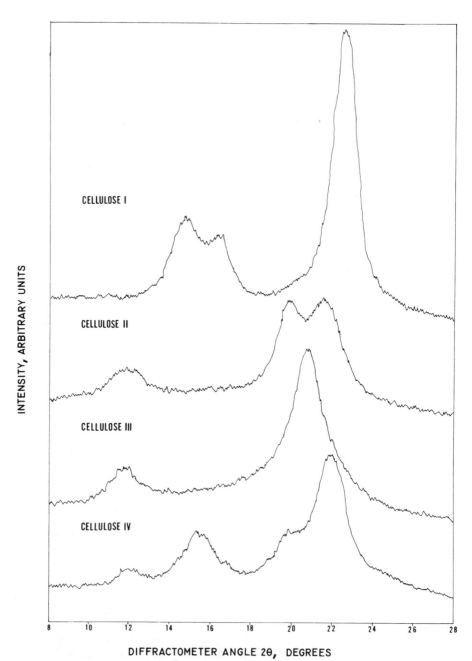

Fig. 1. X-Ray Diffractometer Tracings of Cotton Fiber in the Cellulose I, II, III, and IV Forms (80% Conversion from Cellulose II to Cellulose IV).

prepared from cotton cellulose II, the presence of cellulose II crystallites could be observed on the x-ray diffractometer tracing (Fig. 1). Approximately 80% conversion from cellulose II to IV had occurred. Other workers (8) have also obtained 70-80% conversion from cellulose II to IV when native cellulose has been used as the source of cellulose. Since preparation of cotton cellulose IV involved heat treatment at 270°C, cotton cellulose I was heated under the same conditions (cotton cellulose I-heated). It can be deduced from x-ray diffractometer tracings (Fig. 2) that there was no difference in cell form between cotton cellulose I and cotton cellulose I-heated and that no cellulose IV was present in the heated sample. Cotton in the cellulose II form was annealed to increase the size of the crystallites (cotton cellulose II-annealed). From x-ray diffractometer tracings (Fig. 2), it can be observed that increased resolution of the ($10\bar{1}$) and (002) peaks, and a sharpening of the (101) peak had occurred after annealing indicating the size of the crystallites had increased.

The cotton celluloses with the different cell forms were characterized by sorption ratio, degree of polymerization, copper number, and carboxyl content (Table 1). Ash contents were checked on cotton cellulose I and mercerized cellulose samples and were found to be 0.07% and 0.08%, respectively.

TABLE 1

Characterization of Cotton Cellulose with Different Cell Forms

Cell Form	Sorption Ratio[a]	DP[b]	Copper Number	Carboxyl Content mM/100 g
Cellulose I	1.00	2070	<.01	.43
Cellulose I-(heated sample)	0.87	247	.01	.07
Cellulose II	1.52	1830	.24	.23
Cellulose II-(annealed sample)	1.45	255	.04	.08
Cellulose III	1.31	1920	.13	.32
Cellulose IV[c]	1.26	140	<.01	.93

[a] Sorption ratio = moisture regain of sample/moisture regain of control cotton.

[b] DP = degree of polymerization.

[c] 80% conversion from Cellulose II.

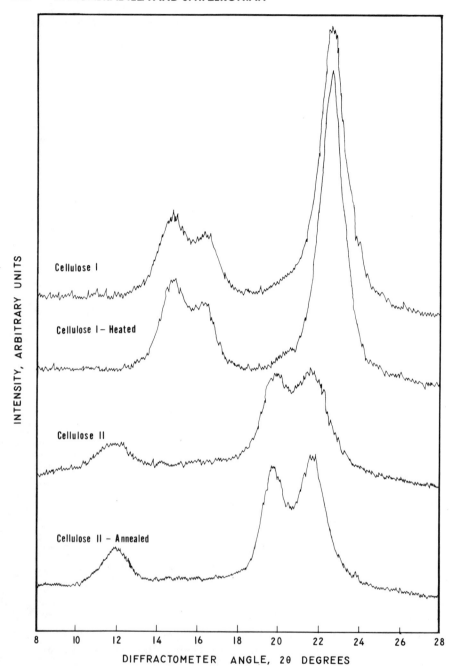

Fig. 2. X-Ray Diffractometer Tracings of Cotton Fiber in the Cellulose I Form; Cotton in the Cellulose I Form - Heated; Cotton Fiber in the Cellulose II Form; and Cotton in the Cellulose II Form - Annealed.

Sorption ratios were used as a measure of crystallinity (11). To calculate the sorption ratio of a sample, its moisture regain is divided by that of untreated cotton measured at the same temperature and relative humidity. A sorption ratio greater than one indicates the sample has a lower crystallinity, or higher accessibility, than cotton cellulose. The fraction of amorphous cellulose (F_{am}) present in a sample can be calculated from its sorption ratio (11). However, the result is less precise if the crystal form in the untreated cotton and the cellulose under examination are not the same (12). In this study, the crystal form of the cellulose was changed, consequently F_{am} was not calculated. To give an indication of the order of magnitude, if the sorption ratio is 0.8, 1.0, 1.2 or 1.5 then F_{am} would be 0.31, 0.38, 0.46 or 0.58, respectively.

The sorption ratios of cotton cellulose II, III and IV were higher than that of cotton cellulose I, indicating that a reduction in degree of crystallinity accompanied crystal transformation (Table 1). Heating cotton cellulose I and annealing cotton cellulose II lowered their sorption ratios compared to cotton cellulose I and II, respectively indicating the heat treatments had increased crystallinity.

Heat treatments sharply lowered the degree of polymerization of cotton cellulose I and II (Table 1). It will be noted however that the reducing power, as indicated by copper number measurements, of cotton cellulose I-heated, cotton cellulose II-annealed and cotton cellulose IV remained low. When cellulose is hydrolyzed by acids, an increase in reducing power occurs as the degree of polymerization is lowered, due to chain scission of the glycosidic linkages increasing the number of hemiacetal groups (13). Thus it appears that as cellulose is degraded by heat, in a non-oxidative environment, reactions occur which eliminate hemiacetal groups at chain ends.

Thermal analysis of fibrous samples. Thermal analysis showed differences between the thermal properties of cotton fibers with the various crystal forms. A typical DTA curve and TG curve for cotton cellulose I are shown in Fig. 3. The DTA curves for cotton cellulose II, III, and IV were similar to that of cotton cellulose I. From the DTA curve for cotton cellulose I it can be seen that a major endothermic decomposition reaction begins at 332°C and peaks at 358°C. The onset and peak temperatures of the major endothermic decomposition reaction were affected by the transformation of the cell form of the cotton (Table 2). Cotton cellulose II, III and IV had lower onset and peak temperatures than cotton cellulose I.

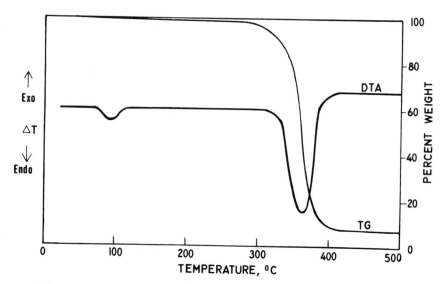

Fig. 3. Typical Differential Thermal Analysis (DTA) and Thermogravimetric (TG) Curves of Cotton Cellulose.

TABLE 2

Thermal Analytical Data of Cotton Cellulose with Different Cell Forms

Cell Form	Differential Thermal Analysis		Thermogravimetry		
	T, onset °C	T, peak °C	T^a °C	residue %	char %
Cellulose I	332	358	337	11.7	8.9
Cellulose I-(heated sample)	300	332	326	9.6	7.1
Cellulose II	316	343	324	20.8	12.5
Cellulose II-(annealed sample)	300	332	315	13.0	9.2
Cellulose III	311	338	318	15.9	12.3
Cellulose IV^b	326	350	333	18.5	14.3

[a] Onset temperature of major weight loss.

[b] 80% conversion from cellulose II.

In the case of cotton cellulose IV, it should be remembered that only 80% conversion from the cellulose II crystal form had been achieved. The residual cellulose II may be partially the cause of the lowered onset and peak temperatures.

Heat treatment of cotton cellulose I (cellulose I-heated) resulted in a large decrease in the onset and peak temperatures, as determined from DTA curves. Annealing cotton cellulose II (cellulose II-annealed) also caused a drop in both onset and peak temperatures when compared to either cotton cellulose I or II. The increase in crystallinity that occurred as the result of the heat treatments may be the cause of the shift of the endothermic reactions to lower temperatures. Further evidence for increased crystallinity being the cause of this shift will be given in a later section.

TG curves were similar for cotton cellulose in the different crystal forms. However, there were differences in the temperature where major weight loss begins, in the percent residue (i.e. weight remaining after thermal decomposition), and in the percent char (i.e. weight remaining at 500°C). The TG curve of cotton cellulose I showed the presence of a very slight weight loss between 100-280°C, a major weight loss due to thermal decomposition between 280-380°C, and a slight weight loss between 380-500°C. The weight loss associated with the major decomposition was 88.3% (11.7% residue). At 500°C, there was 8.9% char. The temperature where major weight loss begins had decreased for cotton cellulose II and III; for cotton cellulose IV, the temperature remained close to that of cotton cellulose I (Table 2). Again, it should be noted that the results for cotton cellulose IV will be affected by the presence of residual cellulose II crystallites. The percent residue and percent char had increased for cotton cellulose II, III, and IV. The percent char for these samples was similar to the percent residue of cotton cellulose I. The amount of weight lost during char reaction was similar for cotton cellulose I, III, and IV: this weight loss is the difference between the percent residue and the percent char.

Heat treatment of cotton cellulose I resulted in a lower temperature at which major weight loss begins when compared to cotton cellulose I, and the percent residue and percent char were slightly lower. Annealing cotton cellulose II (cellulose II annealed) resulted in a decrease in onset temperature, percent residue, and percent char when compared to cotton cellulose II. Comparison of cotton cellulose II-annealed to cotton cellulose I showed that there was a decrease in temperature where major weight loss begins, but the percent residue and percent char were roughly similar for these two samples. It can be concluded from the TG data as well as the DTA data for the heat-treated samples that decomposition of the samples starts at lower temperatures as the degree of crystallinity is increased.

The DP of the cellulose decreased from 2070 to 140 during the preparation of cotton cellulose IV. This lowering of the DP should be considered when examining the thermal analysis data. Samples with which cotton cellulose IV can be compared are cotton cellulose I-heated and cotton cellulose II-annealed, both of which had very low DPs also (Table 1). Differential thermal analysis showed that the major endothermic reaction occurred at a significantly higher temperature for cotton cellulose IV than for the other two samples which had the same onset and peak temperatures for their endothermic decomposition (Table 2). The higher decomposition temperature of cotton cellulose IV should not be due to the higher carboxyl content of this sample (Table 1) since an increase in carboxyl content has been shown to accelerate the thermal decomposition of cellulose (14). There were differences in the TG curves for these three samples which supported the DTA findings. The temperature where weight loss begins increased in the order: cellulose II-annealed, cellulose I-heated, and cellulose IV. The percent residue and percent char increased in the order: cellulose I-heated, cellulose II-annealed, and cellulose IV. When the presence of residual cellulose II crystallites in the cotton cellulose IV is taken into account, this reinforces the difference in thermal properties between cotton cellulose IV and cellulose I-heated or cellulose II-annealed. The presence of the cellulose II crystallites should lower the onset and peak temperatures shown by DTA and the onset temperature, percent residue, and percent char calculated from TG, for cotton cellulose IV.

Levoglucosan analysis of fibrous samples. The amount of levoglucosan formed during vacuum pyrolysis varied for cotton fibers with different crystal forms (Table 3). Cotton cellulose II and IV gave much lower yields of levoglucosan than either cotton cellulose I or III. Heat treatment of both cotton cellulose I and II increased markedly the yield of levoglucosan. Comparison of the heat-treated samples with cotton cellulose IV shows that the levoglucosan content was lower for cotton cellulose IV, although all three samples had received heat treatments and also had very low DPs (Tables 1 and 3). The comparison between cellulose I-heated and cotton cellulose IV is especially noteworthy since in the preparation of cotton cellulose IV the starting sample received the same

TABLE 3

Levoglucosan Content of Cotton Cellulose with Different Cell Forms

Cell Form	Levoglucosan[a] %	Residue After Pyrolysis[a] %
Cellulose I	28.5	13.5
Cellulose I (heated sample)	37.9	8.1
Cellulose II	18.7	13.6
Cellulose II (annealed sample)	31.1	9.7
Cellulose III	32.8	9.1
Cellulose IV[b]	14.0	19.9

[a] Calculated on dry weight of cellulose
[b] 80% conversion from cellulose II

heat treatment as cellulose I-heated. The percent residue formed on pyrolysis was similar for cotton cellulose I and II and was similar for cotton cellulose III and heat-treated samples of cotton cellulose I and II. The percent residue was highest for cotton cellulose IV.

The preparation of cotton cellulose II, III and IV from cotton cellulose I involved an increase in accessibility and a decrease in crystallinity as shown by the sorption ratios of the samples (Table 1). This change in accessibility (or crystallinity) should influence the extent to which the major pyrolytic reaction occurs since reduction of crystallinity in cotton cellulose is known to affect the course of pyrolysis. Examining the present data in this light, the following observations can be made. The reactions of dehydration and char formation had increased in importance when cotton cellulose I was converted to cotton cellulose II since cotton cellulose II produced about 66% of the amount of levoglucosan produced by cotton cellulose I (Table 3). Also, the amount of residue and char, as shown by TG (Table 2), was higher for the former sample. However, in addition to cell form these changes could be due to lower crystallinity. The sorption ratio of cotton cellulose II was 1.52 in comparison to a sorption ratio of 1.00 for cotton cellulose I. There is much additional evidence, including x-ray analysis, that conversion to cellulose II lowers the crystallinity of cotton cellulose (15). Heating cotton cellulose I or annealing cotton cellulose II increased the crystallinity of these samples, as indicated by sorption ratios and raised the amount of levoglucosan produced on pyrolysis. The amount of residue and char as shown by TG (Table 2) had decreased also. These facts support the hypothesis that crystallinity is one of the controlling factors determining the amount of levoglucosan produced during cellulose pyrolysis.

The results obtained with cotton cellulose III indicate, however, that cell form might also be important. Cotton cellulose III produced a greater amount of levoglucosan on pyrolysis than cotton cellulose I (Table 3) even though sorption ratio measurements indicated it had a lower crystallinity than cotton cellulose I (Table 1). Also, the percent residue and percent char, as shown by TG, were higher for cotton cellulose III (Table 2). Thus, the products of the competing reactions were both higher for cotton cellulose III than for cotton cellulose I.

In addition, cotton cellulose IV yielded less levoglucosan than cotton cellulose III on pyrolysis although these samples had approximately similar degrees of crystallinity as indicated by sorption ratios (Table 1). Thus, again, crystal form may be having an effect on the amount of levoglucosan produced.

To separate more clearly the effect of crystallinity or accessibility from that of cell form, microcrystalline samples of the different crystalline forms of cellulose were studied.

Properties of microcrystalline samples. Microcrystalline samples were prepared from cotton cellulose I, II, III and IV by acid hydrolysis which eliminated the amorphous regions of the fibers. Microcrystalline samples are also referred to as level-off degree of polymerization samples. A measure of the crystallite length for a microcrystalline sample can be obtained from its DP. The crystal form of the samples did not change after hydrolysis (16).

The microcrystalline samples were characterized by DP, copper number, and carboxyl content (Table 4). The DP of microcrystalline samples containing the cellulose I crystal form (microcrystalline cellulose I) was at least twice that of the microcrystalline samples containing either the cellulose II, III or IV crystal forms (microcrystalline cellulose II, III or IV, respectively). The copper number reflects the

TABLE 4

Characterization and Thermal Analytical Data of Microcrystalline Cotton Cellulose with Different Cell Forms

Cell Form	DP[a]	Copper Number	Carboxyl Content mM/100 g	Differential Thermal Analysis		Thermogravimetry		
				T, onset °C	T, peak °C	T[b] °C	residue %	char %
Cellulose I	203	6.87	.14	299	331	323	7.3	5.1
Cellulose II	105	11.85	.07	302	339	320	13.8	8.4
Cellulose III	87	12.96	.10	288	319	309	13.0	7.9
Cellulose IV[c]	81	11.00	.95	299	334	321	10.7	7.2

[a] DP = degree of polymerization.

[b] Onset temperature of major weight loss

[c] 80% conversion from cellulose II.

number of reducing groups in a cellulose which would be represented in these samples primarily by hemiacetal groups at cellulose chain ends. Since the DP of microcrystalline cellulose I is approximately twice that of the other samples, the copper number is roughly half that of the others. The carboxyl contents of microcrystalline cellulose I, II and III were lower than those of their fibrous counterparts cotton cellulose I, II and III (cf. Tables 1 and 4). It is interesting to note that the carboxyl content of microcrystalline cellulose IV was the same as its fibrous counterpart, indicating the heat treatment used to induce conversion of the crystal form from cellulose II to IV had caused the formation of carboxyl groups on the surface of the crystalline regions as well as in amorphous regions.

Thermal Analysis of Microcrystalline Samples. The onset and peak temperatures measured from the DTA curves of microcrystalline cellulose I, II, and IV were similar while those of microcrystalline cellulose III were lower (Table 4). Likewise, the temperature at which the major weight loss begins, as shown by TG, was similar for microcrystalline cellulose I, II, and IV but was lower for the cellulose III sample. The percent residue and percent char showed differences due to the cell form. Microcrystalline cellulose II and III had similar residues, which were roughly twice that of the microcrystalline cellulose I. The percent residue of the cellulose IV sample was significantly lower than those of microcrystalline cellulose II and III, but was higher than that of the cellulose I sample.

Levoglucosan analysis for microcrystalline samples. On pyrolysis, microcrystalline cellulose I and III gave higher yields of levoglucosan than did microcrystalline cellulose II and IV (Table 5). The amounts of levoglucosan produced by

TABLE 5

Levoglucosan Content of Microcrystalline Cotton Cellulose with Different Cell Forms

Cell Form	Levoglucosan[a] %	Residue After Pyrolysis[a] %
Cellulose I	41.4	6.7
Cellulose II	27.6	10.4
Cellulose III	39.5	7.3
Cellulose IV[b]	33.8	10.9

[a] Based on dry weight of cellulose

[b] 80% conversion from cellulose II

microcrystalline cellulose I and III were similar. The DP for microcrystalline cellulose I was twice that for the cellulose III sample and the reducing power for microcrystalline cellulose I was half that of the latter sample (Table 4) raising the question whether samples of microcrystalline cellulose I and III would still yield similar amounts of levoglucosan on pyrolysis if the crystallite lengths were the same.

The evidence appears unequivocal that crystal form can affect levoglucosan yield since both microcrystalline cellulose I and III give significantly higher yields of levoglucosan than microcrystalline cellulose II. The DP and copper number of microcrystalline samples cellulose II and III are roughly similar (Table 4) eliminating the need to consider crystallite length as a factor.

4. Cabradilla, K. E. and Zeronian, S. H., in "Thermal Uses and Properties of Carbohydrates and Lignins" (F. Shafizadeh, K. V. Sarkanen and D. A. Tillman, Eds), p. 73, Academic Press, New York, 1976.
5. Shafizadeh, F., in "Advances in Carbohydrate Chemistry" (M. L. Wolfrom and R. S. Tipson, Eds.), Vol. 23, p. 419 Academic Press, New York, 1968.
6. Zeronian, S. H. and Cabradilla, K. E., J. Appl. Polym. Sci., 17, 539 (1973).
7. Atalla, R. H. and Nagel, S. C., J. Polym. Sci., Polym. Lett. Ed., 12, 565 (1974).
8. Hermans, P. H. and Weidinger, A., J. Colloid Sci., 1, 495 (1946).
9. Battista, O. A., Ind. Eng. Chem., 42, 502 (1950).
10. AATCC Technical Manual, Vol. 50, p. 48, American Association of Textile Chemists and Colorists, Research Triangle Park, North Carolina, 1974.
11. Valentine, L., Chem. Ind. (London), 1279 (1956).
12. Jeffries, R., J. Appl. Polym. Sci., 8, 1213 (1964).
13. Nevell, T. P., in "Recent Advances in the Chemistry of Cellulose and Starch" (J. Honeyman, Ed.), p. 89, Heywood and Company, London, England, 1959.
14. Phillip, B., Baudisch, J. and Stohr, W., Cellul. Chem. Technol., 6, 379 (1972).
15. Warwicker, J. O., Jeffries, R., Colbran, R. L. and Robinson, R. N., "A Review of the Literature on the Effect of Caustic Soda and Other Swelling Agents on the Fine Structure of Cotton", Shirley Institute Pamphlet No. 93, The Cotton Silk and Man-Made Fibres Research Association, Manchester, England, 1966.
16. Cabradilla, K. E., Ph.D. Thesis, University of California Davis, 1976.
17. Tripp, V. W. and Conrad, C. M. in "Instrumental Analysis of Cotton Cellulose and Modified Cotton Cellulose" (R. T. O'Connor, Ed.), p. 349, Marcel Dekker, Inc., New York, 1972.
18. Zeronian, S. H., J. Appl. Polym. Sci., 15, 955 (1971).
19. Halpern, Y. and Patai, S., Isr. J. Chem., 7, 673 (1969).

appears therefore that small increases in crystallinity will decrease the thermal stability of cotton cellulose I or II.

When the thermal data is compared for the four crystal forms either as fibrous materials (without further heating) or in the microcrystalline state it appears that samples in the cellulose III crystal form are less thermally stable than samples in the other three forms (cf. Tables 2 and 4). Thus it appears crystal form can influence thermal properties as well as the course of pyrolysis.

CONCLUSIONS

This study shows that crystal form as well as degree of crystallinity can affect the course of pyrolysis and the thermal properties of cellulose. Pyrolysis of microcrystalline samples showed that the levoglucosan yield of samples in the cellulose II crystal form is less than that of samples in the cellulose I, III or IV forms. Also, there is considerable evidence that the thermal stability of samples in the cellulose III crystal form is less than that of samples in the other three crystal forms. Thus when attempting to optimize pyrolysis conditions for conversion of cellulose into useful products, consideration should be paid to the crystal form of the cellulose as well as its degree of crystallinity. Additionally, attention should be paid to the possibility of hydrolyzing cellulose to its microcrystalline state or to heating the sample to induce an increase in its degree of crystallinity since both these treatments cause large increases in levoglucosan production on pyrolysis.

The authors would like to thank Mr. J. J. Creely, Southern Regional Research Center, USDA, New Orleans, Louisiana, for taking the x-ray diffractograms of the cellulose samples.

REFERENCES

1. Golova, O. P., Pakhomov, A. M. and Andriovskaya, E. A., Dokl. Acad. Nauk SSSR, 112, 430 (1957).
2. Basch, A. and Lewin, M., J. Polym. Sci., Polym. Chem. Ed., 11, 3095 (1973).
3. Lewin, M., Basch, A. and Roderig, C., in "Proceedings of the International Symposium on Macromolecules" (E. B. Mano, Ed.), p. 225, Elsevier, New York, 1975.

When cotton cellulose I and II are hydrolyzed to their microcrystalline states, the loss in weight of the samples is only 6% and 8%, respectively (18) but the amounts of levoglucosan formed on pyrolysis increases 45% and 48%, respectively (cf. Tables 3 and 5). It appears therefore that on pyrolysis the amorphous material present in fibrous samples inhibits the tar producing reactions although these reactions occur primarily in the crystalline areas. This might be the explanation for the large increase in the amount of levoglucosan produced by cotton cellulose IV after it has been hydrolyzed to its microcrystalline state (cf. Tables 3 and 5).

When the thermal properties of the fibrous materials are compared with those of the microcrystalline samples several observations can be made. For a given crystal form the onset and peak temperatures as measured from DTA curves and the onset temperature, percent residue and percent char as measured from TG curves are always lower for the microcrystalline sample than for the non-heated fibrous counterpart (cf. Tables 2 and 4). It may be deduced therefore that the greater the crystallinity of the sample the lower its thermal stability, as indicated by DTA or TG. Also the reduction in the amount of residue and char for the more crystalline product indicates that the pyrolysis of such samples is proceeding by the pathway resulting in less char and more tar-forming reactions, supporting the finding that the microcrystalline products produced greater quantities of levoglucosan. The available evidence indicates that when cellulose is pyrolyzed, the non-crystalline regions are the first to be attacked. Rapid cleavage of the chains occurs until the DP of the cellulose falls to 200. The primary degradation to chains of 200 DP is virtually complete before appreciable amounts of volatile products are formed (19). Elimination, or reduction, in the amount of amorphous material in a cellulosic sample before pyrolysis may permit levoglucosan formation to start at a lower temperature.

When cotton cellulose I and cotton cellulose II were heated, they increased in crystallinity as indicated by the sorption ratio results (Table 2). The thermal analytical data for these heated samples is approximately the same as that of microcrystalline cellulose I and II, respectively (cf. Tables 2 and 4). The sorption ratio of microcrystalline cellulose I and II is about 0.88 and 1.14, respectively (Zeronian, S. H., unpublished results) indicating that heated cotton cellulose I (sorption ratio = 0.87) and microcrystalline cellulose I had similar crystallinities but microcrystalline cellulose II had a higher crystallinity than annealed cotton cellulose II (sorption ratio = 1.45). It

Cellulose IV is prepared by heating cotton in the cellulose II form. The levoglucosan yield for microcrystalline cellulose IV was higher than that of microcrystalline cellulose II and presumably would have been higher still if full rather than 80% conversion from cellulose II had been achieved. The percent residue formed on pyrolysis was about 50% higher for microcrystalline cellulose II and IV than for microcrystalline cellulose I and III indicating that dehydration and char-forming reactions were more important for the former samples.

In summary, for crystalline materials it appears on pyrolysis the amount of levoglucosan and residue, or char, produced depends at least in part, on crystal form and thus on the arrangement of the cellulose chains in relation to one another within the crystals. The lattice dimensions are different for each crystal form (17). The unit cells vary from one another in that the planes of the anhydroglucose units of cellulose chains lie at different angles. Thus inter-chain distances and bond strengths vary. It appears these factors may influence the amount of levoglucosan produced on hydrolysis.

Influence of fine structure of cotton fiber on pyrolysis and on thermal properties. The relation of crystal form and degree of crystallinity in determining thermal properties and pyrolysis can be better understood by comparing the results for microcrystalline cellulose I, II, III and IV with those of their fibrous counterparts, cotton cellulose I, II, III and IV which contain amorphous material.

For all cell forms on pyrolysis the yield of levoglucosan increased sharply and the amount of char produced decreased when fibrous samples were hydrolyzed to their microcrystalline state. Since the acid hydrolysis had eliminated noncrystalline material, the degree of crystallinity of the samples does influence the percent levoglucosan formed when cotton cellulose with different crystal forms is pyrolyzed. When cotton cellulose I and II were heated to increase crystallinity, both yielded a greater quantity of levoglucosan on pyrolysis. It should be noted that the quantity of levoglucosan produced was roughly similar to that obtained from microcrystalline cellulose I and II, respectively (cf. Tables 3 and 5). However, irrespective of degree of crystallinity of the fibers, some of the variation in levoglucosan yield between cotton cellulose I or III and cotton cellulose II can be attributed to the crystal form, since there were differences in levoglucosan yield when samples were pyrolyzed in the microcrystalline state.

RECENT DEVELOPMENTS IN THE INDUSTRIAL
USE OF HEMICELLULOSES

R. L. WHISTLER AND R. N. SHAH

Department of Biochemistry
Purdue University
W. Lafayette, Ind.

Recent work on isolation, purification, and structural analysis of hemicelluloses is reviewed. Principal homoglycans such as glucuronoarabinozylans, glucomannans, galactomannans, and arabinoxylans are covered. Attention is given to the use and potential use of hemicelluloses as adhesives, thickeners, coagulants, hardening agents, anti-cancer agents, and as special food additives. Chemically modified hemicelluloses are reviewed and their suggested industrial uses are discussed.

Industrial interest in hemicelluloses has been rekindled because of the rapidly increasing costs of synthetic polymers. Recent and foreseeable prices place the natural polymers in a much more competitive position relative to synthetic polymers. Since hemicelluloses are enormously abundant in the plant world, it is logical to expect their emergence into commercial applications. Present conditions represent a congenial economic environment.

Hemicelluloses constitute 20-30% of annual and perennial plants. They are in largest amount in annual plants and have been rather extensively examined in such agricultural crops as corn stalks, corn cobs, corn seed coat, wheat straw and soy bean hulls. Hemicelluloses are defined as plant cell wall polysaccharides, other than cellulose and pectin, that are extractable by alkaline solutions. Some few hemicelluloses are extractable in hot water but most require mild to concentrated alkaline solution with 10% sodium hydroxide being a common extractant. Lime water has been used satisfactorily in a suggested semi-commercial extraction. After alkaline extraction, one group of hemicelluloses precipitate on neutralization of the solution and are termed hemicelluloses A. These are the linear, or nearly linear, high molecular weight polysaccharides, usually with few, if any, carboxyl groups. Hemicelluloses remaining in the neutral solution are termed hemicellulose B types. They are generally more highly branched and may be lower in molecular weight than the A-type, or they may possess numerous carboxyl groups.

Extraction of hemicellulose, in the laboratory, is usually made from plant material previously freed of lipids and lignin. These two components impede penetration of extraction liquor, while lignin not only encrusts and holds hemicellulose by entrapment but is generally bound to hemicellulose by covalent bonds. Lignin is also, in part, alkali soluble. When present, therefore, some of it may be extracted with the hemicellulose to complicate later purification. Hemicellulose may be bound to other plant components but such binding seems to be by ester linkages that are easily hydrolyzed during even a mild alkaline extraction such as afforded even by lime water.

XYLANS OR XYLOHEMICELLULOSES

The most abundant class of hemicelluloses have a backbone or basic chain of D-xylopyranosyl units linked mainly in $(1 \rightarrow 4)$-β-D fashion. Purely linear, homoglycans are rare. Usually the xylan chain contains other sugar units as single or multiple unit side chains and sometimes the xylan chain is branched with other sugar units attached. The commonest second sugar is in the form of β-L-arabinofuranosyl groups although now and then some D-arabinopyranosyl groups are present. Attachment to the xylan chain is often at C3 of the chain units. α-D-Glucuranosyl units, often as the 4-O-methyl derivative occur as end units on the L-arabinofuranosyl chains or more frequently simply as side chains linked $1 \rightarrow 2$, or to a less extent $1 \rightarrow 3$ to the main xylan chain. D- and L-Galactopyranosyl units are also often attached to the chains particularly in the xylohemicelluloses found in annual plants.

An additional structural feature of many D-xylans, particularly those of hardwoods, is the presence of acetyl ester groups. The amount of O-acetyl groups in hardwood D-xylans ranges from 8-17% corresponding to approximately 3.5-7 groups per ten D-xylose units. All of the ester groups are believed linked to D-xylopyranosyl units of the main chain, predominately at C-2 positions although a small proportion of linkage at C-3 positions has been observed in some D-xylans (1,2).

On the basis of the above structural information, the idealized D-xylan structure as shown in Figure 1 can be given as representing features found in D-xylans from many sources of land plants. Structural details of D-xylans within certain plant groups are often characteristic of that group and can prove useful in classification particularly where questions of proper classification of individual species is concerned.

D-Xylans of algal species, although not as widespread or as thoroughly studied, appear to show considerable structural divergence from those of land plants. A D-xylan (3) from the red seaweed *Rhodymenia palmata*, contains, in addition to the normal (1 → 4)-β-D-linkage seen in D-xylans of land plants, but approximately on-third (1 → 3)-β-D-linkages.

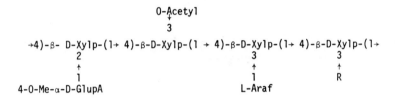

Figure 1. General structure for xylohemicellulose.

Because of their wide occurance, xylohemicelluose (5) are certain to find industrial applications.

An effort was made to develop industrial uses for the abundant hemicellulose present in corn fiber, the seed coat byproduct of the corn wet milling industry. This hemicellulose, extractable by lime water or other mild alkali is, after extraction, a water soluble homogeneous hydrocolloid with viscous and other rheological properties similar to other natural gums frequently used as thickeners, stabilizers and emulsifiers. The viscosity of the hemicellulose is shown in Figure 2. Below 8-10% concentration the viscous characteristics are most like gum arabic, but above 10% concentration they are more like gum karaga.

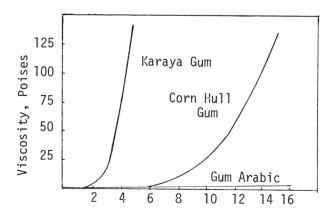

Figure 2. Viscosity versus concentration of corn hull gum, gum arabic, and gum karaya. Brookfield viscometer, No. 4 spindle, 60 r.p.m.

D-GLUCO-D-MANNANS

D-Gluco-D-mannans comprise from 3-5% of the wood of angiosperms and from 3-12% of the wood of gymnosperms where they occur in the cell wall in close association with cellulose and D-xylans. Similar polysaccharides are also found in the tubers of *Amorphophallus* species and in the bulbs and seeds of several plants where they may function as food reserve material. Their removal from the plant material is effected by extraction with potassium hydroxide solution, usually in the presence of borate (5) which complexes with the 2,3-cis-hydroxyl groups rendering the polysaccharide more soluble. Isolation is accomplished by precipitation as the barium or copper complex.

Hardwood D-gluco-D-mannans are pure copolymers which upon hydrolysis yield D-glucose and D-mannose in the approximate ratio of 1:2, respectively. The D-gluco-D-mannans of softwood contain a higher proportion of D-mannose (1:3) and, in addition, contain small amounts of D-galactose residues. Fragmentation and methylation analysis show both hardwood and softwood D-gluco-D-mannans (6-9) to consist of linear chains of (1 → 4)-β-D-linked sugar units. A random, rather than regular, arrangement of sugar units is shown by the isolation following partial hydrolysis of oligomers containing adjacent D-glucose residues and up to six adjacently linked D-mannose residues. The D-galactose units present in softwood D-gluco-D-mannans (10) are linked as single side chain units at C-6 positions of D-mannopyranosyl residues of the main chain and there is evidence for additional branching (11,12) at the C-3 position of some D-glucopyranosyl units.

Although insoluble in water in the native state, the D-gluco-D-mannans frequently show water solubility after isolation. Water solubility tends to increase with increasing D-galactose content, the presence of the side chain units presumably preventing association of the linear chain segments. Solubility in methyl sulfoxide is also observed and this solvent is valuable as an extractant particularly when chemical modification of the native polysaccharide is to be avoided.

D-GALACTO-D-GLUCO-D-MANNANS

This group of polysaccharides was first identified as a constituent of the alkaline extract of sulfite pulp from western hemlock (*Tsuga heterophylla*) wood and has since been found in many varieties of softwoods. Like the related D-gluco-D-mannans they are not water extractable but are removed, usually following chlorite treatment of the plant material, by extraction with potassium hydroxide solution or methyl sulfoxide. Isolation is effected by precipitation as the barium complex.

The three sugars making up this polysaccharide, D-galactose, D-glucose, and D-mannose, are present in the ratio of 1:1:3, the higher proportion of D-galactose serving to differentiate this group from the D-gluco-D-mannans. The basic structural features (13-16) are similar, however, with a main chain consisting of $(1 \rightarrow 4)$-linked β-D-glucose and β-D-mannose residues to which is attached single side chain units of D-galactose at the C-6 positions of both D-galactose and D-mannose. A random arrangement of main chain monomer units is indicated by oligomers obtained following partial hydrolysis.

L-ARABINO-D-GALACTANS

Present in varying amounts in the wood of the Coniferales are a group of water soluble, highly branched polysaccharides (17) composed of L-arabinose and D-galactose.

The polysaccharides are especially abundant in the genus *Larix* (larches) where they may constitute up to 25% of the wood. Similar L-arabino-D-galactans are also present in the sap of the sugar maple.

The L-arabino-D-galactans are exceptional among the hemicelluloses in being water extractable from untreated wood. Extraction is usually carried out from defatted wood with hot water following which the polysaccharide can be isolated in relatively pure, unaltered form by precipitation with alcohol. The interesting properties of this polysaccharide along with its ease of extraction and isolation has led to commercial production and utilization.

Hydrolysis of L-arabino-D-galactans of larch woods yields L-arabinose and D-galactose in the ratio of 1:4 to 1:8 with the usual ratio being near 1:6. Information from classic fragmentation analysis (18,19) show larch wood L-arabino-D-galactan to be composed of a main chain of (1 → 3)-linked β-D-galactopyranose units each of which bears a substituent at the C-6 position. The majority of such side chain substituents appear to be made up of two (1 → 6)-linked β-D-galactopyranosyl residues but with some composed of one or three D-galactopyranosyl units. Other side chains are composed of single L-arabinofuranose units or of 3-O-β-L-arabino-pyranosyl-L-arabinofuranose. Small amounts of D-glucuronic acid residues are also present in L-arabino-D-galactans from tamarack and European larch (20).

It has been suggested that the arabinogalactans from all larch species are similar and that observed differences might be caused by different analytical techniques, as well as varying methods of isolation.

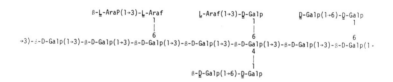

Figure 3. A likely structure of a portion of a molecule of arabinogalactan.

Arabinogalactan, although commercially marketed only recently, has acquired varied uses (Industrial gums). It may be used under the Federal Food, Drug and Cosmetic Act as an emulsifier, stabilizer, binder or bodying agent in essential oils, non-nutritive sweetners, flavor bases, non-standardized dressing, and pudding mixes in the minimum quantity required to product its intended effect.

DEVELOPMENTS IN THE USE OF HEMICELLULOSES 347

The easy availability of this polysaccahride provide a special opportunity for it to find a significant place in industry.

APPLICATIONS AND EXPECTED USES

Hemicellulose in Paper

A great wealth of data fully substantiates the value of hemicelluloses in their action to improve paper properties. Unfortunately, the principal wood pulping processes in commercial practice today have the undesirable feature of removing and degrading most of the hemicelluloses present in the wood. While designed to delignify wood the pulping liquors extensively dissolve and degrade the hemicelluloses. As time goes on, there is no doubt that new pulping methods will evolve that will leave the hemicellulose undissolved and undegraded. Such processes will add to pulp yield, lead to improved paper and make hemicellulose available for broad industrial applications. One pulping process is available today in which the cellulose and hemicelluloses are left intact in their original morphological relation. Such a pulp, termed holocellulose, is usually prepared by the action of chlorine dioxide, an oxidant that attacks and solubilizes lignin, making it easily removable, but does not degrade the polysaccharides. Chlorine dioxide is too expensive for large scale commercial use. None-the-less, other processes will surely evolve. It is conceivable that the new solvents for cellulose, now recommended as economically feasible for rayon production, may show feasibility under appropriate conditions.

It has been known for a long time that well-fibrillated and swollen fibers are needed for the manufacture of high strength paper. Addition of substances to enhance fiber bonding, such as the natural gums and starches has been practiced from the beginning of papermaking. The importance of hemicelluloses during fiber preparation was recognized by Schrolbe in 1927. Since then, it became well recognized that the quality of paper is dependent on the amount and quality of hemicellulose attached to the fibers.

Hemicelluloses with the greatest effect on the strength of paper may be those with xylan backbones. In general, hemicelluloses improve breaking stress, modules of elasticity, yield point stress and work-to-rupture of fibers and bursting strength, tensile strength and fold endurance (21-25) of paper. Holocellulose with a 20% hemicellulose content produces a paper with a very high bursting strength (25). Pulp, from birch wood, containing 8% hemicellulose has shown a 20-30%

increase in bursting strength and a 3-fold increase in breaking strength (26).

FURFURAL

It is well known that steam distillation of pentose sugars in the presence of acids such as 12% hydrochloric or sulfuric acid produces a high yield of furfuraldehyde. In fact, the process was once the standard analytic process for quantitative estimation of pentose sugars, and mainly for those pentose sugars occuring in the polysaccharides of plant material. Expanded to a commercial scale the process produced large tonnage of furfural from hemicelluloses in oat hulls and in corn cobs. Nylon was originally produced commercially from furfural starting material. Although commercial furfural production from plant material has almost ceased in the United States, a large plant has just gone into production in Kenya and revival of furfural as a source of competitive organic chemicals can be anticipated as petrochemical prices esculate. Competitive yields of furfural can only be obtained from plant material rich in polysaccharides of the xylan-type. Fortunately, these xylan-type hemicellulose are the principal hemicelluloses of annual plants and therefore are abundantly available in agricultural residues such as corn stalks, corn cobs, wheat straw and soybean hulls. Hardwoods contain less hemicellulose than annual plants but hardwood residues are a potential source of furfural (27,28).

XYLOSE-XYLITOL

Xylan-type hemicellulose predominates in annual crops and hardwoods such as birch wood, making these raw materials sources for D-xylose sugar or for products derived from D-xylose of which the most important today is D-xylitol. D-Xylitol has attracted industrial attention because its sweetness is equivalent to sucrose and because xylitol is not cariogenic. Xylitol has a high endothermic heat of solution. Chewing gum with it as sweetner feels cool on initial chewing. Unfortunately, xylitol is completely metabolized so that it is not useful as a low caloric sweetner replacement. It is considered as a conditional food additive by the Food and Drug Administration. Xylitol is made in Finland from birch wood chips by hydrolysis of the hemicellulose, crystallization of xylose and hydrogenation. It is marketed by the Finnish Sugar Company of Helsenki. Xylitol has been tested in a variety of food products (29). Xylitol is a recognized normal constituent of the glucuronate-xylulose cycle (30), has no insulin requirement and has been used for intravenous infusion. However, investigators in South Australia find that its use tends to induce crystals of calcium oxalate in body tissue and more work has been recommended before xylitol is further applied as an intraveneous nutrient (32).

D-Xylose is used to the extent of about 400 tons per year in the United States as an ingredient in media to increase the production of isomerase, used for the commercial production of high fructose syrup from D-glucose.

A potential source of D-xylose in the United States is plant residues, especially corn cobs. When corn cobs are hydrolyzed with dilute acid such as 0.1-0.2% sulfurous acid (33,34,35) they produce a good yield of D-xylose, not contaminated appreciably with D-glucose from cellulose hydrolysis. Pretreatment of the cobs with water at 140° for 90 minutes removes ash, soluble sugars and reduces the protein to low levels. Such removal of contaminant increases the yield of crystalline D-xylose. In another acid hydrolysis (36), a yield of 15% D-xylose was obtained with a purity of more than 94%. Apricot shells (37) have also been found to give a high yield of D-xylose. Xylonase has been recommended (38,39) as superior catalysts for hydrolysis.

Xylo-hemicelluloses have been examined in various clinical and biochemical applications. A variety of hemicelluloses, but especially wheat straw hemicellulose has shown antitumor action (40-43).

The polysulfate esters (44,45) have antiplasmin activity (45), antithrombic effects (46,47), anti-inflammatory action (48-52), decreased histamine (52,53), inhibited hyaluronidase (54), and reduced serum lipids (55-62). Phosphates have also been prepared (63).

Carboxymethylxylohemicelluloses have been made (64,65) and show some of the properties of CMC with potential application in detergent building. These derivatives may also be useful for dispersing inorganic or organic pigments (66) or for coating paper (67). Michael addition with acrylonitrile(68) produces derivatives suitable for thickeners of organic solvents. Acetates (69-71) butyrates (72) and benzoates (72) of xylohemicelluloses have been made on numerous occasions as well as have the higher fatty acid exters. Flocculents and adhesives (73) are prepared by reacting hemicellulose with epichlorohydrin. Dithiobisthioformates (75) have been suggested for rubber reinforcements and as chelating agents. Nitrite esters (76) have been made and used to solubilize hemicellulose in dimethylforamide with recovery of the hemicellulose by mild acid denitrition.

Xanthates (77-82) have been made and examined. Hemicellulose xanthates in viscose have poor stability but though they do not detract from viscose their presence is undesirable from a handling point of view.

Hemicellulose have been grafted with acrylate (83-85) to form interesting products.

The $\underline{O}$-acetyl derivatives have interesting gum properties (86) and form cellophane-like films. Birch $\underline{O}$-acetylhemicellulose with 9% acetyl content forms a film with a tensile strength of 590 kg/cm^2 while the arabinoglucuronoxylan of white pine forms a film with a tensile strength of 885 kg/cm^2.

HEMICELLULOSE IN BAKING

The presence of hemicelluloses in wheat flours improves the water binding capacity (87), mixing quality (88,89), reduces the energy requirement for dough mixing (88), and in the incorporation of added protein (89,90), and improves leaf volume (91).

All wheat flours contain water soluble and water insoluble hemicelluloses (92-95). They are often given the old fashioned and incorrect name of pentosans. The commonest and perhaps most abundant water soluble hemicellulose is an arabinoxylan consisting of a linear xylan with $\underline{L}$-arabinofuranosyl units attached principally at C-3 position. Some soluble hemicelluloses have galactose and $\underline{D}$-glucuronic acid units. Other polysaccharides are present such as glucofructan extracted from *Triticum durum* (96). A large part, perhaps as much as 75% of the hemicelluloses in soft and hard wheat flour are insoluble in water but are extractable in alkali. These may be crosslinked or joined to other flour components perhaps by ferulic ester linkages (97,98).

An interesting observation that requires further examination stems from the finding (99) that addition of plant hemicellulose to wheat flour results in the bread remaining fresh, some three times as long as bread baked without the added hemicellulose. While the loaf volume is increased substantially by the hemicellulose, the action of the hemicellulose on protein or more likely starch would be the cause of the reduced staling.

COMPONENTS OF DIETARY FIBER

The recent interest in dietary fiber is largely a result of observations of Dennis Burkitt and Hugh Trowell (100,101). They were impressed that rural Africans, whose diets were composed principally of whole corn, sorghum or millet were free from numerous intestinal disorders that plague people in more advanced societies. They attributed the beneficial effects to the whole grain diet contributing a high level of dietary fiber. As a consequence, it is now common to add fiber to breads and other foods, such as breakfast cereal. Wheat brand and refined cellulose are principally used with another likely source being the fibrous parts of the corn kernel (102). By the fiber method of Van Soest (103) it is 97% fiber or neutral detergent residue. The latter consists of 22% cellulose and 77% hemicellulose. Like the hemicellulose in wheat endosperm cell walls and bran, the hemicellulose in corn fiber may be crosslinked by ester linkages and perhaps diferulic acid (104). Although ferulic acid has not been reported in pericarp cell walls, such a crosslink would account for the easy removal of hemicellulose from corn fiber by lime water. Corn fiber absorbs about 4 times its weight of water. Water absorbtion is influenced, in part, by the hydrophylic nature of the hemicellulose.

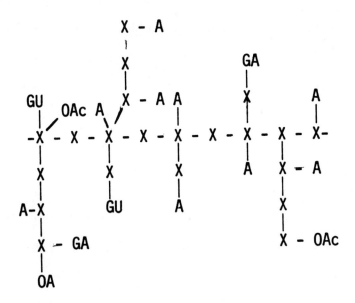

Figure 4. One possible structure (105-107) of hemicellulose of corn fiber, A is arabinosyl, Ac is acetyl, GA is galactosyl, GU is glucuronosyl, X is xylosyl.

Dietary fiber is not digested in the small intestine but is attack by the microflora of the large intestine with the hemicelluloses being digested 40-60% and perhaps the cellulose degraded and digested to a maximum of 10% although higher digestion values may be found. There is need for more accurate determination of hemicellulose and cellulose digestibility. Dietary fiber binds inorganic molecules, or ions, probably in the most part by way of reaction with the hemicellulose. Absorption of bile acids and cholesterol by dietary fiber has created interest with respect to a role in serum cholesterol control (108-111,112).

REFERENCES

1. Bouveng, H. O., Garegg, P. J., and Lindberg, B., Acta Chem. Scand. 14, 742 (1960).
2. Bouveng, H. O., Acta Chem. Scand. 15, 87 (1961), Acta Chem. Scand. 15, 96 (1961).
3. Percival, E. G. V., and Chanda, S. K., Nature, 166, 787 (1950).
4. Mackie, I. M., and Percival, Elizabeth, J. Chem. Soc. 1151 (1959).
5. Timell, T. E., Svensk Papperstidn. 63, 472 (1960); Tappi, 43, 844 (1960).
6. Timell, T. E., Adv. Carbohydr. Chem. 19, 247 (1964); Adv. Carbohydr. Chem. 20, 409 (1965).
7. Jabbar, Mian A., and Timell, T. E., Svensk. Papperstidn. 63, 884 (1960).
8. Jones, J. K. N., Merler, E., and Wise, L. E., Can. J. Chem. 35, 634 (1957).
9. Aspinall, G. O., Begbie, R., and McKay, J. E., J. Chem. Soc., 214 (1962).
10. Koolman, P., and Adams, G. A., Can. J. Chem., 39, 886 (1961).
11. Meier, H., Acta Chem. Scand. 12, 1911 (1958).
12. Croon, I., and Lindberg, B., Acta Chem. Scand. 12, 453 (1958).
13. Schwarz, E. C. A., Timell, T. E., Can. J. Chem. 41, 1381 (1963).
14. Hamilton, J. K., Partlow, E. V. and Thompson, N. S., J. Am. Chem. Soc., 82, 451 (1960).
15. Mills, A. R., and Timell, T. E., Can. J. Chem. 41, 1389 (1963).
16. Meier, H., Acta. Chem. Scand. 14, 749 (1960).
17. Adams, M. F., and Douglas, C., Tappi, 46, 544 (1963).
18. Bouveng, H. O., and Lindberg, B., Acta. Chem. Scand. 10, 1515 (1956).

19. Bouveng, H. O., Acta Chem. Scand. 13, 1869 (1959); Acta Chem. Scand. 13, 1877 (1959).
20. Urbas, B., Bishop, C. T., and Adams, G. A., Can. J. Chem. 41, 1522 (1963).
21. Spiegelberg, Harry L., Tappi, 49 (9), 388 (1966).
22. Fineman, M. N., Tappi, 35 (7), 320 (1952).
23. Monograph No. 6, "Nature of the Chemical Components in Wood", T.A.P.P.I., New York, p. 92 (1948).
24. Swanson, J. W., Tappi, 33 (9), 451 (1951).
25. Andreev, V. I., and Zimina, E. S., Sb. Tr., Tsent. Navch-Issled. Inst. Bum., No. 6, 147 (1971).
26. Filatenkov, V. F., Bum. Prom. 5, 4 (1974).
27. Eisner, K., Drevo, 21(3), 79 (1966).
28. Konovalov, V. K., Savinykh, A. G., Potopalskaya, A. A., and Oleinikova, V. E., Khim. Pererab. Drev. No. 16, 12 (1967).
29. Russo, J. R., Food Eng. 48(4), 77 (1976).
30. Touster, O., Fed. Proc. 19, 977 (1960).
31. Kaller, U., and Froesch, E. R., Diabetologia 7, 349 (1971).
32. Thomas, D. W., Edwards, J. B., and Edwards, R. G., Sugars Nutr., (Pap. Int. Conf.) 567-590 (1974) Ed. Sipple, H. L. and McNutt, K. W.
33. Dunning, J. W., and Lathrop, E. C., U.S.Pat. 2,450,586 (1948).
34. Yoshi, Kamiyama, Yasuhiko, Hirabayashi, Yoshio, Sakai and Tatsuyoshi, Kobayashi, Hakko. Kogaku Zasshi 52, (9), 669 (1974).
35. Isao, Kusakabe, Tsuneo, Yasui, and Tatsuyoshi, Kobayashi, Hakko Kogako Zasshi 53 (3), 135 (1975).
36. Renato, Maspoli, Ger. Offen. 2,413,306 (1974).
37. Hermann, Friese, Fr. pat. 2,070,377 (1971).
38. Jeffreys, Roy A., and Bryan, E., Brit. Pat. 928,591 (1963).
39. Yoshi, Kamiyama, Yasuhiko, Hirabayashi, Yoshio, Sakai and Tatsuyoshi, Kobayashi, Hakko Kogaku Zasshi, 52(9) 669 (1974).
40. Yoshio, Sugihara and Fumio, Araki, Csch. Med. Sci., 21(4) 115 (1972).
41. Takeo, Hashimoto, Japan Kokai, 75,100,100 (1975).
42. Barry, Vincent C., Conalty, M. L., McCromick, Joan E., McElhinney, R. S., and O'Sullivan, J. F., Proc. Roy. Irish Acad., Sect. B 64 (18), 335 (1966).
43. Whistler, Roy L., Bushway, Afred A., Singh, Prem P., Nakahara, Waro, and Tokuzen, Reiko, Adv. Carbohyd. Chem. Biochem. 32, 235 (1976).
44. Masakiyo, Sakurai, Yoshimasa, Fujita, and Masao, Nomoto (Seikagaku Kogyo Co., Ltd.) Japan 7,343,100 (1973).
45. Coccheri, S., Pretolani, E., and Gennari, P., Boll. Soc. Ital. Biol. Sper. 39, 184 (1963).

46. Raynaud, Robert, Brochier, Mireille, Griguer, Paul, and Raynaud, Philippe, Therapie, 20(5), 1259 (1965).
47. Henk, W., and Neboisis, G., Wien, Klin. Wochenschr. 80(10) 191 (1968).
48. Mauricio, Rocha e Silva, Cavalcanti, Rejanne Q., and Reis, Marina L., Biochem. Pharmacol. 18(6), 1285 (1969).
49. Kalbhen, Dieter A., Clausen, H., and Domenjoz, R., Arzneim-Forsch. 20(12), 1872 (1970).
50. Giroud, J. P., and Timsit, J., Therapie, 28(5), 889 (1973).
51. Kalbhen, D. A., Pharmacology 9(2), 74 (1973).
52. Kalbhen, D. A., and VanHeek, H. J., Arzneim-Forsch 21(10), 1608 (1971).
53. Mordelet-Dambrine, Madeleine and Parrot, Jean L., J. Pharmacol. 4(3), 317 (1973).
54. Kalbhen, D. A., Karzel, Karlfried, Dinnendahl, V., and Domenjoz, R., Arzneim-Forsch. 20(10), 1479 (1970).
55. Deysson, Guy, Compt. Rend. Soc. Biol. 159(1), 125 (1965).
56. Raveux, Roger, Ravold, J., Segal, V., and Brunaud, M., Progr. Biochem. Pharmacol. 2, 451 (1967).
57. Brunaud, Marcel, Segal, V., Ravold, J., and Raveux, R., Progr. Biochem. Pharmacol. 3, 393 (1967).
58. Raynoud, Robert, Brochier, Mireille, Griguer, Paul, and Raynoud, Philippe, Therapie 20(5), 1249 (1965).
59. Henk, W., and Nebosis, G., Wien. Klin. Wochenschr. 80(10), 191 (1968).
60. Liberti, R., Cicconetti, C. A., and Gentili, G., Riforma Med. 87(7), 260 (1973).
61. Mieko, Hashi, and Takayasu, Takeshita, Mokuzai Gukkaishi 19(2), 101 (1973).
62. Mieko, Hashi, and Takayasu, Takeshita, Agr. Biol. Chem. 39(3), 579 (1975).
63. Dudkin, M. S., Shakantova, N. G., and Lemle, N. A., Zh. Prikl. Khim. 47(10), 2319 (1974).
64. Schmorak, J., and Adams, G. A., Tappi 40, 378 (1957).
65. Dudkin, M. S., Kogan, E. A., and Grinshpun, S. I., Latvijas PSR Zinatnu Akad. Vistis; Khim Ser. 5, 633 (1964).
66. Naoterv, Ohtani, Takashi, Takeuchi, and Tsutomu, Koyama Ger. Offen. 2,523,161 (1975).
67. Kruger, Emilio, U.S. Pat. 3,540,480 (1971).
68. Nordgren, Robert, Ger. Offen. 2,064,810 (1971).
69. Millett, M. A., and Stamm, A. J., J. Phys. and Colloid. Chem. 51, 134 (1947).
70. Hibbert, H., Can. J. Research 15B, 490 (1937).

71. Carson, J. F. and Maclay, W. D., J. Am. Chem. Soc. 68, 1015 (1946); J. Am. Chem. Soc. 70, 293 (1948).
72. Husemann, E., J. Prakt. Chem. 155, 13 (1940).
73. Pulkkinen, Erkki, J., Reintjes, Marten, and Starr, Laurence D., U.S. Pat. 3,833,527 (1974).
74. Kunz, Frederick L. G., Can. Pat. 947,280 (1970).
75. Trimnell, Donald, Shasha, Baruch S., and Doane, William M., U.S. Publ. Pat. 391,184 (1975).
76. Schweiger, Richard G., J. Org. Chem. 41(1), 90 (1976).
77. Heuser, E., and Schorsch, G., Cellulosechemie 9 (93), 109 (1928).
78. Burkart, Phillip, Horst, Dautzenberg, and Willi, Schmiga, Faserforsch. Textiltech. 20(12), 537 (1969).
79. Croon, I., and Donetzhuber, A., Tappi, 46(11), 648 (1963).
80. Matthes, A., Faserforschung und Textiltechn. 3, 128 (1952).
81. Philippi, B., Faserforschung und Textiltechn. 8, 21 (1957).
82. Jerzy, Skoracki, and Henryka, Drozdz, Polimery 18(1), 31 (1973).
83. Church, John A., J. Polymer Sci., part A-1, 5(12), 3183 (1967).
84. O'Malley, James J., and Marchessault, R. H., J. Phys. Chem., 70(10), 3235 (1966).
85. Garbuz, N. I., Mikhailov, G. S., Zhabankov, R. G., and Livshits, R. M., Vysokomol. Soedin., Ser. B., 11(7), 533 (1969).
86. Cafferty, P. D., Glaudemans, C. P. J., Coalson, R., and Marchessault, R. H., Svensk. Papperstidn. 67(12), 845 (1964).
87. D'appolonia, B. L., and MacArthur, L. A., Amer. Assoc. Cereal Chem. 52, 230 (1975).
88. Jelaca, S. L., and Hlynka, I., Amer. Assoc. Cereal Chem. 48, 211 (1971).
89. Jelaca, S. L. and Hlynka, I., Amer. Assoc. Cereal Chem. 49, 489 (1972).
90. D'appolonia, B. L., and Gilles, K. A., Amer. Assoc. Cereal Chem. 48, 427 (1971).
91. D'appolonia, B. L., Gilles, K. A., and Medcalf, D. G., Cereal Chem. 47, 194 (1970).
92. See Review: Bechtel, W. G., Gedds, W. F., and Gilles, K. A., "Carbohydrates in Wheat Chemistry and Technology", Ed. Hlynka, I., Am. Assoc. Cereal Chem. monograph III, p. 277-352 (1964).
93. Cole, E. W., Cereal Chem. 46, 382 (1969).
94. Perlin, A. S., and Suzuki, S., Cereal Chem. 42, 199 (1965).
95. Lintas, C., and D'appolonia, B. L., Amer. Assoc. Cereal Chem. 49, 731 (1972).
96. Medcalf, D. G., and Cheung, P. W., Amer. Assoc. Cereal Chem. 48, 1 (1971).

97. Geissmann, T., and Neukom, H., Amer. Assoc. Cereal Chem. 50, 414 (1973).
98. Neukom, H., Geissmann, T., and Painter, T. J., Bakers Digest, 41, 52 (1967).
99. Casier, Joris, Ger. Offen. 2,047,504 (1971).
100. Burkitt, D. P., Walker, R. P. and Painter, N. S., J. Am. Med. Assoc. 229, 1068 (1974).
101. Trowell, H., Am. J. Clin. Nutr. 29, 41 (1976).
102. Walson, S. A., "Structures, Composition and Fermentability of the Fibrous Components of Corn Kernel", 61th Ann. Mtg., Am. Assoc. Cereal Chem., Oct. 1974.
103. Van soest, P.J., and McQueen, R. W., Proc. Nutr., 32, 123 (1973).
104. Markwalder, H. U., and Neukom, H., Phytochem. 15(5), 836 (1975).
105. Whistler, R. L. and BeMiller, J. N., J. Am. Chem. Soc. 78, 1163 (1956).
106. Whistler, R. L., and Corbett, W. M., J. Am. Chem. Soc. 77, 6328 (1955).
107. Feather, M. S., and Whistler, R. L., Arch. Biochem. Biophys. 98, 111 (1962).
108. Birkner, H. J. and Kern, F., Jr., Gastroenter 67, 237 (1974).
109. Balmer, Joanne and Zilversmit, D. B., J. Nutr. 104, 1319 (1974).
110. Story, Jon A., and Kritchevsky, David, J. Nutr. 106, 1292 (1976).
111. Raymond, T., Connor, W. E., Tin, O., and Connor, S., Circulation 54, 550 (1976).
112. Hashi, Mieko and Takeshita, Takayasu, Agr. Biol. Chem. 39, 579 (1975).

Index

A

Acetylation of
  cellulose, 5–7, 107, 151, 285–287
  cotton, 15, 16, 289
  paper sheets, 8
  pulp fiber, 8
Alkali cellulose, 25, 26, 29, 30, 33
Ammonia, liquid, 303–318, 324
Anhydroglucose, 4, 16, 17, 28, 137, 142, 287, 298
Annealing, cellulose, 323, 327–329
Antibacterial agents, 260–266

B

Bacteriostats, 260–262, 270, 278
Benzoin ethyl ether, 206, 208, 210, 214, 219, 220
Bioassay, fish, 244, 254
BOD (biological oxygen demand, 241, 243, 244, 248, 249, 255

C

Carboxyl groups, 295, 325, 327, 332, 334, 341
Cellobiose, 287, 288
Cellulase, 285, 287, 291, 296, 298
Cellulose,
  accessibility, determination of, 117, 136, 137, 142, 349
  acetylation of, *see* Acetylation of cellulose
  acid hydrolyzed, 119, 120, 127, 162
  amorphous fraction, 118, 137, 138, 329
  ball-milled, 127, 128, 130, 131, 136
  bromine accessibility, 141, 142
  char from, 331, 333, 337
  crystallinity index for, 117, 131, 130, 133

crystallinity, determination of, 117–143, 329
degree of cystallinity of, 3, 6, 126, 128, 130, 136, 143, 310, 322
degree of substitution of, 3, 7, 8, 16, 28, 35, 135, 136, 286, 288, 289, 293, 295, 297
esterification, 3, 4, 11, 14, 15, 107, 289
etherification, 3, 11, 14, 16, 28, 75, 110–112, 149, 151, 234, 285–289
films derived from, 6, 23, 26, 28, 60, 65, 74, 75, 109, 135
gauze from, 15
infrared spectrum, 117, 133
iodine sorption on, 117–119, 141, 142
molecular weight, 149, 289–291
morphology, 25, 26, 32, 304
oxidation of, 3, 11, 14, 15, 121, 288, 295, 296
paracrystalline, 129, 130
penetration of, 148
plastics derived from, 6, 26, 35, 60, 65, 74, 75, 96–99, 109
reactivity of, 14, 147–165
reagents, 148–151, 164
solvents for, 4, 27, 106, 109, 347
unit cell of, 123–124
x-ray analysis of, 117–132, 156, 305–314, 325–328
Cellulose I, 122, 124, 126, 127, 130, 133, 134, 152, 157, 160, 306–314, 321, 325–339
Cellulose II, 122, 123, 125–127, 130, 133, 134, 152, 159, 160, 304, 306–314, 321, 323, 325–339
Cellulose III, 304, 306–314, 321, 324–339
Cellulose IV, 304, 321, 324–339
Cellulose acetate
  fibers, 7, 100, 107, 112, 265
  properties, 6, 8, 75, 139, 143
  reactivity, 4, 104, 107, 135
  solubility, 4, 5

357

358 INDEX

Cellulose acetate (*continued*)
  solvents, 4, 5
  uses, 6, 32, 34, 95, 96, 99, 104, 112
Cellulose derivatives
  acetate, 109, 137, 285, 286, *see also* above
  carboxymethyl-, 16, 29, 60, 82, 96, 110, 111, 285, 289–295
  cuprammonium, 34, 126, 322
  cyano-, 29
  dithiocarboxylic esters, 235
  ethyl-, 28, 60, 110
  fatty acids with long chain, 293
  hydroxyethyl-, 29, 30, 82, 87, 110, 285, 286, 289–291
  hydroxypropyl-, 31, 110
  mercerized, *see* Mercerized cellulose
  methyl-, 28, 29, 60, 110, 136, 285, 286, 289, 291
  nitrate, 26, 27, 75, 109, 110, 289
  sulfate, 27
  tetrahydropyranyl derivatives, 285–291
  vinyl ethers, 285–289
  xanthate, 7, 33, 109, 280
Chlorine dioxide, 347, 255, 347
COD (chemical oxygen demand), 243, 244, 248, 249
Copper number, 325, 327, 334
Cotton
  consumption, 39, 50, 53
  geographic distribution of, 43
  processing, 39, 47–49
  production, 39, 40, 44–47, 87
  reactions with, 7, 15, 296, 303, 322
  tagging accessible regions in, 162
  technological prospects for, 39, 60–62
  users, 50–52
Cotton liners
  output and use, 55–57, 75, 97, 98, 110
  processing, 57–60
Crosslinking, 3, 17, 31, 32, 82, 90, 163, 164, 289, 291–296, 350, 351
Crystallite size, 123, 135, 309, 336

D

Declourization, of mill effluent, 242–255
Decrystallized cellulose, 133, 148, 152, 154, 155
Deuterated cellulose, 117, 119, 121, 137–139
Differential thermal analysis, 325, 330, 332, 335
Dissolving pulp
  composition of, 75, 98

  demand for, 71–74, 95
  processes, 74–75, *see also* Rayon, regeneration processes for
  production, 71–77, 98
  yield, 13
Dyeing, 14, 50, 61, 62, 88, 107, 317
Dyes, 90, 260, 317

E

Elementary fibrils, 13, 156, 157, 159, 163, 164
Endothermic decomposition, 322, 329
Enzymatic treatment, 62, 285, 288, 290–292, 298
Exothermic decomposition, 322

F

Fabric
  ammonia treatment, effect of, 316, 317
  flammability, 11, 18, 26, 87, *see also* Flame retardants
  moisture absorbency, 14, 135
  shrink resistant, 17
  treatments, 14, 50
  water repellent, 11, 16, 17, 51
  wrinkle resistant, 7, 17, 18, 32
Fibers
  cellulose acetate, *see* Cellulose acetate
  corn, 99, 343, 351
  cotton, 7, 12, 13, 23, 39–42, 73, 81, 88, 92, 102–105, 117, 121, 127, 131, 139, 143, 276, 325, *see also* Cotton
  crimped, 35, 90, 91, 107
  dietary, 350–352
Fortisan, 7, 127, 139, 161
  man-made, 54, 95, 97, 99, 100, 107
  mechanical properties of, 7, 8
  price index for, 108
  prima, 90–92
  pulp, 8, 241, 265, *see also* Pulp
  ramie, 129, 131, 143, 160, 322
  rayon, 73, 81, 95, 98–106, 112, 129, 131, 143, 309, *see also* Rayon
  recycling of, 253–255
  silk, 101
  textile, 39, 57, 96–108
  viloft, 88–89
  vulcanized, 35
  water absorbent, 8, 87, 89, 285, 294, 295, 298
  water repellent, *see* Fabrics, water repellent
  water retention of, 89, 246, 285, 294, 295, 298, 299

INDEX 359

Fibers, antibacterial, 259–279
 aesthetic uses, 267, 272–275, 277
 durability to laundering, 261, 262, 266, 267, 278
 durability to dry cleaning, 267, 278
 federal regulations, 278, 279
 for apparel, 272, 275, 277
 for paper products, 261, 264, 266–269, 272, 278
 for textiles, 261, 263, 269, 272–274, 277, 278
 hygienic uses, 262, 266, 269, 272–275, 277
 marketability, 276, 278
 medical uses, 261, 263, 264, 271–275, 277, 278
 moisture transport with, 276
 selective toxicity of, 276
Fibers, antimicrobial activity
 controlled release mechanism of, 268–269
 permanent barrier mechanism for, 269
 test methods for, 269–271
 with dermtophytic fungi, 265, 269, 272, 275
  bacteria, 265, 269–275
  parasitic worms, 275
  protozoa, 275
  yeasts, 265, 269, 275
Fibers, antimicrobial agents for, 265
Finishing agents, 15–19, 259–267
Finishing of fibers, 11, 61, 84, 303
Finishing techniques
 insolubilization, 14, 267
 ion-exchange, 267
 metastable bonds, 267, 268
 polymerization, 14, 268
 regeneration principle of, 268
 with thermosetting agents, 267
Flame retardants, 18, 51, 82, 87, 107, 142, 262, 311, *see also* Fabrics, flammability of
Forest land area, 66
Fungistats, 260
Furfural, 348

G

Gamma irradiation, 292–298, *see also* Graft copolymerization, initiation with gamma rays
Gel permeation chromatography, 279, 285, 290, 297
Germicidal activity, 261, 262
Glass transition temperature (Tg), 318, 332, 334
Graft copolymers, 31, 260, 262, 268, 296, 297, 350

Graft copolymerization
 by electron transfer, 171, 172
 by the xanthate method, 227–239
 molecular weights, 229
 with mixed solvents, 200–203, 215, 216
Graft copolymerization, initiation with
 ceric ions, 229, 234–236
 dyes, 198
 ferric ions, 198
 gamma rays, 197, 198, 200, 220–223, 285
 manganese complexes, 108, 171, 178–193
 peracetic acid–hydrogen peroxide, 171–178, 193
 ultraviolet light, 171, 197–223
Graft copolymerization, polymer substrates
 amylose, 178
 cellophane, 176
 cotton, 14, 259
 cotton linters, 190, 193
 ferrated cellulose xanthates, 172, 227
 nylon, 176, 178, 267, 272
 polyethylene terephthalate, 176, 178
 pulp, 190, 191, 227, 241
 wood pulps, 171, 193, 227
Graft copolymerization, vinyl monomers
 acrylamide, 182, 184, 218, 227, 230–233, 235
 acrylonitrile, 31, 171, 182, 185, 186, 189–193, 227, 228, 230–232, 234, 235
 amino, 239, 255
 2-dimethyl aminoethyl methacrylate, 241–243, 264
 ethyl acrylate, 227, 230–233, 235–238
 methyl methacrylate, 171, 173–178, 182, 187, 191, 193, 199, 227, 230–232, 235–236
 methacrylic acid, 227, 230–232, 235, 237–239
 n-butyl acrylate, 182, 190, 191, 230–232, 235, 239
 styrene, 197–200, 203–216, 230–232, 235, 237, 238, 295
 vinyl acetate, 227, 230–232, 235, 237–238
 4-vinyl pyridine, 173, 176–178, 193
 with monomer mixtures, 236–238
Grafted pulp, mechanical properties, 237
Grafting
 acridine in, 208, 209, 218
 hydroquinone in, 209
 organic acids in, 204–207
 thiourea in, 208, 209
 with β-propiolactone, 82
Grafting, nonwetting solvents in
 benzene, 203, 206–208, 215, 220
 hexane, 203, 206–208, 215

Grafting, wetting solvents in
  acetic acid, 204–207
  acetone, 198, 204–206, 215, 229, 235
  butanol, 200–202, 215, 218, 221
  dimethyl formamide, 204–206, 215
  dimethyl sulfoxide, 204–206
  dioxan, 198, 204–206, 215
  ethanol, 200, 201
  methanol, 198–218, 297
  octanol, 200–203, 218
  n-propanol, 200
Grafting parameters, 109, 176, 229

H

Hemacetal groups, 329
Hemicelluloses
  arabinogalactans, 345–347
  galactoglucomannans, 341, 345
  glucomannans, 76, 99, 341, 344, 345
  in pulps, 74–77, 99, 110
  xylans, 76, 99, 341–343, 347, 348, 350
Hemicelluloses, uses
  as adhesives, 341, 349
    anti-cancer agents, 341
    coagulants, 341
    emulsifiers, 346
    food additives, 341, 346, 351
    hardening agents, 341
    stabilizers, 343, 346
    thickeners, 341, 343, 349
  in baking, 350
  paper, 347
Hemicellulose derivatives, 349, 350
Homopolymerization during grafting, 173, 182, 193, 197, 199, 203, 211–213, 221, 227, 229, 268
Homopolymer removal, 173, 199
Hydrocellulose, 122, 135, 159, 160
Hydrogen bonding, 8, 32, 110, 137, 149, 154–159, 165
Hydroxyl groups
  availability, 5, 16, 30, 82, 137, 147, 153–165, 203, 285, 289, 298
  rate of substitution, 154
  reactivity, 13, 14, 28, 110, 147, 150–153, 287

I

Ion-exchange, 29, 30, 243, 246–248

L

Levoglucosan content
  influence of crystal form, 321–322, 332–334, 337
  influence of fine structure, 321–322, 332–334, 337
Lignin, 74, 76, 171–173, 177, 193, 291, 342

M

Mercerization, 26, 99, 121, 127, 130, 149, 150, 228, 237, 315, 323
Mercerized cellulose, 26, 117, 120, 122, 123, 125, 126, 129, 131, 133–135, 139, 161, 262, 324, 325
Microcrystalline cellulose, 4, 30, 321, 334–337
Modified cellulose, growth forecast, 112

N

Natural gums, 24, 343, 347
Nonwovens, fiber use in, 33, 50, 62, 87, 89, 95, 104, 317
Nylon, 99, 107, 108, 262, 348

P

Paper
  mechanical properties, 8, 347
  reaction with, 8, 24
Photosensitizers, for grafting, 197–199, 202–216
Plasticizers, 7, 26, 314–316
Polyelectrolytes, 242
Polyester, 16, 34, 81, 83, 84, 91, 96, 99, 102–105, 109
Polymorphism, 304
Propylene oxide, 31
Pulp
  composition and properties, 75–78, 267, 293
  dissolving, *see* Dissolving pulp
  fibers, *see* Fibers, pulp
  grafted, *see* Grafted pulp, Graft copolymerization, substrates
  in nonwovens, 105, *see also* Nonwovens
  kraft, 75, 98, 227, 231, 243
  production, 71
  sulfite, 75, 132, 173, 345
  thermomechanical, 285, 291–293
  with amino groups, 248

Pulping process
  chemical byproducts, 76, 77
  prehydrolysis kraft, 74, 98
  spent liquors, 15, 76, 347
  sulfite, 74–76, 98
Pulpmill
  effluents, 241–244, 255
  locations, 72
Pulpwood
  consumption, 71
  demand, 70, 71
  projections, 71
  supply, 70
Pyrolysis
  effect of crystal form, 321, 334–335, 337–339
  effect of fine structure, 321, 334–335, 337–339

R

Radial scavengers, 208–210
Rayon
  definition of, 82
  energy requirements, 105
  flame proof, 87, *see also* Flame retardants, Fabrics flammability
  grafted, 175–174, 262
  Mechanical properties, 83–85, 89–92, 106, 293, 316, 318
  new fibers, 81, 88–92, 103, 104
  regeneration processes for, 106–108, 251, *see also* Regenerated cellulose
  tire cord, 13, 54, 83, 105, 126, 127
Reaction of cellulose
  with ethylene oxide, 152
  -N,N-diethylaciridinium chloride (DAC), 152, 157
Regenerated cellulose, 7, 34, 83, 99, 107, 117, 126, 176

S

Schiffli process, 16
Solvent exchange, 121, 122
Sulfuric acid treatment, 27, 35, 232
Swelling of cellulose, 87, 119–122, 148, 157, 303, 310

T

Textile fibers
  consumption, 101
  uses, 7
  world demand, 86
  world production, 100
Thermal properties
  effect of crystal form, 321, 325–332, 337–339
  effect of fine structure, 321, 325–332, 337–339
Thermogravimetry, 325, 330
Timber inventories, 66
Timber supply, 65, 69
Timberland
  commercial, 66–69
  ownership, 66–69
Tromsdorff effect, 204, 205, 215, 218, 221
Twinning mechanism, 312–314

U, V

Uranyl nitrate, 204, 206, 207, 209–213
Viscose, 7, 34, 81, 87, 88, 98, 103, 105–107, 126, 132, 133, 137, 139, 172, 228, 316, 318, 350

W

Wood, dimensional stabilization of, 9, 32
Wood cellulose, 23, 65
  area by ecosystem, 67
  outlook, 65
  productivity, 69, 71
  projections, 73
  supply, 69–70, 97
Wool, 73, 88, 100, 101, 267
World demand
  cellulosic fibers, 85
  man-made fibers, 97
  textile fibers, 86

X, Z

Xylitol, 348, 349
Zelan process, 16

STAFFORD LIBRARY COLUMBIA
668.44 MOD c.1

Modified cellulosics

3 3891 00028 8327

Dinosauring
Contents

BOOK 1
Fantasy Close to Home

Jumanji
- Vocabulary 1
- Selection Comprehension 3

The Mysterious Girl in the Garden
- Vocabulary 4
- Selection Comprehension 5

Theme Outcome/Supporting Skill
- Making Inferences 6

The Shrinking of Treehorn
- Vocabulary 8
- Selection Comprehension 9

BOOK 2
They Walked the Earth

Tyrannosaurus
- Vocabulary 10
- Selection Comprehension 11

Theme Outcome/Supporting Skill
- Identifying Main Idea and Supporting Details 12

Wild and Woolly Mammoths
- Vocabulary 14
- Selection Comprehension 15

Theme Outcome/Supporting Skill
- Identifying Main Ideas Across Texts 16

Strange Creatures That Really Lived
- Vocabulary 18
- Selection Comprehension 19

BOOK 3
Battle of Wits

The Boy of the Three-Year Nap
- Vocabulary 20
- Selection Comprehension 21

Farmer Schulz's Ducks
- Vocabulary 22
- Selection Comprehension 23

Theme Outcome/Supporting Skill
- Comparing Solutions 24

The Sign in Mendel's Window
- Vocabulary 26
- Selection Comprehension 27

BOOK 4
Laura Ingalls Wilder

Little House on the Prairie
- Vocabulary 28
- Selection Comprehension 29

Farmer Boy
- Vocabulary 30
- Selection Comprehension 32

Theme Outcome/Supporting Skill
- Identifying Characteristics of Historical Fiction 33

On the Banks of Plum Creek
- Vocabulary 35
- Selection Comprehension 36

BOOK 5
The Mystery Hour

Meg Mackintosh and the Case of the Missing Babe Ruth Baseball
- Vocabulary 37
- Monitoring Comprehension 38
- Selection Comprehension 40

Theme Outcome/Supporting Skill
- Noting Details About Clues in a Mystery 41

Paddington Turns Detective
- Vocabulary 43
- Selection Comprehension 44

The Case of the Missing Roller Skates
- Vocabulary 45
- Selection Comprehension 46

BOOK 6

The Dreamers

Walt Disney: Master of Make-Believe
Vocabulary 47
Selection Comprehension 48

Marian Anderson
Vocabulary 49
Selection Comprehension 50

Theme Outcome/Supporting Skill
Understanding Influences on
People's Lives 51

Roberto Clemente
Vocabulary 53
Selection Comprehension 54

BOOK 7

Dear Diary

Teacher's Pet
Vocabulary 55
Selection Comprehension 56

Making Room for Uncle Joe
Vocabulary 57
Selection Comprehension 58

Theme Outcome/Supporting Skill
Making Inferences About
Characters 59

Justin and the Best Biscuits in the World
Vocabulary 61
Selection Comprehension 62

Chasing After Annie
Vocabulary 63
Selection Comprehension 64

Information Skills

Unfamiliar Word Meanings 66
Parts of a Book 67
Using an Index 68
Following Directions 69
Card Catalog 70
Using the Library 71
Using an Encyclopedia 72
Reading a Street Map 73
Reading Diagrams 74
Bar Graphs and Line Graphs 75
Tables 76
Locating Information Quickly ... 77
Outlines 78
Summarizing Information
Graphically 79
K-W-L 80
SQRRR 82
Taking Tests 84

Language and Usage Lessons

The Sentence
Statements and Questions 85
Commands and Exclamations 87
Complete Subjects and
Complete Predicates 89
Simple Subjects 91
Simple Predicates 93
Combining Sentences: Subjects
and Predicates 95
Correcting Run-on Sentences 97

Nouns
Common and Proper Nouns 99
Singular and Plural Nouns 101
Nouns Ending with *y* 103
More Plural Nouns 105

Verbs
Action Verbs 107
Main Verbs and Helping Verbs .. 109
Present, Past, and Future 111
Making Subjects and
Verbs Agree 113
Irregular Verbs 115
The Special Verb *be* 117

Adjectives
What Is an Adjective? 119
Adjectives After *be* 121
Using *a*, *an*, and *the* 123
Making Comparisons 125
Comparing with *more* and *most* ... 127

Pronouns
What Is a Pronoun? 129
Subject Pronouns 131
Object Pronouns 133
Using *I* and *me* 135
Possessive Pronouns 137
Pronouns and Homophones 139

Adverbs
What Is an Adverb? 141
Comparing with Adverbs 143
Using *good* and *well* 145
Negatives 147

Capitalization, Punctuation, and Usage Guide

Abbreviations 149
Titles 150
Quotations 150
Capitalization 151
Punctuation 152
Problem Words 153
Adjective and Adverb Usage 155
Negatives 155
Pronoun Usage 156
Verb Usage 156

*Overview of Language and
Usage Lessons* 157

*Modeling of Language and
Usage Lessons* 158

Jumanji

Study the picture carefully. Then use the words in the box to answer the questions.

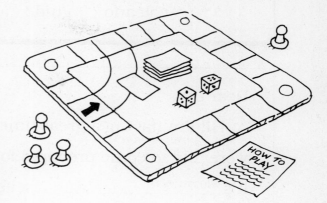

> select   dice
> board game
> shortcut   instructions

1. What does the picture show? __board game__
2. What do you read to find out how to play? __instructions__
3. What do you do before starting the game? Before starting the game, you __select__ a playing piece.
4. What can you roll to see how many spaces to move? __dice__
5. What can you do if you land on the space with an arrow? You can take a __shortcut__ across the board.

Now write about a board game you enjoy playing with your friends. __(Answers will vary.)__
_____
_____
_____
_____
_____

Use the words in the box to complete the letter.

| tsetse flies | monsoon | stampede | jungle |
| volcano | guide | python | |

Dear Jill,

Hi! I'm having a great time on Danger Island. We were told to come prepared for wet weather, because it's __monsoon__ season here. So far, though, we've had a lot of sunshine. The day we arrived, we heard — and felt — a __volcano__ rumble, but it did not erupt.

Today a __guide__ led us into the __jungle__ to see the animals. We saw a giant __python__ wrapped around a tree branch, waiting for its dinner to come by. Then we almost got caught in a __stampede__ of wart hogs. Luckily, we got out of the way just in time. We were warned about being bitten by __tsetse flies__, but I don't think we saw any.

I'm getting sleepy right now, so I'll close this letter. I'll try to write again soon.

Your friend,
Tony

VOCABULARY Book 1 Fantasy Close to Home

The Shrinking of Treehorn

Read each question, paying attention to the underlined words. Then write your answer on the lines. **(Sample answers)**

1. If you dropped your mother's favorite vase <u>on purpose</u>, what <u>privileges</u> might you lose?
   You might not be able to watch television or have dessert.

2. Why would it be <u>strange</u> for an elephant to <u>disappear</u>?
   An elephant is so large that it would be unusual if one disappeared.

3. What would your teacher say if you <u>shirked</u> your homework?
   He or she might tell you to stay after school.

4. What could you do if you wanted to <u>pretend</u> that you were <u>shrinking</u>?
   You could get on your knees or wear extra large clothing.

5. What could you do to stop your shoes from <u>shuffling</u>?
   You could stuff newspaper in them or else buy smaller shoes.

VOCABULARY — Book 1 Fantasy Close to Home

grumpy because he couldn't get the shell off the telephone.

Then something odd happened. As soon as the words "dumb old Fantastaplex" came out of Randy's mouth, the two children on the screen suddenly dropped the controls of their space-age car. The looks on their faces changed from delight to disbelief. "What does he mean, 'dumb old Fantastaplex'?" the girl asked the boy.

Then a man in jeans and a T-shirt rushed into view. He turned out to be the owner of the friendly voice, but now he didn't sound very friendly.

"Just who do you think you are, young man?" he demanded, pointing his finger into the TV camera. "If you think you can get away with calling Fantastaplex 'dumb,' you've got another think coming!"

The man and the children, all three, glared out of the TV — straight at Randy!

(Sample answers)

1. Which events in the story are realistic?

   <u>Randy is watching cartoons and is taking apart an old telephone. The</u>

   <u>cartoon has been interrupted by a toy commercial.</u>

2. At what point did you know for sure that this story is a fantasy?

   <u>The children in the commercial stop playing with the car, and the girl asks,</u>

   <u>What does he mean? in response to something Randy says.</u>

3. Think about the rules the author has set for this fantasy. Then finish these sentences:

   One rule is that TV characters can <u>see Randy and speak to him.</u>

   Another rule is that Randy can <u>talk to the people in TV commercials.</u>

4. Based on the rules for this fantasy, what do you think might happen next?

   <u>(Answers will vary, but students' responses should be based on the rules</u>

   <u>above.)</u>

SKILLS SUPPORT **Book 1 Fantasy Close to Home**

Making Inferences

As you read this story beginning, think about which events are realistic, which are fantastic, and what the rules for this fantasy are. Then follow the directions after the story.

Fantastaplex

"Are *you* the kind of kid who likes to build things and take things apart?" asked a very friendly voice from the TV set.

"Yep," said Randy, without looking up. In fact, at that moment he was taking apart an old telephone his mother had given to him. At the same time he was watching — or at least listening to — his favorite Saturday morning cartoon.

"Do you just *love* wheels, gears, motors, and electric switches?" continued the friendly voice.

"Sure," answered Randy. The commercial interested him, but he couldn't look up because he was trying to unscrew the last tiny screw that held on the telephone's plastic shell. Then he'd be able to lift it off and look inside.

"If this sounds like *you*," said the voice, "then you'll have *fantastic fun* with FANTASTAPLEX — the toy that lets *you* be the builder!"

Oh — Fantastaplex, thought Randy. Big deal. He had hoped that the commercial was for a toy he had never heard of. He had played with a Fantastaplex set at Eric's house. There was nothing fantastic about it. All the parts (the ones that Eric hadn't lost) were cheaply made, and the moonwalker they had put together fell to pieces before it had taken two steps.

The voice was still talking about the wonders of Fantastaplex. Randy glanced up. A space-age car was zooming across someone's living room while the boy and the girl working the controls looked as if they were having the greatest time of their lives.

"Yeah, right!" said Randy. "Good luck making a car like that with dumb old Fantastaplex." He was feeling

SKILLS SUPPORT   Book 1   Fantasy Close to Home

The Mysterious Girl in the Garden

Think about the story *The Mysterious Girl in the Garden.* Complete the sentences below. (Sample answers)

1. Terrie was upset about spending the summer in England because she had no one to play with and she wanted to spend the summer with her grandmother.

2. Tuesday turned out to be a special day for Terrie because she met Charlotte and her dog, Lioni, at Kew Botanic Gardens.

3. Terrie felt that she was like Charlotte in several ways because they both had no one to play with and both were told what to do by their families.

4. Terrie was shocked when she saw the museum portrait of the princess and her dog because they looked exactly like the girl and the dog she had met the day before.

5. Terrie stared at the portrait for a long, long time because she wasn't sure if Charlotte was playing a trick on her or if Charlotte really was a princess from long ago.

Terrie thought Charlotte was both pathetic and obnoxious. Answer these questions about Charlotte.

What made Charlotte seem pathetic? She was lonely and sad because her parents were fighting over her.

How was Charlotte obnoxious? She was rude and uppity and ordered Terrie around.

SELECTION COMPREHENSION   Book 1 Fantasy Close to Home

Jumanji

Below are five events from the story *Jumanji*. Write why each event is important.   (Sample answers)

1. Peter and Judy find the game JUMANJI in the park.
   They don't know it yet, but the game is soon to bring them more excitement and adventure than they could have imagined.

2. Judy reads aloud the last instruction: "ONCE A GAME OF JUMANJI IS STARTED IT WILL NOT BE OVER UNTIL ONE PLAYER REACHES THE GOLDEN CITY."
   Peter and Judy have to continue the game until the end regardless of what happens.

3. Peter lands on the space that says, "Lion attacks, move back two spaces."
   When a lion appears, Peter begins to take the game seriously.

4. Judy reaches the Golden City on the game board and quickly yells, "Jumanji!"
   Judy ends the game, saving them from the snake, lion, and other dangers.

5. Peter and Judy watch as Danny and Walter Budwing run through the park with a long thin box.
   Peter and Judy know that the same frightening events will probably happen to the two boys.

SELECTION COMPREHENSION   Book 1   Fantasy Close to Home

The Mysterious Girl in the Garden

Write a sentence that means the opposite of each of the sentences below. Change the underlined words. The first one has been done for you.   (Sample answers)

1. The newspaper printed the scandal about the mayor.
   The newspaper printed a story that praised the mayor.

2. That's a pathetic-looking dog.
   That's a very healthy-looking dog.

3. Usually Melanie is a very obnoxious person.
   Usually Melanie is a very sweet, nice person.

4. Kenny told his friends gossip about his brother.
   Kenny told his friends good things about his brother.

5. It's impossible to finish this in time.
   It's possible to finish this in time.

6. Christina is very impatient when she has to wait.
   Christina is very patient when she has to wait.

7. I'm going to ignore you.
   I'm going to pay attention to you.

8. Bill acts uppity around other people.
   Bill acts very humble around other people.

9. I detest Brussels sprouts.
   I love Brussels sprouts.

VOCABULARY — Book 1 Fantasy Close to Home

The Shrinking of Treehorn

Think about *The Shrinking of Treehorn*. From the box, choose the name of the character who made each statement below, and write the name on the blank. Then tell why the person made the statement. The first one has been done for you.

Moshie	Principal	Father	Teacher
Bus Driver		Treehorn	

1. __Father__ "Do sit up, Treehorn. I can hardly see your head." (Page 57) **Treehorn was getting small, but his father thought he wasn't sitting up straight.**

2. __Moshie__ "How come you can't mail it yourself, stupid?" (Page 62) **He could not figure out why Treehorn could not reach the mailbox.**

3. __Bus Driver__ "First time I ever heard of a family naming two boys the same name." (Page 64) **Treehorn was so small that the bus driver thought he must be Treehorn's younger brother.**

4. __Teacher__ "We don't shrink in this class." (Page 64) **She did not want any disturbances in her classroom.**

5. __Principal__ "We can't have any shirkers here, you know." (Page 66) **He thought Treehorn had written "shirking" instead of "shrinking."**

6. __Treehorn__ "If I don't say anything, they won't notice." (Page 71) **Treehorn noticed that he was green but thought his parents wouldn't notice it if he didn't mention it.**

SELECTION COMPREHENSION   Book 1 Fantasy Close to Home

Tyrannosaurus

Read each group of words below. Draw a line through the word that does not belong in the group. Then write why the other three words belong together. The first one has been done for you. (Sample answers)

1. tracks   footprints   trail   ~~hoof~~
   All except *hoof* are signs left behind by a creature that has moved through an area.

2. catastrophe   ~~health~~   disease   disaster
   All except *health* name something harmful.

3. armor   weapons   shield   ~~feathers~~
   All except *feathers* name something used for fighting or in battle.

4. ~~hunter~~   prey   food   victim
   All except *hunter* name something that could be hunted and eaten.

5. suddenly   quickly   ~~gradually~~   swiftly
   All except *gradually* mean "fast."

6. mighty   strong   ~~weak~~   powerful
   All except *weak* describe something that has great strength.

7. ~~slept~~   roamed   wandered   traveled
   All except *slept* relate to movement across an area.

8. cow   pig   ~~dinosaur~~   sheep
   All except *dinosaur* are animals that are still alive today.

Tyrannosaurus

Think about the selection *Tyrannosaurus*. Then answer each question.   (Sample answers)

1. What were some of the ways that dinosaurs protected themselves? Some ran from trouble. Some moved together to form a wall. Some had armor on their bodies and horns on their heads.

2. How was the body of the Tyrannosaurus "designed for hunting"? It had strong back legs for chasing its prey, three sharp claws on each foot, and six-inch teeth with sharp edges.

3. What strange problem did the Tyrannosaurus's small, weak arms create for it? If it lay down, it probably had a hard time getting back up again.

4. What do tracks and other evidence suggest to scientists about the family life of the Tyrannosaurus? It probably traveled alone or in pairs. It probably laid eggs.

5. What other questions do scientists still hope to answer about the Tyrannosaurus? How did Tyrannosaurus mothers care for their young? What caused it and other dinosaurs to die out 65 million years ago?

SELECTION COMPREHENSION   11   Book 2 They Walked the Earth

Identifying Main Idea and Supporting Details

Read the article below. Think about the main idea, the most important idea the author presents. Also look for details that support this main idea. Then follow the directions on the next page.

The Mangrove Trees of South Florida

Imagine that you are flying in a plane over southern Florida. As you fly over the part nearest the sea, you see thousands of tiny islands that look like green puzzle pieces scattered on a shiny mirror. These are mangrove islands.

Mangrove islands are formed by mangrove trees. Mangroves are very unusual and useful trees. Unlike most trees, they can live in seawater, which is salty. Their long roots act like nets in the water and trap dirt and rocks. Over time, the dirt and rocks build up on the roots, forming islands around the trees.

Mangrove trees are useful to the many kinds of animals that live on the mangrove islands. Tiny worms and crabs eat the leaves of mangrove trees. In turn, the worms and crabs provide food for raccoons, snakes, and other larger animals that hunt among the roots of the mangroves. Birds build their nests in the high branches of mangrove trees. Underwater, the mangrove roots make good homes for shrimp and other marine life.

Mangrove trees also help protect the land. During storms, mangrove trees act as a barrier between the sea and the land. The sea hits the mangroves hard, but the mangroves break the power of the waves, keeping the land behind and around them from being washed away.

1. Read the sentences below. Decide which sentence best tells the main idea of "The Mangrove Trees of South Florida." Then write the sentence on the lines.

    A. Mangrove roots make good homes for marine life.
    B. Mangroves are very unusual and useful trees.
    C. Mangrove trees act as a barrier between the sea and the land.

    Mangroves are very unusual and useful trees.

2. Write three details that support the main idea. (Sample answers)

    A. **Unlike most other trees, mangrove trees can live in salt water.**

    B. **The leaves of mangrove trees provide food for tiny worms and crabs.**

    C. **During storms, mangrove trees protect the nearby land from high waves.**

    Now use the main idea and supporting details you wrote to write a brief summary of the article.

    (Answers will vary.)

SKILLS SUPPORT — Book 2 They Walked the Earth

Wild and Woolly Mammoths

Read each sentence. Find a word in the box that means almost the same as the underlined word or words. Write the word on the line.

> trunk   tusks   ruins   mammoths
> extinct   enemies   climate

1. Thousands of years ago woolly <u>animals that looked like elephants</u> lived on the earth.   **mammoths**

2. On the sides of its face, this animal had two <u>long teeth</u> that it used for digging.   **tusks**

3. It had a <u>long nose</u> that it used to breathe and smell with and to carry food and water to its mouth.   **trunk**

4. This animal's long, hairy coat was good protection against the very cold <u>weather of the area</u> where it lived.   **climate**

5. These animals had few <u>creatures that wanted to cause them harm</u> — except for saber-toothed tigers and Stone Age people.   **enemies**

6. Scientists have found clay figures and bone carvings of these animals among the <u>remains of buildings</u> where Stone Age people once lived.   **ruins**

7. Today this animal is <u>no longer living on the earth</u>, but it lives on in books, museums, and our imaginations.   **extinct**

VOCABULARY   Book 2   **They Walked the Earth**

Wild and Woolly Mammoths

Think about the selection *Wild and Woolly Mammoths*. Decide whether each statement below is true or false, and circle that answer. Then write a reason for your answer. The first one has been done for you.
(Sample answers)

1. Woolly mammoths looked like cows.   True   (False)
They looked like furry elephants with two curved tusks, a long hairy trunk, and a heavy coat.

2. Scientists can only guess about the habits of woolly mammoths because none of these creatures have ever been found.   True   (False)
Scientists have found frozen mammoths, including one that still had food in its stomach.

3. Woolly mammoths were reptiles like dinosaurs.   True   (False)
They were mammals. They were warm-blooded, had hair, and nursed their young.

4. Archaeologists have learned a great deal about how Stone Age people lived.   (True)   False
By studying villages, caves, and carvings, scientists know much about these people.

5. Stone Age hunters figured out skillful ways to trap and kill woolly mammoths.   (True)   False
They used fire to scare mammoths down steep cliffs. They covered deep pits with branches.

6. Stone Age people hunted woolly mammoths only for their meat.   True   (False)
They also used mammoth bones and tusks for tent frames, jewelry, fuel, and musical instruments.

SELECTION COMPREHENSION   Book 2   They Walked the Earth

Identifying Main Ideas Across Texts

Read passage 1. Then read passage 2. Ask yourself what topic both passages share.

1. The Everglades in southern Florida is home to some of the most beautiful and unusual creatures in the world. The graceful egret and the great white heron are among the many birds that can be seen in the skies. White-tailed deer, bobcats, and the rarely seen Florida panther live in the woods.

2. Many of the creatures in the Everglades were once in danger of becoming extinct. The manatees — large, slow-moving mammals that live underwater — were almost wiped out by speeding motor boats and people who hunted them for their flesh, hide, and oil. Certain kinds of turtles were hunted for their shells or killed for food. At one time, nearly all the egrets in the Everglades were killed for their valuable feathers.

In the warm, swampy waters of the Everglades, crocodiles, alligators, giant turtles, and snakes can be found. Another creature that makes its home in the water is the manatee, or sea cow. These large mammals can eat more than one hundred pounds of water plants in one day!

In recent times, successful efforts have been made to save endangered wildlife in the Everglades. Manatees are now protected in parts of the Everglades set aside for their safety. Turtles now live unharmed in turtle preserves. And many once-endangered birds such as the egret are now protected by law. Once again, the Everglades has become a place where many kinds of wildlife can live safely.

Now follow the directions below. (Sample answers)

What topic do both of these passages tell about? __creatures that live in the Everglades__

Fill out the chart below. Write a sentence that tells the main idea in each passage. Then write two details that support each main idea.

Passage 1

Main Idea: Some of the most beautiful and unusual creatures in the world live in the Everglades.

Details:

1. Deer, bobcats, and panthers live in the woods of the Everglades.

2. Alligators, snakes, and manatees can be found in the warm, swampy waters of the Everglades.

Passage 2

Main Idea: Many creatures of the Everglades that were once in danger of becoming extinct are now protected by law.

Details:

1. Manatees are now protected in some parts of the Everglades.

2. Egrets and other once-endangered birds are now protected by law.

Use what you wrote on the chart above to write a paragraph that sums up the information from both passages.

Many beautiful and unusual creatures live in the Everglades. Egrets, bobcats, panthers, and white-tailed deer live there. Manatees, alligators, and giant turtles can be found in the warm waters. Many of these creatures, which were once in danger of becoming extinct, are now protected by law.

SKILLS SUPPORT Book 2 They Walked the Earth

Strange Creatures That Really Lived

Read each question below. Write your answers on the lines. **(Sample answers)**

1. Why do scientists go on **expeditions**?
   They go to do research. They go to find answers to questions about the past.

2. What do you think a **tar pit** looks, smells, and feels like? A tar pit might look black and bubbly, smell like tar, and feel very sticky.

3. Why might a **tar pit** be a good place for an **expedition**? Animals of long ago might have become stuck in a tar pit. A scientist might be able to find their remains there.

4. A **lizard** is one kind of **reptile**. What are some other reptiles? Alligators, turtles, snakes

5. Some reptiles, such as lizards, have **scaly** skin. What do you think scaly skin feels like? Dry, hard, rough

6. When an animal dies, its flesh **decays**. What happens to flesh when it **decays**? It rots and falls away from the bone.

Now imagine that you are a scientist about to go on an expedition. Where are you going and what are you looking for?

VOCABULARY   Book 2   They Walked the Earth

Strange Creatures That Really Lived

Look back at *Strange Creatures That Really Lived*. Write two or more details from the selection to explain why each creature below seems strange. (Sample answers)

1. **pteranodon:** It looked like a huge bat. It had leathery wings and a long, pointed bill.

2. **archelon:** It was a twelve-foot-long sea turtle. It weighed 6,000 pounds. It once swam in an inland sea that covered what is now South Dakota.

3. **archaeopteryx:** It looked like a small dinosaur with feathers. It had a tail and rounded wings.

4. **uintatherium:** It had six horns on its head. It had sharp teeth even though it was a plant-eater.

5. **baluchitherium:** It looked like a rhinoceros but had no horns. It was the size of a small house. It could stretch out its long neck like a giraffe.

6. **dodo:** It was the size of a turkey. It waddled like a duck. It had wings but could not fly.

BOOK 3

The Boy of the Three-Year Nap

Read each group of words below. Draw a line through the word that does not belong in the group. Then write why the other three words belong together. **(Sample answers)**

1. samurai warrior   patron   guard   ~~merchant~~
   All except *merchant* can provide protection.

2. ~~gentle~~   fierce   violent   wild
   All except *gentle* could describe an angry person.

3. frown   sneer   ~~smile~~   scowl
   All except *smile* are angry facial expressions.

4. church   ~~storehouse~~   temple   shrine
   All except *storehouse* are buildings where someone might worship.

5. a command   an order   ~~a request~~   a demand
   All except *a request* are orders to do something.

6. debate   bargain   ~~listen~~   convince
   All except *listen* involve arguing or discussing.

7. decreed   ordered   demanded   ~~pleaded~~
   All except *pleaded* mean "told to do something."

8. ~~refuse~~   allow   consent   agree
   All except *refuse* mean "to give permission."

VOCABULARY   Book 3   Battle of Wits

The Boy of the Three-Year Nap

Think about the folktale *The Boy of the Three-Year Nap.* Then answer the questions below. **(Sample answers)**

1. Why was Taro known as "The Boy of the Three-Year Nap"? <u>He was so lazy that the villagers said he could sleep for three years at a time.</u>

2. Why was Taro so impressed with his new neighbors? <u>They lived in a large mansion with a lovely garden, pond, and teahouse. They wore fine clothes.</u>

3. How did Taro prepare to put his plan into action? <u>He dressed himself in a black kimono and priest's hat. He painted scowl lines on his face. He waited by the shrine for the merchant.</u>

4. How did Taro fool the merchant? <u>He pretended to be the ujigami and convinced the merchant that Taro must wed his daughter.</u>

5. How did Taro's mother fool the merchant? <u>She would not let Taro marry the man's daughter until the house had been fixed up so his daughter could live in comfort.</u>

6. How did Taro's mother outsmart Taro? <u>She made the merchant promise to give Taro a job so Taro would have to work to get what he wanted.</u>

SELECTION COMPREHENSION Book 3 Battle of Wits

Farmer Schulz's Ducks

Use the words in the box to complete the article.

> accelerated   concussion
> wreckage   frustration
> impatience   semitrailer
> swerve   contented

CRASH KEEPS SCHOOL COOL

A freak accident occurred yesterday when a __semitrailer__ loaded with ice cream was chugging up Route 99. A sports car came speeding up from behind, the driver honking his horn with __impatience__. He __accelerated__ to pass the truck, but had to __swerve__ to avoid an oncoming car. Both the truck and the sports car ended up in a ditch.

The ice cream landed in front of nearby Happy Valley School. Children ran from class, attacking the ice cream with plastic spoons. Both drivers sat near the __wreckage__ of their vehicles, eating butterscotch ice cream. Asked if he felt any __frustration__ over the accident, the truck driver said, "Nope! These kids have the right idea!"

The sports car driver complained of a headache and was treated for a mild __concussion__. The __contented__ children returned to class.

VOCABULARY 22 Book 3 Battle of Wits

Farmer Schulz's Ducks

Think about *Farmer Schulz's Ducks*. Then read each problem below and explain how the problem was solved. (Sample answers)

	PROBLEM		SOLUTION
1.	Farmer Schulz's ducks liked to swim on the Onkaparinga River, but after dark, the river became too dangerous for them to stay.	1.	Each night the ducks returned to Farmer Schulz's yard, where it was safer for them.
2.	As the city grew, more and more drivers began to race back and forth on the road. Sometimes they did not stop for the ducks.	2.	Farmer Schulz nailed up a sign that read "Ducks Crossing" where the drivers could see it.
3.	Not all cars stopped for Farmer Schulz's "Ducks Crossing" sign. Finally, a car crashed into the ducks.	3.	Farmer Schulz built a bridge over the road with safety fences and ramps for the ducks to use.
4.	A semitrailer hit the duck bridge, destroying it. Several ducks were hurt or killed.	4.	Farmer Schulz decided to put in a pipe so the ducks could travel underneath the road.
5.	Farmer Schulz wanted to avoid any trouble with the government about building a duck pipe.	5.	He wrote a letter to the government, asking for official permission to build a pipe.
6.	Farmer Schulz needed to teach his ducks how to use the duck pipe.	6.	He built mesh flaps to guide the ducks into the tunnel. His family shooed the ducks into the tunnel.

SELECTION COMPREHENSION

Comparing Solutions

Read both stories. As you read, think about the problem that has to be solved in each story. Also think about how the characters solve their problems.

The Fishers and the Greedy Duke

There once was a selfish Duke who seized a fishing boat that was the home of a clever old couple.

The next day the old couple appeared at the Duke's castle. "We have no place to live and no way to make a living," they told the keeper of the castle. "So we have come here to work." They were hired at once.

The Duke was afraid of being robbed, so he had guards at the castle gate to search everyone going out. At mid-morning the old couple was stopped at the gate. They were leading four of the Duke's mules with forty bags strapped to their backs out of the castle.

"What's in those bags?" asked the guards.

"Twigs to make the Duke's brooms," the couple replied.

The guards emptied every bag but found no stolen gold, so they let the couple go.

At noon, the couple came to the gate again with forty bags on four more mules. They told the guards, "We're carrying straw to make the Duke's bricks."

The guards searched every bag but found no stolen gold, so they let the couple go.

At mid-afternoon, the couple came to the gate again with forty bags on four more mules. They said, "We're carrying only feathers to make the Duke's pillows."

The guards took every feather out of every bag but found no stolen gold, so they let the couple go.

At the end of the day, the guards looked for the old couple, but they never came back. They had sold the Duke's twelve fine mules, bought a new fishing boat, and sailed far away.

Yours, Mine, and His

The three Choy boys usually got along well, but sometimes they quarreled.

"Hey, that's my sweat shirt you have on," said ten-year-old John.

"No, it's mine," said his twin, James. "You left yours at school."

Or twelve-year-old Matthew would snap, "You ran off with my notebook this morning."

One day after such a quarrel, their father made a suggestion. "Why don't you put your names on your things?"

"Oh, Dad," they groaned, "only *little* kids have their names on everything."

"Well then," Mr. Choy said, "I'll count on you boys to come up with your own solution."

The next day Matthew came home carrying a bag. He called his brothers together and emptied the bag. Out came three markers — blue, red, and yellow.

"Pick a color," he said to the twins. We can mark the labels of our things with these colors. That way we'll always know whose things are whose."

And that is what they did.

Answer these questions about the two stories. (Sample answers)

1. What is the problem in "The Fishers and the Greedy Duke"? <u>The selfish Duke took the old couple's fishing boat where they lived and worked.</u>

2. How was the problem solved? <u>The old couple tricked the Duke by leading twelve of his mules out of the castle. Then they sold the mules and bought a new boat.</u>

3. What is the problem in "Yours, Mine, and His"? <u>The three Choy brothers argued about their clothes and other belongings.</u>

4. How was the problem solved? <u>Each boy chose a different color to mark his things.</u>

With a partner, compare the ways in which the characters in the two stories solved their problems. Which characters used a clever trick? Which used common sense?

The Sign in Mendel's Window

Read the story below, paying attention to the underlined words.

In the village of Rivka, there lived a ¹poor ²tenant named Grinkov who decided to run for mayor. He gave speeches in the marketplace, telling everyone what he would do if elected. Although Grinkov was a very ³awkward speaker, the villagers were not ⁴disappointed by what he had to say. Everyone who knew Grinkov said he was a ⁵humble man. He was also regarded throughout the village as ⁶an honest person. The people of Rivka believed these qualities would make Grinkov a good mayor, and so they voted for him.

Below is the same story, with blanks in place of the underlined words. On each blank, write a word from the box that means nearly the opposite of the word with the same number in the first story. The first one has been done for you.

In the village of Rivka, there lived a ¹ **prosperous** ² **landlord** named Grinkov who decided to run for mayor. He gave speeches in the marketplace, telling everyone what he would do if elected. Although Grinkov was a very ³ **eloquent** speaker, the villagers were not ⁴ **impressed** by what he had to say. Everyone who knew Grinkov said he was a ⁵ **braggart**. He was also regarded throughout the village as a ⁶ **scoundrel**. The people of Rivka believed these qualities would *not* make Grinkov a good mayor, so they did not vote for him.

> scoundrel
> landlord
> impressed
> prosperous
> eloquent
> braggart

VOCABULARY Book 3 Battle of Wits

The Sign in Mendel's Window

Think about *The Sign in Mendel's Window*. Then complete each sentence below. (Sample answers)

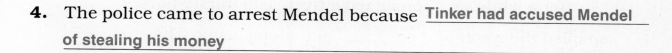

1. Mendel and Molly's neighbors were worried about the FOR RENT sign in the butcher shop window because <u>they were afraid that Mendel and Molly were sick or leaving town</u>.

2. At first Mendel was impressed by Tinker because <u>Tinker was eloquent, well-dressed, and humble</u>.

3. Tinker encouraged Mendel to count the shop's earnings aloud because <u>Tinker wanted to know exactly how much money Mendel had</u>.

4. The police came to arrest Mendel because <u>Tinker had accused Mendel of stealing his money</u>.

5. Simka proved that Tinker was lying because <u>everyone in town knew how much money was in the box</u>.

6. Molly dumped the coins into a pot of boiling water because <u>she wanted to show that they were covered with fat, which proved they belonged to a butcher</u>.

7. Indeed, Tinker's biggest mistake was in coming to Kosnov because <u>the neighbors were very close and always knew what was going on in town</u>.

SELECTION COMPREHENSION   Book 3   Battle of Wits

Little House on the Prairie

Use the words in the box to complete the diary entry below. Write one word on each line.

| groves | ford | cliffs | prairie |
| bluffs | trotted | lurched | |

August 29, 1880

Dear Diary,

Another boring day of rolling along in the covered wagon! I was so glad to see even small __groves__ of trees after seeing nothing but the wide, flat __prairie__ day after day!

After a while we approached a river. The thirsty horses smelled the water and __trotted__ along faster. The wagon __lurched__ from side to side on the uneven ground. Soon we came to a __ford__ where the river was not deep and we could cross in safety.

On one side of the river, tall, steep __cliffs__ rose up from the flat land. Beyond that were smaller, sloping __bluffs__. Father said that is where our new home will be. The long journey is almost over!

Now write a few sentences telling what you know about the *prairie*.
(Answers will vary.)

Little House on the Prairie

Think about the story *Little House on the Prairie*. Decide whether each statement below is true or false, and circle that answer. Then explain your answer.

1. Laura and her family traveled across Kansas in a fast-moving train.

    True   (False)

    They traveled in a covered wagon that moved very slowly.

2. The Ingalls family had no trouble crossing the creek.

    True   (False)

    The water was very deep, and Pa had to help the horses cross. The family lost Jack during the crossing.

3. At the end of the day, the Ingalls family was too tired to prepare and eat supper.

    True   (False)

    They made coffee, cornmeal cakes, and salt pork. Then they ate in front of the fire.

4. The wolf Laura thought she saw turned out to be their dog, Jack.

    (True)   False

    Jack had made his way to their camp. When Laura saw his green eyes shining in the dark, she thought he was a wolf.

Write at least one way Laura Ingalls' life was different from yours and one way her life was similar to yours.

Different: (Answers will vary.)

Similar: _____

SELECTION COMPREHENSION   Book 4   Laura Ingalls Wilder

Farmer Boy

Study the picture carefully. Write each word from the box by the place or thing in the picture that it names.

| buggy | colt | haymow | pantry | parlor | pasture |

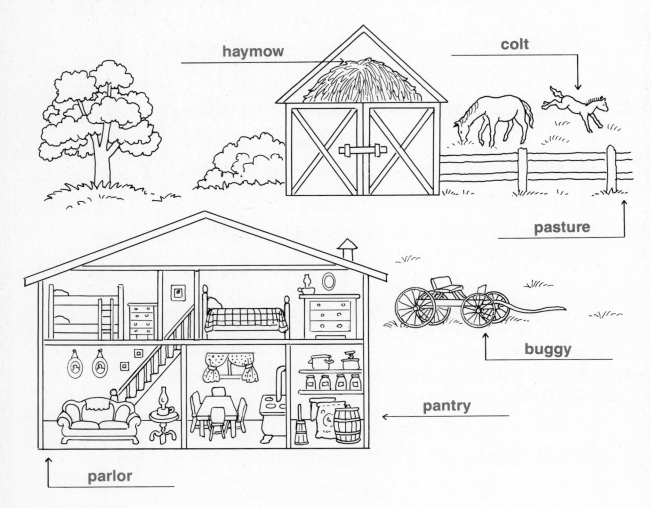

Now write a few sentences telling what might happen if the *colt* got out of the *pasture*.

(Answers will vary.)

VOCABULARY Book 4 Laura Ingalls Wilder

Read each sentence. Find the meaning of the underlined word in the list of meanings below. Write the word on the line before its meaning.

1. Luke's big sister was always trying to be the boss and tell him what to do.
2. He found this very aggravating and he sometimes got angry with her.
3. One day she told him to churn the butter, even though that was really her chore.
4. "I don't have to obey you!" he shouted. "You're not my mother."
5. "If you don't," she said, "I'll tell Father to give you a licking you won't forget!"
6. Luke did not feel anxious because he knew his father wouldn't do that.
7. Still, the milk would be ruined if they let it sit in the churn while they argued.
8. Luke hesitated just a bit, and then he began to do the churning.

licking	a beating or whipping
ruined	spoiled
boss	a person who controls or manages
anxious	worried; concerned
aggravating	annoying; irritating
hesitated	paused or stopped for a short while
obey	follow the orders of
churn	to beat milk or cream to make butter

Now write about something you have found *aggravating*.
(Answers will vary.)

VOCABULARY   Book 4 Laura Ingalls Wilder

Farmer Boy

Below are some incomplete sentences about the story *Farmer Boy*. Fill in the blanks with the name of the character who performed each action. Then answer the questions that follow.

The Wilders:
Eliza Jane
Royal
Almanzo
Alice

1. **Royal** scolded Almanzo for bothering the colts in the pasture. Why? **Father Wilder had told the children not to bother the colts.**

2. **Almanzo** fed molasses candy to Lucy the pig. What happened to poor Lucy? **Lucy's teeth became stuck together, and she could not eat, drink, or squeal.**

3. **Almanzo** threw the blacking brush at the wall-paper. What happened next? **He hid in the haymow and later felt so sick that he couldn't eat.**

4. **Alice** admitted to Mother that they had eaten almost all the sugar. What was Mother's response? **She said that she wouldn't scold because the children had been so good.**

5. **Eliza Jane** patched over the splotch on the parlor wall-paper. Why? **She was sorry that she had been so aggravating to Almanzo, so she wanted to help him.**

Identifying Characteristics of Historical Fiction

Read the story. Think about which details are based on historical fact and which are probably from the author's imagination. Then follow the directions after the story.

The Pinto

It was a clear morning in 1842 when Monterey was still the capital of California, and California still belonged to Mexico. The excitement of the upcoming rodeo stirred the air as groups of cowhands gathered at the Customs House. They joked and boasted about their horses, tied in a long line.

The cowhands wore shiny leather boots or deerskin shoes embroidered with silver. They were dressed in velvet pants with gold braid and silver buttons and wore their wide, flat-topped hats. These people lived in the saddle. They roped cattle at full gallop. It was easy for them to pick up a coin, a handkerchief, or a trailing lasso from their galloping horses.

As Hernando, a cowhand, stepped out of the Customs House, he nearly bumped into his younger brother Antonio.

"Azul is gone!" Antonio cried. Anger and pain mixed in his voice.

Hernando circled his brother's shoulders with one arm. "I'm sorry, Antonio," he said, "but horses are easy to come by. You know that."

"Azul was special!" Antonio insisted.

Antonio loved the large brown spots on his pony's shiny white coat. He loved the flecks of blue in the pony's eyes. That was why he called his pony Azul — the Spanish word for blue.

"Without Azul, I can't ride in the rodeo," Antonio said. "I must find him!"

"I see," Hernando said, "but he might have joined a wild herd. He will not be easy to find. The herd may be in the valley by now, and that's a day's journey."

➡

SKILLS SUPPORT   33   Book 4   Laura Ingalls Wilder

Antonio had not thought of this. He was not thinking clearly.

"There will be other ponies and other rodeos," Hernando said, trying to comfort Antonio.

"There is no pony like Azul!" Antonio said. "With Azul, I was ready for the rodeo. I have worked with him all year. We were partners. We *are* partners!"

Suddenly Antonio had an idea. "Will you help me track Azul, Hernando?" he asked.

Hernando gave his younger brother a long look. He had watched Antonio and Azul practice turns and roping day after day. For a year now, hardly a day had passed without Antonio riding out into the pasture to work on his roping.

"All right, Antonio," Hernando sighed. "We will look for this little pinto of yours. Tomorrow morning we will leave before sunrise. Now I have work to do."

1. Look back at the story. Find details that are probably based on historical fact, and write them on the lines below. Hints: Where and when does the story take place? What did the cowhands wear? **(Sample answers)**

   In 1842, Monterey was still the capital of California, and California still

   belonged to Mexico.

   The cowhands wore shiny leather boots or deerskin shoes embroidered with

   silver.

2. Look back at the story again. Find details that are probably from the author's imagination, and write them below. Hints: What do the characters say to each other? What happened to Antonio's pony? How does Antonio feel? **(Sample answers)**

   "Azul is gone!" Antonio cried. Anger and pain mixed in his voice.

   "There will be other ponies and other rodeos," Hernando said, trying to

   comfort Antonio.

SKILLS SUPPORT   Book 4   Laura Ingalls Wilder

On the Banks of Plum Creek

Use the words in the box to complete the story below. Write one word on each line.

| blizzard | whirling | swiftly | suddenly | bitter |
| swirled | hauled | stagger | frantically | |

Amy and her little sister were walking home from school when, without warning, a winter storm __suddenly__ hit. The air became so __bitter__ cold that Amy's nose hurt. At first, the snowflakes __swirled__ in circles in the air. Then the wind blew in a __whirling__ blast against their faces. Amy gripped her sister's hand and fought her way against the heavy snow and strong wind. She knew that this storm was a real __blizzard__!

The girls ran for home as __swiftly__ as they could. The wind made them __stagger__ and almost fall. When they reached home, snow was piled against the door. Amy tugged __frantically__ at the door until it finally opened. She fell inside and __hauled__ her sister in after her.

Now describe a *blizzard* or other storm you have been in.
(Answers will vary.)

VOCABULARY — Book 4 Laura Ingalls Wilder

On the Banks of Plum Creek

Think about what happened in the story *On the Banks of Plum Creek.* Then fill out the story map below.
(Sample answers)

Setting
Time: winter, late 1800s
Place: near Walnut Grove, Minnesota

Characters:
Pa and Ma Ingalls, Mary, Laura, and Carrie

Problem:
While Ma and Pa are away, a sudden blizzard hits.

Action/Events:
Ma and Pa go to town for the day, leaving the girls alone at home. A blizzard unexpectedly hits. Mary and Laura, afraid of freezing to death, bring all the wood from the woodpile into the house.

Resolution/Ending:
Ma and Pa return home safely and laugh at the sight of all the wood in the house.

SELECTION COMPREHENSION   Book 4   Laura Ingalls Wilder

BOOK 5

Meg Mackintosh and the Case of the Missing Babe Ruth Baseball

Read each group of words below. Draw a line through the word that does not belong in the group. Then write why the other three words belong together.

1. fake   false   ~~authentic~~   untrue
   All except *authentic* mean "not real."

2. investigate   detect   examine   ~~ignore~~
   All except *ignore* are part of a detective's job.

3. ~~difference~~   resemblance   similarity   likeness
   All except *difference* mean "ways in which things are alike."

4. deduction   ~~confusion~~   conclusion   solution
   All except *confusion* are answers to a problem or mystery.

Read the story below. Find a word in the box that completes the meaning of each sentence. Write the word in the blank.

> solved
> proof
> decode
> deduce
> telltale

Detective Snoop examined the note carefully for clues to help him __deduce__ who had left it — and why. Snoop was convinced that it was a secret spy note, although he had no __proof__ that spies were involved.

"This __telltale__ reddish stain on the note makes me think that whoever wrote it was in a struggle with another spy," Snoop said to himself. "And this unusual green scribbling could be a secret code. I must __decode__ it!"

Just then, Baby Buster came into the room holding a hot dog with ketchup in one hand and a green crayon in the other. The case of the mysterious note was __solved__.

VOCABULARY   Book 5   The Mystery Hour

Meg Mackintosh and the Case of the Missing Babe Ruth Baseball

This is your Detective's Note Pad. Use it to help you solve the "Case of the Missing Babe Ruth Baseball."

> **Words You Should Know** Be on the lookout for these words as you read this mystery:
>
> decipher   deduction
> resemblance   telltale
>
> You can figure out what the words mean by looking for familiar word parts and by thinking about what meanings make sense in the sentence. If you need more help, look up these words in the Glossary.

Before You Read Think about the prediction you wrote on Journal page 121. Keep your prediction in mind as you read this mystery.

Read from page 275 to the end of page 278. Find out what Meg does to unravel the first clue. Then answer the questions below.

What was Clue One? <u>Not a father / Not a gander / Take a look / In her book</u>

Which book did Meg reach for, and why? <u>She reached for *Mother Goose* because its title fit the clue: a mother is not a father, and a goose is not a gander.</u>

Read from page 279 to the end of page 288. Then answer the following questions.

What was Clue Two? <u>Little Boy Blue with the cows in the corn / Whatever you do / Don't blow this ? </u>

MONITORING COMPREHENSION 38 Book 5 The Mystery Hour

What horn did Alice mean? an old powder horn on the mantle

When Clue Three was unscrambled, what did it say?
little bo peep lost her ___?

Where did this clue lead Meg? to an old painting of sheep

What was Clue Four? rub a dub/three men in a ___?

Where did this clue lead Meg? to an old washtub in Gramps's toolshed

Read to the end of the selection. Then fill in the rest of this note pad. See if you can solve the mystery before Meg does!

How did Meg know that the next clue was a fake? It looked different from the other clues and was numbered Clue #4, even though Meg had already found the fourth clue. Also, it had nothing to do with Mother Goose rhymes.

What was the next **real** clue? The little dog laughed

Where did Meg discover the hidden baseball? inside an old stuffed dog

After You Read Answer these questions to help you piece together the parts of the puzzle.

What did all the clues have in common? They all had to do with Mother Goose rhymes and were hidden in Gramps's house. They were written on unlined paper in Alice's handwriting. Also, all of the clues were found in old things.

How was Meg able to solve the mystery? by paying attention to small details and figuring out the meaning of each clue; by thinking logically and making a list of deductions; by being patient and observant

MONITORING COMPREHENSION Book 5 The Mystery Hour

Meg Mackintosh and the Case of the Missing Babe Ruth Baseball

Think about *Meg Mackintosh and the Case of the Missing Babe Ruth Baseball*. Read each clue below. Write how Meg used the clue to figure out where to look for the next clue. **(Sample answers)**

CLUE

1. Not a father
   Not a gander
   Take a look
   In her book

2. Little Boy Blue
   with the cows in
   the corn
   Whatever you do
   Don't blow this ___?___

3. ucle reeth
   tillet ob epep
   stol reh ___?___

4. rub a dub
   three men in a ___?___

HOW MEG FIGURED IT OUT

1. Not a father is a mother. Not a gander is a goose. Meg decided to look for the next clue in the Mother Goose book.

2. Meg figured out that the answer was *horn*. At first she looked in the bugle. Then she looked for a clue in the old powder horn.

3. When unscrambled, clue three said "little bo peep lost her ___?___." *Sheep* was the missing word. Meg looked behind a painting of sheep.

4. Meg decided that the missing word was *tub*. She found the next clue scratched into the bottom of the old metal bathtub.

SELECTION COMPREHENSION — Book 5 The Mystery Hour

Noting Details About Clues in a Mystery

Read this story. See if you can figure out who the thief is. Then answer the questions on page 42.

The Mystery of the Missing Matsuko

"I'm so glad you came quickly, Detective Yee," Mr. Daley sighed miserably. "My prize Matsuko painting has vanished. It's worth thousands of dollars!"

Detective Yee took out her notebook. "Suppose you tell me what happened."

"Well, the painting was here an hour ago, just before I went out," Mr. Daley said. "I usually bring my lunch and eat in the back room, but today it's so hot that I decided to eat by the pond in the park. I carefully locked up the gallery, as I always do. Then five minutes ago I came back — and the painting was gone."

Detective Yee was exploring the small gallery. "It doesn't look as if anyone broke in. Does anyone else have a key?"

"Yes," said Mr. Daley. "I rent three small rooms upstairs to three artists. Each artist has a key. They've all been here since about nine."

"Hmmmm," said Detective Yee thoughtfully. "It is quite possible that one of them came through here on an errand, saw that you were gone, and stole the painting. I'd better find out what they've all been doing in the last hour."

Detective Yee went upstairs and knocked on the first door. It was opened by a woman with bushy white hair. She put down her half-eaten sandwich and gave her name as Jennie Mahoney. In the room, Officer Yee could see an unfinished oil painting of a vase of flowers. On a paint-stained table were a crumpled lunch bag and an orange. Jennie Mahoney said that she had not been out of her room since ten.

The second door had the name Dwight Magruder on it. It was opened by a thin young man with a stringy beard. In answer to Detective Yee's questions, he pointed to a nearly finished drawing. "I was so wrapped up in my work that I haven't been out of this room since I got here at nine." →

SKILLS SUPPORT   Book 5   The Mystery Hour

Behind Magruder Detective Yee saw only art supplies and a half-empty paper cup of lemonade with ice cubes still floating in it.

The third door was opened by a cheerful man wearing a green apron smeared with clay. He told Detective Yee that his name was Andy Gooden, and he showed her a statue of a fox he'd been working on all morning. He'd gone out once at around eleven to talk to Mr. Daley, but he hadn't been out since. Detective Yee didn't see any signs of lunch in the room and asked whether he had gone out to eat. "Oh, I always go out late — at around three — to get a bite," he told her.

Detective Yee went back into the hallway and nodded to herself. She was sure now that she knew who the thief was. Then she knocked again on one of the doors. When it opened, she said, "Since you clearly didn't tell the truth, you'd better come down to the station with me. I have a lot more questions for you now, Mr. Magruder."

(Sample answers)

1. What was the most important thing that Detective Yee tried to find out from each of the artists?

   whether they had gone out in the last hour

2. What two things did Detective Yee especially notice in the room of Dwight Magruder?

   A. his art supplies

   B. a cup half-full of lemonade and ice

3. What did Detective Yee see in Mr. Magruder's room that told her that he must have gone out on an errand?

   the ice in the cup

4. How did Detective Yee know that Mr. Magruder had not brought this item in with him at nine o'clock?

   It was hot, and by noon the ice would have melted.

5. What do you know from real life that helped you figure out the mystery?

   Ice melts quickly on a hot day.

Paddington Turns Detective

Imagine that you are interviewing for a job with the Dogstar Detective Agency. Read the questions below, paying attention to the underlined words. Then answer the questions. Use the underlined words in your response. The first one has been done for you. **(Sample answers)**

1. "Your application looks strong, but I have a few questions. First, what special characteristics do you think a detective needs to <u>solve</u> a mystery?"

   "To solve a mystery, a detective needs to have a good eye for details and a good mind to understand clues."

2. "Good! What signs make someone look <u>suspicious</u>?"

   "Someone looks suspicious if that person seems to be in a hurry or gets very nervous when I ask questions."

3. "Excellent! How would you <u>track</u> down a criminal?"

   "I would carefully follow the clues left behind by the criminal until I had tracked the criminal to his or her hideout."

4. "Very good! What could you do to <u>recognize</u> criminals who often <u>disguise</u> themselves?"

   "To recognize criminals who disguise themselves, I'd look for things that may not change, such as voice, height, and weight."

5. "Perfect. Now, what would you do before you <u>accused</u> someone of a crime?"

   "Before I accused anyone, I would be sure that I had enough proof."

6. "Hmmmmm. Finally, what would you use to <u>secure</u> a criminal to make sure he or she doesn't escape?"

   "I would secure a criminal with this special rope I always carry with me."

"I am impressed. When can you start working?"

VOCABULARY   Book 5 The Mystery Hour

Paddington Turns Detective

Think about *Paddington Turns Detective*. Decide whether each statement below is true or false, and circle that answer. Then write a reason for your answer. **(Sample answers)**

1. Mr. Brown's prize squash was missing. (True) False
   It had disappeared sometime Wednesday night. Mr. Brown thought that someone had stolen it.

2. Paddington did not think that the flashing light in the garden was connected to the missing squash. True (False)
   He saw the light the same night the squash had disappeared, so he thought the person with the light must be the thief.

3. Paddington wanted to catch the thief. (True) False
   He disguised himself with a beard. He got rope and batteries for his flashlight. He hid in the greenhouse that night.

4. Mr. Curry was amused by Paddington's claims that he was a burglar. True (False)
   He said that Paddington would be sorry for accusing him. He demanded that the bear be punished and that Mr. Curry be paid damages.

5. Mrs. Brown cooked the prize squash by mistake. (True) False
   She hadn't realized it was the prize one and had served it for dinner.

6. Paddington was unhappy about the trouble he had caused. (True) False
   He slipped away, packed his bags, and decided to return to darkest Peru.

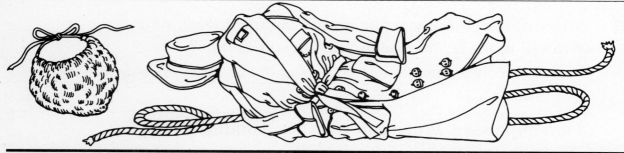

SELECTION COMPREHENSION Book 5 The Mystery Hour

The Case of the Missing Roller Skates

Read the news brief below.
Then answer the questions.

"We interrupt this program to bring you a special report: A valuable statue of the founder of Chocodonuts has disappeared from the lawn of City Hall. We go to Carmen Perez, reporting live. Carmen?"

"Thank you, Douglas. With me here at City Hall is Detective Jones. Detective Jones, I understand that you became aware of this theft when the lawn sprinkler system at City Hall went off unexpectedly. Do you think that whoever set off the sprinklers also stole the statue?"

"Carmen, I can't say. We have no leads, except some doughnut crumbs by the spot where the statue had stood. Perhaps the thief wasn't working alone. The statue would be too heavy for anyone except a weightlifter to lift."

"Thank you, Detective. Citizens of Centerville, please contact City Police if you have any clues."

1. Where was the **scene of the crime**? the lawn of City Hall

2. What was Detective Jones able to report with **certainty**? He could report that the statue was gone, that the sprinkler system had gone off unexpectedly, that there were doughnut crumbs on the spot where the statue had once stood, and that the statue was too heavy for one person.

3. How did the detective **reason** that the thief might have had a **partner**? Detective Jones said that the statue was too heavy for one person to carry, unless that person was a weightlifter.

4. What clues might the police use to **trip** up a **suspect**? The doughnut crumbs; the fact that only a weightlifter could lift the statue.

VOCABULARY Book 5 The Mystery Hour

The Case of the Missing Roller Skates

Think about *The Case of the Missing Roller Skates.* Complete each sentence below.
(Sample answers)

1. Although Encyclopedia Brown was at the scene of the crime, he did not see who took the skates because <u>he was having his tooth pulled by the dentist</u>.

2. Encyclopedia reasoned that he should not look for a grown-up thief because <u>a grown-up was not likely to steal an old pair of skates. A grown-up would be too hard to catch in the busy medical building</u>.

3. Encyclopedia checked *every* office in the building to see if a boy or girl had been there that morning because <u>he assumed that a child had taken the skates. He wanted to gather all the information he could about any suspects</u>.

4. Billy Haggerty became Encyclopedia's number one suspect because <u>Billy had been the only other child in the building that morning</u>.

5. When Billy was asked about Dr. Wilson, he gave himself away because <u>he referred to the doctor as *he*, which meant that Billy must have seen the doctor before or had been to his office</u>.

6. When Billy said, "I had a sprained wrist, not a toothache," he made another mistake because <u>this meant that Billy knew Dr. Wilson was a dentist</u>.

SELECTION COMPREHENSION Book 5 The Mystery Hour

BOOK 6

Walt Disney: Master of Make-Believe

Read each group of words below. Draw a line through the word that does not belong in the group. Then write why the other three words belong together.

1. animator ~~author~~ cartoonist artist
   All except *author* are people who draw.

2. cartooning tracing ~~typing~~ painting
   All except *typing* are part of an animator's job.

3. ~~scissors~~ celluloid airbrush colored inks
   All except *scissors* are tools or materials an animator uses.

4. competitors opponents rivals ~~friends~~
   All except *friends* are people who are against one another.

5. series ~~single~~ several multiple
   All except *single* mean "more than one."

Use the words in the box to complete the story below.

At last *Mavis Meets the Swamp Monster* was completed. The filmmakers had spent months on the __production__ of the film. They were sure they would make a huge __profit__ on the movie.

> profit
> projected
> production
> contract

Opening night arrived. The filmmakers had signed a __contract__ to show their film at the Capitol Theater. From the moment the first scene was __projected__ on the screen, the audience started laughing. Unfortunately, the movie wasn't supposed to be a comedy.

VOCABULARY — Book 6 The Dreamers

Walt Disney: Master of Make-Believe

Think about the biography of Walt Disney. Write an answer to each question below. (Sample answers)

1. How did Walt Disney become interested in cartoon animation? **He got a job at the Kansas City Film Ad Company making cartoon ads. He liked the job so much that he learned as much about animation as he could.**

2. In what ways was Walt a good salesman? **He talked people into lending him money for his business or into buying his work.**

3. How did Walt improve his animated cartoons? **Instead of photographing cutout figures, he photographed a series of drawings to create more lifelike figures. Later he added sound to his cartoons and used storyboards.**

4. Why was Walt usually so short of money in the years before he created Mickey Mouse? **Starting a new business was expensive. Walt was always using any profits he made to improve his cartoons.**

5. What lesson did the Oswald Rabbit series teach Walt? **After Mintz had copyrighted Oswald in his own name, Walt vowed to never again work for anyone else.**

6. How did Walt get the idea for Mickey Mouse? **He sketched a mouse based on his old pet mouse, Mortimer, but his wife suggested that he change the name to Mickey.**

SELECTION COMPREHENSION — Book 6 The Dreamers

Marian Anderson

Read the advertisement below, paying attention to the underlined words. Then find the meaning of each underlined word in the list of meanings. Write the word on the line next to its meaning.

Are you looking for a career in music? Do you have musical talent, the desire to be a success, and, most important, a spirit of cooperation and hard work?
The Bay City Orchestra and Chorus is looking for you. The first concert of the season is scheduled for late October. This group of professional musicians, which has received many honors from critics across the nation, has openings for the following positions:

- An assistant conductor to lead the orchestra rehearsals
- An accompanist to play the piano with guest singers
- A qualified tenor and soprano to sing duets
- Singers with experience in performing spirituals

MEANINGS

1. religious songs
2. person who plays an instrument as the singers sing
3. ability
4. person who directs an orchestra
5. signs of respect or high regard
6. someone who has done well
7. songs sung by two singers
8. paid and trained
9. a musical performance
10. way of earning a living
11. one's nature or character

WORDS

spirituals

accompanist

talent

conductor

honors

success

duets

professional

concert

career

spirit

VOCABULARY　　　　　Book 6　The Dreamers

Marian Anderson

Think about *Marian Anderson.* Use details from the biography to complete each sentence below. (Sample answers)

1. Even as a child in the church choir, Marian's voice stood out because it was rich and velvety and she could sing any part of a song.

2. Marian was able to begin taking private music lessons because the people of the Union Baptist Church believed in her talent and held a concert to raise money for her.

3. Marian decided to go to Europe to study and perform because her career was standing still. She was rarely asked to perform in important theaters at home.

4. Although they could not always understand English, people the world over loved Marian's singing because she poured her heart and soul into her music. They loved her voice and they loved her.

5. Marian gave an outdoor concert at the Lincoln Memorial on Easter Sunday because she had not been allowed to sing in Constitution Hall. She wanted to do whatever she could to gain justice for her people.

6. When Marian sang with the Metropolitan Opera Company, she "opened a door for her own people" because she was the first black person to sing such an important part in an opera. She sang the difficult part successfully.

SELECTION COMPREHENSION Book 6 The Dreamers

Understanding Influences on People's Lives

Read this biographical article. Think about the influences that made José Clemente Orozco a great artist. Then answer the questions that follow.

Clemente Becomes an Artist

When José Clemente Orozco was a young boy, his family moved to Mexico City. Across the street from their apartment was a printing shop.

Clemente watched for hours at a time as an artist and several helpers made beautiful pictures. He was fascinated by what he saw.

The artist first drew a picture on a metal plate, using a sharp tool instead of a pencil or paintbrush. Next, the artist used the plate to print copies of the picture he had drawn. When they came off the printing press, the pictures were black and white. Then the helpers went to work, coloring in the pictures. Plain prints became colorful works of art.

"This was my awakening to the world of painting," Clemente would later write.

Clemente decided then and there that he would become an artist. He begged his mother to let him study drawing. Finally she agreed. She took him to meet the director of an art school.

"Why, this little child is hardly old enough to be in school!" the director exclaimed. But when he saw some of Clemente's pictures, he let the boy visit a drawing class.

Clemente soon showed the teachers at the school that he had real artistic talent. The director called to tell Clemente's parents that their son was a better drawer than any of the older students in his class. Clemente was then allowed to become a regular student at the art school. He was certain he could become an artist.

When he finished school, Clemente was given an award of money to study further. But the award wasn't given to study art — it was to study farming! Clemente's father was happy. After all, his family was poor. If Clemente learned how to help farmers, he would soon be able to earn money.

Clemente was not happy. He wanted to be an artist, but he did go on to study farming. The only part he liked, however, was drawing maps of farms.

Soon after that, an event happened that would change Clemente's life. One day Clemente and a friend were trying to make fireworks. Suddenly there was a huge explosion! Clemente's left hand was so badly burned that a doctor had to operate and remove it. Clemente's eyes were injured too. He would wear thick glasses for the rest of his life.

But Clemente did not let the accident ruin his life. In fact, it made him more determined than ever to pursue his dream of being an artist. He began taking art classes at night and taking odd jobs as an artist.

At art school Clemente's favorite teacher was Dr. Atl. Dr. Atl suggested that Clemente paint some pictures showing important events in Mexico's history. Clemente took Dr. Atl's advice. When the paintings were shown, people loved them. This was only the beginning of José Clemente Orozco's fame as one of the great Mexican painters of the twentieth century.

Think about these actions. What does each one show about Clemente? Write your answer in the second column.
(Sample answers)

Action	What It Shows About Clemente
1. He spent hours watching the printers make beautiful pictures at the print shop across the street.	1. He was curious and had a lot of patience. He loved art.
2. He asked his mother if he could go to art school.	2. He knew at a young age that he wanted to be an artist. He knew he needed to work at his talent.
3. After his terrible accident, Clemente went back to art school.	3. He didn't allow the accident to ruin his life. He was determined to become an artist.

Roberto Clemente

Use the words in the box from the selection *Roberto Clemente* to complete this biography about a baseball player named Lou Gehrig.

> stadium
> championships
> inning
> equals
> outfield
> major leagues
> desire
> treated
> patient

Even as a young man, Lou Gehrig was a powerful hitter. Many professional baseball teams in the __major leagues__ wanted him to play for them. When Lou's father needed money for an operation, it was Lou's greatest __desire__ to help. So he signed to play with the New York Yankees.

Lou didn't get to play right away, but he was __patient__. After two years, he got to bat in the ninth __inning__ of a game. He hit the ball far into the __outfield__, but it was caught. However, after that he never missed playing in another game.

Lou became one of the all-time great hitters, once hitting four home runs in one game. With Lou, the Yankees won many __championships__, including the World Series seven times. He was also known as a gentleman who __treated__ others with respect and as __equals__.

Sadly, Lou found out that he had a disease, which forced him to retire. The Yankees set aside a game to honor him. The __stadium__ was packed with fans. As Lou stood in the infield one last time, he told the crowd, "Today I consider myself the luckiest man on the face of the earth." Two years later, he died.

Today, Lou Gehrig lives on at the Baseball Hall of Fame and in the hearts of all who knew him.

VOCABULARY   Book 6   The Dreamers

Roberto Clemente

Think about *Roberto Clemente*. Write two or more details from the biography to support each statement below.   (Sample answers)

1. As a boy, Roberto loved to play baseball.
   He and his friends made their own baseballs out of string and played baseball after school. He ran track and threw the javelin in high school to improve his baseball skills.

2. Roberto's efforts to become a good player soon paid off.
   He was picked to play for a team in Santurce before he finished high school. He played on the Brooklyn Dodgers' farm team when he was nineteen.

3. Clemente played outstanding baseball for the Pittsburgh Pirates.
   He was a tremendous fielder. He got a hit in every game of the 1960 and 1971 World Series. He won many batting titles. He was a member of the All-Star team many times. He made 3,000 hits.

4. Clemente was not always happy in the United States.
   He felt that he and other Spanish-speaking players were not always treated fairly. He was often sick or injured. After each baseball season, he returned to Puerto Rico.

5. Although Clemente died in 1972, he is still honored today.
   He was elected to the Hall of Fame. A sign was placed on the door of the room where he had lived. His dream of a sports center for Puerto Rican children became a reality.

SELECTION COMPREHENSION Book 6   The Dreamers

BOOK 7

Teacher's Pet

Read each statement below, paying attention to the underlined words. Then answer the question that follows the statement.   (Sample answers)

1. It is the first day of school, and Mrs. Bartles's students want to enhance the blank bulletin board in their classroom. What can they do to enhance it?
   They could put up posters or paint a mural on it.

2. Stanley hopes to impress Mrs. Bartles on the first day. What can he do to impress her?
   He might raise his hand if he has a question. He might be in his seat when the bell rings.

3. Mrs. Bartles does not have a very good memory and cannot remember the names of all her students. What can she do to help her memory?
   She could make a seating chart or have the students wear name tags.

4. Claudia has always been the best student in science, but now that Matthew has joined the class, she has finally met her match. How do you think she feels?
   She might be determined to study harder or to do work for extra credit.

5. José is trying to convince his classmates to vote for him for class president. How can he convince them?
   He might tell them how honest and hardworking he is.

6. Shirley put much effort into writing a report, but her puppy ate it. How do you think she felt afterward?
   She was probably angry that she had to start all over again after spending so much time on the first report.

VOCABULARY

Teacher's Pet

Think about the story *Teacher's Pet*. Read each event below, and choose the word from the box that best tells how Cricket felt. Write the word on the line. Then write a reason for your choice. **(Sample answers)**

| stunned | eager | strange |
| foolish | proud | |

How Did Cricket Feel When . . .

1. . . . she realized that Zoe was as good a student as she was? **strange**
   Cricket felt strange not being the best student in class, but at the same time, the competition made her learn more.

2. . . . Mrs. Schraalenburgh said that everyone had to write a book report once a month? **eager**
   Cricket liked writing book reports and hoped to prove to the teacher that she was the best student.

3. . . . she completed her book report? **proud**
   She had copied it over neatly and had made a special cover for it. She thought the book report would impress her teacher.

4. . . . she saw a B– written at the top of her book report? **stunned**
   Cricket was certain she had done A+ work.

5. . . . Mrs. Schraalenburgh wrote Beverly Cleary's name on the chalk board? **foolish**
   Cricket realized she should have taken more care in writing her book report, instead of wasting time making a fancy cover.

SELECTION COMPREHENSION   Book 7   Dear Diary

Making Room for Uncle Joe

Use the words in the box to complete the two journal entries below.

responsible	routine	embarrassing
appreciates	nuisance	encouraged

Dear Journal,

My friend Gil went on vacation and left his dog, Mayonnaise, with me. That dog is always causing trouble — he's a real __nuisance__. Yesterday I went to the park to play ball, and Mayonnaise followed me. I petted him and __encouraged__ him to behave himself, but every time someone hit the ball, that crazy mutt chased it and ran off with it. Finally we had to stop the game. The guys actually thought that dog was mine! It was really __embarrassing__. I sure hope Gil __appreciates__ what I'm doing for him.

Dear Journal,

Sorry I haven't written in a while, but I've been busy taking care of you-know-who. I'm __responsible__ for feeding him in the morning and walking him three times a day. That's been our __routine__ all week. At night he sleeps beside my bed. I think I'm going to miss that old Mayonnaise when Gil comes home. I'm sure glad they live next door.

VOCABULARY — Book 7 Dear Diary

Making Room for Uncle Joe

Think about the story *Making Room for Uncle Joe.* Complete the chart below by explaining why the family members felt as they did at different points in the story. (Sample answers)

When Uncle Joe First Came

1. Beth was upset because she was embarrassed to have Uncle Joe around the house. She didn't want any of her friends to come over with Joe there.

2. Amy was excited because she thought she could help Uncle Joe and be his friend.

3. Dan was worried because he thought Uncle Joe might be a nuisance. His friend Ben had made fun of retarded people.

4. Uncle Joe felt out of place because he missed his old friends. He tried to be helpful around the house but didn't always succeed.

After They Got to Know One Another

1. Beth felt happy because she gave Uncle Joe piano lessons. She enjoyed being a teacher.

2. Amy was thrilled because she had found a friend who listened to her read and tell stories.

3. Dan felt relieved because he and his friends learned to enjoy being with Joe. They had fun bowling, playing, and working together.

4. Uncle Joe felt that he belonged because he found ways to be helpful and he got along with everyone.

SELECTION COMPREHENSION — Book 7 Dear Diary

Making Inferences About Characters

Read this story. Try to understand what Marty is like by paying attention to what he says, thinks, and does.

First Day

Marty's classmates were listing the products of Texas while he was listing his good points to himself: "I'm nice. People say I have a good sense of humor. I'm a good shortstop ... and I'm pretty smart." He stopped. Marty wasn't sure whether being smart would be okay with all the students at his new school.

"Jason!" said Mrs. Thornsby sharply. "You seem to have lots to say. Perhaps you can tell me: What is Texas's leading crop?"

Marty already knew which of his new classmates was Jason. He was the one who came into class that morning wearing his baseball cap backwards, clowning around.

But at this moment Jason's face was a blank wall. "Corn?" he responded loudly.

Mrs. Thornsby shook her head. "Marty, can you answer the question, please?" she said.

Marty had been slouching in his chair, trying to look invisible. But now he sat up and looked around the room. "Ummmm ... cotton?" he said in a quiet voice.

"Very good," Mrs. Thornsby said.

Marty hunched over his open book and blushed.

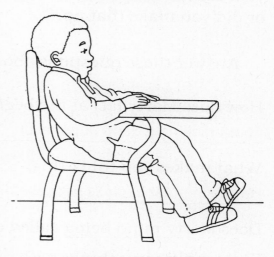

Just before noon, Mrs. Thornsby said to the class, "I introduced Marty Chapman to you this morning. We have a few minutes before lunch. Marty, would you tell us a little about yourself?"

Marty stood up slowly, then said, "I'm from Virginia." He glanced around the room.

"Where in Virginia?" asked Mrs. Thornsby.

"Just outside Norfolk. There's a huge base there and my father just retired from the Navy, so we lived there until last year —"

There was much chatter at this remark. Had he said something wrong? To Marty's relief, the noon bell rang.

Marty tried to slip past Jason and the rest of his classmates without being noticed. Then, to his horror, Marty heard his name called in a loud voice, and then: "Was your father really in the Navy, or did you make that up?"

Marty froze. He recognized that clowning voice behind him. He turned around. "Yes, it's true," he replied carefully.

"Wow!" Jason exclaimed. "Have you ever been on a battleship?"

Marty let out a sigh. "Yeah, my dad was an officer and one time our whole family got to sail on . . ." he began as they walked together toward the lunchroom. He had passed the first test of the new school year.

Answer these questions about the story. (Sample answers)

1. How does Marty feel at the beginning of the story? __scared, nervous about making a good impression__

    What makes you think this? __He lists what he thinks are his good points. He keeps quiet in class because he doesn't want to seem like a showoff.__

2. Does Marty mind being called on by Mrs. Thornsby? __yes__

    What makes you think this? __Marty is slouching in his chair to avoid attention. He hunches over and blushes after he has answered and is praised.__

3. Why does Marty freeze when he hears Jason call his name? __He isn't sure whether Jason is being friendly or mean. He's nervous about talking to Jason for the first time.__

4. How does Marty feel after he hears Jason's question about being on a battleship? __relieved, at ease__

    How do you know this? __He lets out a sigh and starts chatting with Jason.__

Justin and the Best Biscuits in the World

Read each group of words below. Draw a line through the word that does not belong in the group. Then write why the other three words belong together.

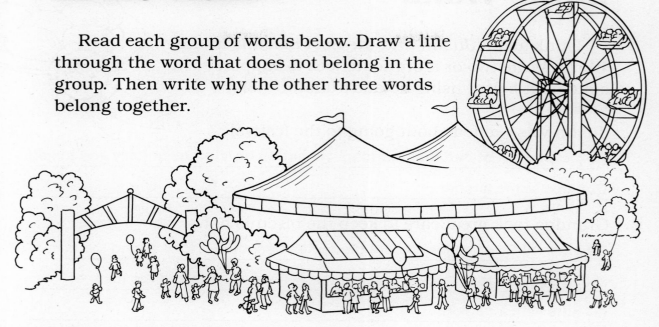

1. festival   party   carnival   ~~ordeal~~
   All except *ordeal* are kinds of celebrations.

2. fairgrounds   pavilion   ~~ranch house~~   competitions
   All except *ranch house* are part of a festival.

3. entry rules   categories   prizes   ~~ingredients~~
   All except *ingredients* are part of a contest.

4. ~~buried~~   displayed   showed   exhibited
   All except *buried* mean "set out for viewing."

5. rolls   ~~pumpkins~~   buns   biscuits
   All except *pumpkins* are types of bread.

6. determine   ~~hesitate~~   conclude   decide
   All except *hesitate* mean "to make up one's mind."

7. eagerness   ~~boredom~~   enthusiasm   excitement
   All except *boredom* mean "great interest in something."

8. declare   announce   proclaim   ~~conceal~~
   All except *conceal* mean "to make known."

VOCABULARY   Book 7   Dear Diary

Justin and the Best Biscuits in the World

Think about *Justin and the Best Biscuits in the World*. Write two or more sentences to support each statement below, using details from the story.
(Sample answers)

1. Justin was excited about going to the festival.
   He jumped out of bed that morning. He put on his clothes quickly and joined Grandpa in the kitchen.

2. Grandpa took great care to keep the biscuits hot.
   He planned to arrive just before the contest deadline. He wrapped the hot skillet of biscuits in towels. Once at the festival, he covered the plate of biscuits with a napkin.

3. Justin was a good sport about losing the pie-eating contest.
   He praised the winner. He was glad he hadn't eaten any more.

4. Grandpa took great care to get ready to return to the festival that night.
   He took a long time to shower and get dressed. He wore a sharp-looking suede vest with fringe. He put on a nice fragrance.

5. By the time the judges were ready to announce the winner for the best biscuits, Justin was worried.
   He wondered why the announcer didn't hurry up. His stomach felt weak, and his hands were cold.

6. Grandpa was thrilled about winning first prize in the biscuit-baking competition.
   He smiled and rushed up to receive his prize. He celebrated later by dancing with all the ladies.

SELECTION COMPREHENSION

Chasing After Annie

Read the story below, paying attention to the underlined words. Then find the meaning of each underlined word in the list of meanings. Write the word on the line next to its meaning.

The scrawny kitten was meowing in its cage. Its previous owner had abandoned it. Luckily, someone had found the little ball of fur in an alley and brought it to Joan's animal shelter.

Joan Steiner, director of the shelter, had seen lots of animals that had been abandoned, but it was always a shock when it happened. Joan was a true animal lover. Everyone who knew her said she wasn't a phony who ran the shelter only to make money.

Joan liked to brag that she could immediately tell someone's personality just by the way he or she acted toward animals. "You can't pretend with me," she liked to say. She made sure her animals were placed only with people who had a genuine love of animals.

MEANINGS	WORDS
a person's nature	personality
fake	phony
act in a false way	pretend
a great surprise	shock
real	genuine
boast	brag
a place for stray animals	animal shelter
earlier	previous

VOCABULARY — Book 7 Dear Diary

Chasing After Annie

Think about the story *Chasing After Annie.* Answer each question below. **(Sample answers)**

1. Why didn't Annie like Richie at first?
   He bragged all the time about everything he did.

2. Why did Richie *think* Annie liked him?
   Richie thought Annie was impressed by his muscles, his good grades, and his fish scrapbook. He also thought she liked him because he could play chess, spell *microgroove,* and high dive.

3. What did Annie and Richie do to try to find Annie's dog?
   Richie and Annie looked all over the neighborhood for Fritz. Richie drew a picture of Fritz and put it up at school. He also went to the animal shelter to look for Fritz.

4. What made Richie give the dog from the animal shelter to Annie?
   He thought she needed any Fritz she could get. He wanted to make a good impression on her.

5. How did Annie feel toward Richie when he gave her the dog?
   She began to like Richie more. When she thought of Richie, she no longer thought of bugs and itches and liver. She smiled at him in school, and she bought him a book about high diving.

6. Why did the new Fritz seem strange to Annie?
He smelled funny, had to be trained again, had two different-colored eyes, and had a funny bump on his leg.

7. What made Richie feel like a phony?
He had played a dirty trick on Annie by pretending to find Fritz. Now Annie was so grateful that she wanted to give him a present, which he knew he didn't deserve.

8. Why did Richie suddenly run from Annie's door just as he was about to tell her the truth about the dog?
He saw the real Fritz returning and was embarrassed that he had been caught in a lie.

9. How did Annie feel toward Richie when she found out the truth about the two dogs?
She was mad that he had tricked her. She hated Richie more than bugs, itches, and liver.

10. What made Annie finally decide that there were worse people in the world than Richie Carr?
She saw how he treated Duchess and realized he was a dog person like herself.

SELECTION COMPREHENSION

Unfamiliar Word Meanings

Read the paragraphs below, paying attention to the underlined words. Write what you think each underlined word means. Then write any clues that helped you understand the word.

Frank was so hungry that he <u>consumed</u> two helpings of meat loaf and mashed potatoes for dinner. His mother had to <u>restrain</u> him from eating the cherry pie afterward, because she was saving it for company later. Frank didn't want to wait, but his mother promised to give him a big piece if he was patient. **(Sample answers)**

1. meaning of <u>consumed</u>: ate

    clues: so hungry, meat loaf and mashed potatoes

2. meaning of <u>restrain</u>: hold back

    clues: saving it for later, didn't want to wait

At first Luisa was <u>hesitant</u> about writing to a pen pal, but later she was glad she had decided to do it. Luisa found that she and her pen pal had many things in common. They enjoyed reading each other's letters so much that they <u>corresponded</u> at least once a week.

3. meaning of <u>hesitant</u>: not certain

    clues: at first, but later, glad she had decided

4. meaning of <u>corresponded</u>: wrote

    clues: reading each other's letters

Now use a dictionary to check the meanings that you wrote.

INFORMATION SKILLS

Using an Encyclopedia

Look at the picture of the set of encyclopedias. Then read each topic listed below. Write the key word or words you would use to find information about the topic and the volume number in which you could find it. The first one has been done for you.

Topic	Key Word or Words	Volume
1. Explorer Thor Heyerdahl's journeys	Heyerdahl, Thor	9
2. The history of the United Nations	United Nations	20
3. The crops grown in North Dakota	North Dakota	14
4. How motion pictures are made	motion pictures	13
5. The habits of koalas	koalas	11
6. The parts of the solar system	solar system	18
7. Rachel Carson's work for wildlife	Carson, Rachel	3
8. Why Yorktown, Virginia, is famous	Yorktown	21

Now answer these questions about guide words.

9. Would information about koalas be on, before, or after the pages with the guide words *Knoxville* and *Korea*? **on**

10. Would information about North Dakota be on, before, or after the pages with the guide words *North America* and *North Carolina*? **after**

Now use an encyclopedia to find out some facts about one of the above topics. Write the facts on a separate sheet of paper.

INFORMATION SKILLS

Using the Library

The top part of four card catalog cards are shown below. Study each card and decide in which section of the library — **Fiction, Nonfiction,** or **Biography** — you would look to find the book. Write the name of the section. Then write where in that section you would look to find the book. **(Sample answers)**

362.42	Bergman, Thomas
Be	Finding a common language: children living with deafness, by Thomas Bergman

1. Section: __Nonfiction__
2. Where to Find Book: __The call number tells me this book would be on the 300's shelf.__

Hurwitz, Johanna

Hurray for Ali Baba Bernstein, by Johanna Hurwitz; illustrated by Gail Owens

5. Section: __Fiction__
6. Where to Find Book: __It would be on the Fiction shelves under H for Hurwitz.__

B	Wilbur & Orville Wright
W	Rowland-Entwistle, Theodore Wilbur & Orville Wright, by Theodore Rowland-Entwistle; illustrated by W. Francis Phillipps

3. Section: __Biography__
4. Where to Find Book: __The book would be under W for Wright on the Biography shelves.__

ELECTRICITY — EXPERIMENTS

537.07	Markle, Sandra
Ma	Power up: experiments, puzzles, and games exploring electricity, by Sandra Markle

7. Section: __Nonfiction__
8. Where to Find Book: __It has a call number and would be on the 500's shelf.__

INFORMATION SKILLS

Card Catalog

There are three kinds of cards in the card catalog. Look at the sample card catalog cards below. Then use what you know about the card catalog to answer each question.

```
598.61   Patent, Dorothy Hinshaw           Author card
Pa         Wild turkey, tame turkey, by
           Dorothy Hinshaw Patent. Photos

598.61   Wild turkey, tame turkey           Title card
Pa         Patent, Dorothy Hinshaw
           Wild turkey, tame turkey, by

         TURKEYS                            Subject card
598.61   Patent, Dorothy Hinshaw
Pa         Wild turkey, tame turkey, by
           Dorothy Hinshaw Patent. Photos
           by William Muñoz.
           New York, Clarion Books, 1989
           52 p. illus.
             Relates the history of the native
           North American turkey and compares
           that proud bird with its farmyard cousin.
```

1. Which kind of card would you use to find out what books the library has by Patricia McKissack? **author**

2. Which kind of card would you use to find out if the library has any books on hockey? **subject**

3. Which kind of card would you use to find out if the library has the book *Stone Fox*? **title**

4. What letter would the title card for *The Hundred Dresses* be listed under? **H** Why? **because the first important word in the title is *hundred***

5. What letter would the author cards for books written by Daniel Pinkwater be listed under? **P**

The next time you go to the library, use the card catalog to find the title of one book about hockey, the title of one book by Daniel Pinkwater, and the author of *Stone Fox*. Write your answers on a separate sheet of paper.

INFORMATION SKILLS

Following Directions

Read the directions for making homemade clay. Then answer the questions that follow.

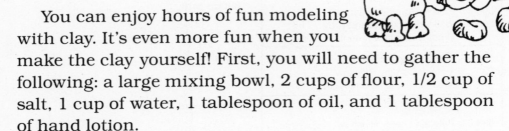

You can enjoy hours of fun modeling with clay. It's even more fun when you make the clay yourself! First, you will need to gather the following: a large mixing bowl, 2 cups of flour, 1/2 cup of salt, 1 cup of water, 1 tablespoon of oil, and 1 tablespoon of hand lotion.

Mix the flour and the salt together in the mixing bowl. Next, add the oil and water. Then work the mixture with your hands until it is doughlike. If the clay is too stiff, add more water. If it is too wet, add more flour. When the clay is a thickness that you like, knead in the hand lotion.

After you have made the clay, you can store it in the refrigerator or make something with it right away!

1. What is the first thing you do after gathering the items needed? __Mix the flour and salt in the mixing bowl.__

2. What key word tells you when to add the oil and water? __next__

3. What do you do if the clay is too stiff? __Add more water.__

4. When might you need to add more flour to the mix? __if the clay is too wet__

5. When do you knead in the hand lotion? __when the clay is a thickness that you like__

6. What two things can you do with the clay after you have made it? __Put the clay in the refrigerator or make something with it right away.__

INFORMATION SKILLS

Using an Index

The index below is from a book about the Old West. Use the information in the index to answer each question.

> Games: indoor, 21; outdoor, 22
> Gold: discovery of, 28–29; gold rush, 34–35; mining of, 40
> Goodnight, Molly, 51, 53
>
> Hickok, Wild Bill, 46–48, 50
> Hispanic settlers, 4, 17–19
> Homes: building of, 8, 10, 15–16; furnishings for, 13–14; types of, 11–12
> Homesteaders, 7, 14, 27

1. How many pages tell about Wild Bill Hickok? **four**
2. Which pages might you look at to find out who the homesteaders were? **7, 14, 27**
3. What are the subtopics listed under the main topic **Games**? **indoor, outdoor**
4. Under which main topic can you find out about how gold was mined? **Gold**
5. Which pages give information about Hispanic settlers? **4, 17–19**
6. How many pages give information about Molly Goodnight? **two**
7. On which pages would you expect to find information about log cabins, sod houses, or other types of homes? **11–12**

Use the index of a social studies book to find the main topic **Indians** or **Native Americans**. On a separate sheet of paper, list a few of the subtopics found under that topic.

INFORMATION SKILLS

Parts of a Book

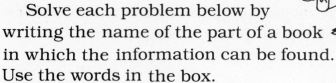

Solve each problem below by writing the name of the part of a book in which the information can be found. Use the words in the box.

> title page   copyright page   table of contents   glossary

1. Becky has six books about space, and she wants to read the one with the most up-to-date facts. Which book part will tell her when each book was published?
   copyright page

2. Rick wants to know if his book titled *Great Baseball Players* has a chapter on Babe Ruth. Where can he look?
   table of contents

3. Carlos is writing a book report on the book he just read. He wants to list the title, the author, and the publisher in his report. Where can he find this information?
   title page

4. Brenda came across the word *fuse* in her science textbook and wants to find out what it means. What part of the book can she use to find the meaning?
   glossary

5. Gaby wants to know how many stories there are in her new book of folktales. Where should she look?
   table of contents

Now use a textbook of your own to find the date the textbook was published, the name of the publishing company, and the title of the first chapter or section.

INFORMATION SKILLS

Reading a Street Map

Study the street map of Fairy Tale Village. Use the information on the map to answer the questions that follow.

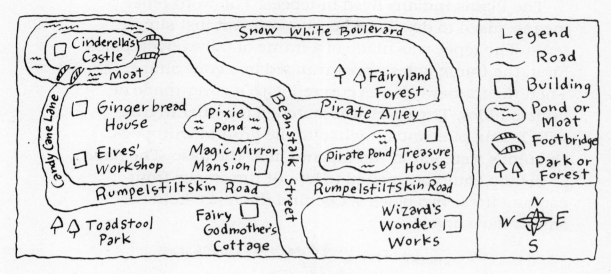

1. In the legend, what does ☐ stand for? __Building__

2. How many parks or forests are there in Fairy Tale Village? __two__

3. If you wanted to go from Elves' Workshop to Gingerbread House, in which direction would you travel on Candy Cane Lane? __north__

4. What intersection is near the Fairy Godmother's Cottage? __Rumpelstiltskin Road and Beanstalk Street__

5. How would you go from Magic Mirror Mansion to Cinderella's Castle? __Go north on Beanstalk Street, turn west on Snow White Boulevard, and cross over the footbridge.__

6. How would you go from Toadstool Park to Fairyland Forest? __Go east on Rumpelstiltskin Road, turn north on Beanstalk Street, and turn east on Pirate Alley.__

INFORMATION SKILLS

Reading Diagrams

Read the paragraph below and study the diagram. Then answer each question.

The Plains Indians lived in tepees. The word *tepee* means "used to dwell in." Tepees were not just simple tents. The tepee was made of a frame of tall wooden poles. First, the frame poles were arranged in a cone shape. Then the framework was covered with a cover made of buffalo hides. The cover was fastened together in front with lacing pins and attached to the ground with pegs. Openings were left at the top and near the base. The flaps at the top could be opened to let out the smoke from cooking fires. Two poles set behind the tepee held the smoke flaps open.

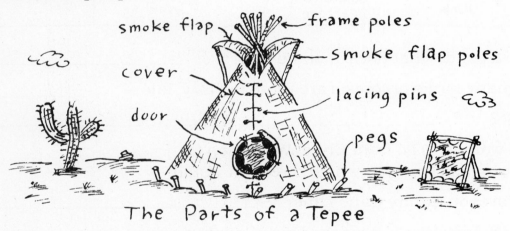

The Parts of a Tepee

1. What does the diagram above show? __the parts of a tepee__
2. How was the cover fastened together in front? __with lacing pins__
3. Why was an opening left at the top of the tepee? __so smoke from cooking fires could be let out__
4. How many poles held the smoke flaps open? __two__
5. What were used to attach the cover to the ground? __pegs__
6. If you wanted to make a model of a tepee, how might this diagram help you? __(Sample answer) It shows what parts are needed and how they fit together.__

INFORMATION SKILLS

Bar Graphs and Line Graphs

Study the bar graph and the line graph. Use the graphs to answer the questions below.

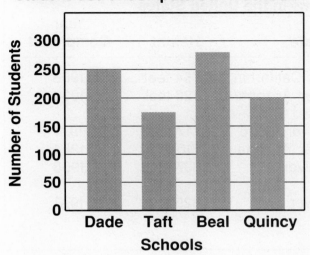

1. In which school do the most students use the computer? __Beal__

2. How many students at the Dade School use computers? __250__

3. In which school is the student use of computers greater, Taft or Quincy? __Quincy__

4. How many more students use the computer at Dade than use the computer at Quincy? __50__

5. In what month did the most fourth grade students at Beal use computers? __January__

6. How many Beal fourth graders used computers in October? __15__

7. How did the number of fourth graders who used the computer change from September to October? __It went up by 5.__

8. During what two months did the number of fourth graders using computers stay the same? __November and December__

INFORMATION SKILLS

Tables

Study the table below. Use the information on the table to answer the questions.

Six Highest Dams in the United States			
Dam	River and State	Height	Year Completed
Oroville	Feather River, California	754 feet	1968
Hoover	Colorado River, Arizona/Nevada	726 feet	1936
Dworshak	Clearwater River, Idaho	717 feet	1974
Glen Canyon	Colorado River, Arizona	710 feet	1964
New Bullards Bar	North Yuba River, California	637 feet	1968
New Melones	Stanislaus River, California	625 feet	1978

1. Which U.S. dam is the highest? **Oroville Dam**
2. Which of the six dams is the newest? **New Melones Dam**
3. Which of the six dams is the oldest? **Hoover Dam**
4. Where is Dworshak Dam located? **on the Clearwater River in Idaho**
5. Which of the dams listed are located on the Colorado River? **Hoover Dam, Glen Canyon Dam**
6. Which of the dams listed are located in California? **Oroville Dam, New Bullards Bar Dam, New Melones Dam**
7. Which one of the two dams completed in the 1970's is higher? **Dworshak Dam**
8. How many of these dams are over 700 feet high? **four**

INFORMATION SKILLS

Locating Information Quickly

Skim the article by reading the title, headings, and first paragraph. Try to get an idea of what the article is about in general. Then answer the questions that follow.

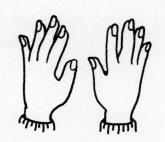

Fingernail Factories

How do fingernails grow? Not by accident! They are designed like factories to make nail growth possible.

How Fingernails Grow

The main part of your fingernail is the nail plate, which protects the nail bed underneath. The nail plate is not like skin. No blood flows into the plate.

Look at your thumbnail. Can you see the half-moon at the base? The half-moon, or *lunula,* is the "nail-plate factory." It helps make new nail. When new nail is added to the old nail plate, the whole nail is pushed up and out. That is how nails grow.

How Fast Nails Grow

How fast nails grow depends on several things — your age, the time of year, your health and eating habits, and which hand you use the most. Nails grow more slowly as you get older. For an adult, a nail grows about one-eighth of an inch a month. Nails grow more slowly in winter than in summer. Being ill or not eating the right foods may slow nail growth. The nails on the hand you use most usually grow faster than those on your other hand.

1. What is the article about in general? <u>how fingernails grow</u>
2. What helped you figure this out? <u>the opening paragraph and the two subtitles</u>

Now scan the article to answer the following questions.

3. What part of the nail is the "nail-plate factory"? <u>half-moon or lunula</u>

4. What things affect how fast your nails grow? <u>your age, the time of year, your health and eating habits, which hand you use the most</u>

INFORMATION SKILLS

Outlines

Read the following article. As you read, think about the main ideas and the important details. Then complete the outline of the article.

President's Day

President's Day is a national holiday that falls on the third Monday in February. This holiday celebrates the birthdays of two important United States presidents — George Washington and Abraham Lincoln.

The holiday has changed over the years. People first celebrated the holiday in the late 1700's in honor of George Washington's birthday on February 22. Later, people decided to also honor Abraham Lincoln, whose birthday was on February 12. Finally, the two holidays were combined into one day to honor these great men.

On President's Day many places are closed, just as they are on other national holidays. In most states, government and business offices are closed on this day. There is no mail delivery because post offices are closed, and schools and banks are closed, too.

Title: President's Day

   I. What President's Day celebrates

      A. George Washington's birthday

      B. Abraham Lincoln's birthday

  II. How the holiday has changed

      A. Used to be for Washington's birthday only

      B. Lincoln's birthday was added later

 III. Which places are closed on President's Day

      A. Government and business offices

      B. Post offices, schools, and banks

INFORMATION SKILLS

Summarizing Information Graphically

Read the article about camels. Then summarize the information graphically by completing the chart.

Camels

Why Camels Are Useful

Camels are best known for carrying heavy loads across the deserts of Asia and Africa. On long trips, a large camel can carry four hundred pounds on its back. Also, the hair, meat, and milk of camels are important to desert people. The camel's soft hair is woven into cloth, while its milk and meat are used as food.

The Camel's Body

A camel's body is well suited for desert travel. Its feet are broad and padded. Its hoof is only at the tip of the toes, which helps the camel to walk on loose sand. Fat is stored in its hump. When there is not much food, the camel's body draws on the fat. Besides the water stored in the fat, water may be stored in the camel's stomach. Camels can go three days without water.

Complaints About Camels

Camels are considered bad tempered, ungraceful, and ugly. No matter how well they are treated, they never seem to even notice their masters. Also, it is costly to keep camels. They can find some grasses for themselves, but they still must be fed hay and grain.

(Sample answers)

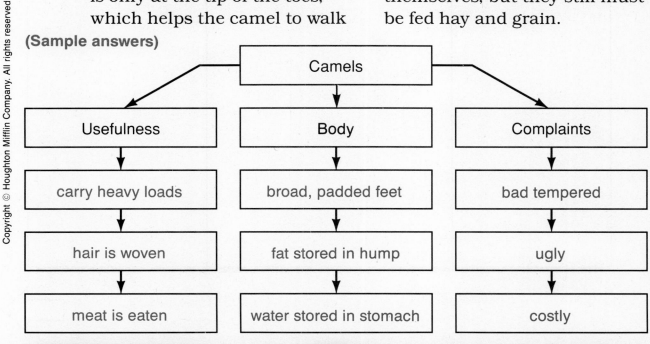

INFORMATION SKILLS

K-W-L

Use the K-W-L chart to help you read the article on the next page. First, look at the article and identify what the topic is. Write the topic on the chart.

Next, write in the "Know" column what you already know about this topic. In the second column, write what you want to know about this topic.

Then read the article. When you finish reading, return to the chart and write in the third column what you learned.

Topic: Paul Revere's Many Skills		
What I Know	**What I Want to Find Out**	**What I Learned**
(Answers will vary.)	(Students' notes will vary but will probably indicate that they want to learn what skills Paul Revere had.)	(Sample answer) Paul Revere was a famous silversmith. He also made false teeth. He was a leader of the Boston Tea Party.

INFORMATION SKILLS

Paul Revere: A Man of Many Skills

For many people, the name Paul Revere brings to mind the image of a man on a galloping horse on his way to warn everyone that the British were coming. Revere's famous midnight ride has become a legend and the subject of poetry and art. Although he is best known today for his ride, the people of Revere's day knew him for many other reasons.

Paul Revere's Trade

Paul Revere was an outstanding silversmith who learned the trade from his father. When the senior Revere died, Paul worked as a silversmith to support the rest of the family. He made a wide range of items, from silver trays and teapots to more unusual things, such as a silver dog collar and a baby rattle. He was considered to be the best silversmith in Boston.

Revere's Other Occupations

When hard times came to Boston and people had less money to spend on items made from silver, Revere switched to other activities to earn a living. He made prints of political cartoons from engraved copper plates. These prints are collectors' items now, although Revere was not thought to be a great artist. Revere also worked for a while putting false teeth in place for people and fixing them with wires.

Revere the Patriot

Paul Revere was active in the colonial efforts against the British. He was a leader of the Boston Tea Party, and after the cases of tea had been dumped into Boston Harbor, Revere rode to New York and Philadelphia to spread the news. During this trip, Revere covered 800 miles in eleven days. He often went on long rides to carry important messages to other parts of the colonies. He did carry out his famous midnight warning ride, but it was only one of his many contributions to the American Revolution.

INFORMATION SKILLS

SQRRR

Use the SQRRR study method to help you read the article on the next page. Follow the directions below.
(Sample answers)

S 1. Survey the text. Look at the title, headings, picture, and caption to get an idea of what the text is about. Write what you think the text is about.

what giant pandas look like, where they live, what they eat, and why

they are lucky

Q 2. Read the first heading and turn it into a question. Write your question below.

Heading 1: What do giant pandas look like?

R 3. Read to answer your question. Write the answer to your question below.

Pandas are partly black and partly white. They are very large.

R 4. Now recite in your own words the answer to your question.

Repeat steps 2 – 4 for Headings 2, 3, and 4. Write the question and answer it.

Heading 2: Where do pandas live?

Answer: They live in bamboo forests in China.

Heading 3: What are the panda's eating habits?

Answer: It picks bamboo shoots, then sits down to eat them slowly.

Heading 4: Why are pandas lucky?

Answer: They are thought of as "treasures," are protected, and get special attention.

R 5. Review the headings. Try to answer each question from memory.

INFORMATION SKILLS

Pandas

What animal is black and white, fat and furry, cute and comical? You guessed it — the panda! Pandas have always lived in the far-off forests of western China. Nowadays, a few pandas also live in zoos.

What Giant Pandas Look Like

The Chinese name for the panda is *beishung,* a word that means "white bear." However, giant pandas are white *and* black. They have black ears and eye patches. They look as if they are wearing a short jacket with a black collar and black sleeves, and high black socks. Giant pandas are very large. An adult panda can weigh up to 350 pounds!

Where Pandas Live

Pandas living in the wild in China almost never leave their forest home. They move about in the thick jungle of bamboo trees, far from people. Most adult pandas live on their own. In bad weather, they may go inside caves or hollow tree trunks. Most of the time, though, they spend the day eating, sleeping, and playing.

The Panda's Eating Habits

Because pandas are so large, they need a lot of food. Usually the only kind of food they eat is the leafy shoots, or stems, of bamboo trees. The panda stands on its hind legs and takes the bamboo shoot in its front paw. The panda's paw is the perfect "tool" for this job. Below its five claw pads is a thumblike pad, designed to grasp the bamboo shoot and pull it down. The panda gets several shoots and sits down to eat. It uses its strong jaws to chew the tough bamboo.

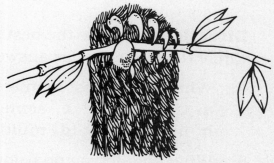

A panda's "thumb" allows it to grasp and pull down its food.

Why Pandas Are Lucky

Pandas might be called "lucky." These animals are rare, but efforts are being made to save them from dying out. The Chinese government has said that pandas are a "national treasure," so they are protected by law. The pandas that live in zoos are considered "treasures," too. These lovable animals get lots of special care and attention.

Taking Tests

Complete this page as if you were taking a test. Pretend that you have fifteen minutes. Look quickly through the entire test before you begin.

Directions: Complete each sentence by writing the correct answer in the blank.

1. You should look over the entire test __before__ you answer the questions.

2. If there's a question you can't answer, you should __skip it__ and go on to the next one.

3. When you finish the test, you should __check__ your answers.

Directions: Choose the best answer to each question. Circle the letter of your answer.

4. What kind of test item is this question an example of?
   a. true–false   c. sentence completion
   b. matching   (d.) multiple choice

5. What should you do before you write the answer for this kind of test item?
   a. go on to the next question   c. time yourself
   (b.) read all the answer choices   d. ask for help

Directions: Write **T** in the blank before the statement if the statement is true. Write **F** if it is false.

__F__ 6. This kind of test item is an example of a multiple-choice question.

__F__ 7. It is not important to read the directions for a test carefully.

__T__ 8. Before you begin a test, you should find out how much time you have to answer all the questions.

INFORMATION SKILLS

Statements and Questions

Different kinds of sentences have different jobs. A sentence that tells something is a **statement**. A statement ends with a **period** (.). A sentence that asks something is a **question**. A question ends with a **question mark** (?).

A sentence always begins with a capital letter.

Statements	**Questions**
The airport was crowded.	Was the airport crowded?
Her plane landed on time.	When did her plane land?
Carlos bought a ticket.	Did Carlos buy a ticket?

GUIDED PRACTICE You will need to guide students through the Guided Practice activity, providing support as necessary.

Is each sentence a statement or a question? What end mark should follow each sentence?

Example: The flight attendant welcomed all the passengers
    statement   period

1. I pushed my small brown bag under the seat S .
2. Have you fastened your seat belts Q ?
3. The takeoff was very smooth S .
4. Can you see out the window Q ?
5. How high will the plane climb Q ?
6. We will land in about an hour S .
7. Is this your first flight Q ?

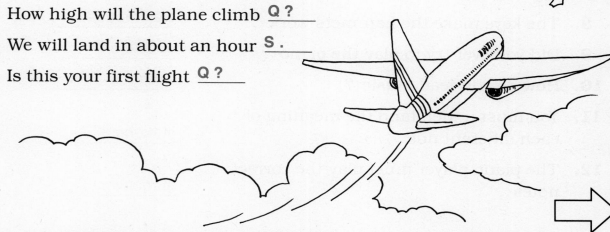

LANGUAGE AND USAGE   85

REMINDER

▶ A **statement** tells something. It ends with a **period** *(.)*.

▶ A **question** asks something. It ends with a **question mark** *(?)*.

Statements	Questions
The concert is today.	When is the concert?
I know that tune.	Do you know that tune?

INDEPENDENT PRACTICE

Circle the end mark for each sentence. Write *statement* if the sentence tells something. Write *question* if the sentence asks something.

Example: The music is wonderful(.)   statement

1. What instrument is being played(?)   question
2. It sounds like a piano(.)   statement
3. A piano has eighty-eight keys(.)   statement
4. How does the piano make sounds(?)   question
5. There are wires inside the piano(.)   statement
6. Felt-covered hammers hit the wires(.)   statement
7. The piano player strikes the keys on the keyboard(.)   statement
8. The keys make the hammers work(.)   statement
9. Did you ever try to play the piano(?)   question
10. How do you read music(?)   question
11. You must understand the meaning of each different note(.)   statement
12. The piano player must play the correct notes(.)   statement

LANGUAGE AND USAGE

Commands and Exclamations

You have learned about two kinds of sentences called statements and questions. Now you will learn about two other kinds of sentences.

A sentence that tells someone to do something is a **command**. A command ends with a period. A sentence that shows strong feeling such as surprise, excitement, or fear is an **exclamation**. It ends with an **exclamation point** (**!**).

Remember to begin every sentence with a capital letter.

Commands	Exclamations
Please wait at the bus stop.	**T**he bus finally arrived**!**
Meet me at Page's Bookstore.	**W**hat a great store it is**!**
Take the subway home.	**H**ow fast the train travels**!**

GUIDED PRACTICE

Is each sentence a command or an exclamation? What end mark should be put at the end of each sentence?

Example: Planning a trip is so exciting   *exclamation*   **!**

1. Apply for your passport   C .
2. Please answer all questions carefully   C .
3. Have your picture taken   C .
4. We're leaving at last   E !
5. My dream is coming true   E !

LANGUAGE AND USAGE

87

REMINDER

> ▸ A **command** tells someone to do something. It ends with a **period** *(.)*.
>
> ▸ An **exclamation** shows surprise, excitement, or fear. It ends with an **exclamation point** *(!)*.
>
> **Commands**
> **P**lease bake some bread.
> **F**ind a wooden spoon.
>
> **Exclamations**
> **W**hat a terrific idea!
> **H**ow easy this is!

INDEPENDENT PRACTICE

Circle the end mark for each sentence. Label each sentence *command* or *exclamation*.

Example: Get all the ingredients ready(.) — **command**

1. Look in the refrigerator for butter and eggs(.) — command
2. What large eggs these are(!) — exclamation
3. Place all the ingredients on the table(.) — command
4. Take out a large mixing bowl(.) — command
5. How carefully you crack the eggs(!) — exclamation
6. Finally we can add the yeast to the batter(!) — exclamation
7. Stir the mixture in the bowl(.) — command
8. Knead the dough for five minutes(.) — command
9. This is so much fun(!) — exclamation
10. Ask Charles to turn on the oven(.) — command
11. How hot the oven gets(!) — exclamation
12. Put the loaf pans on the middle rack(.) — command
13. I can't wait to taste the bread(!) — exclamation
14. The bread is finally done(!) — exclamation
15. How delicious this wheat bread is(!) — exclamation

LANGUAGE AND USAGE

Complete Subjects and Complete Predicates

A sentence must have two parts to tell a complete thought. The **subject** tells *whom* or *what* the sentence is about. The **predicate** tells what the subject *does* or *is*.

All the words in the subject make up the **complete subject**. All the words in the predicate make up the **complete predicate**. A complete subject or a complete predicate may be one word.

Complete Subjects	Complete Predicates
Angela Kelly	is the captain of the boat.
We	waited at the dock.
The red ferryboat	stops.
Passengers	get off the boat.

You can find the subject of a sentence by asking *whom* or *what* the sentence is about. You can find the predicate by asking what the subject *does* or *is*.

GUIDED PRACTICE

What are the complete subject and the complete predicate of each sentence?   **Subj. underlined once; pred., twice.**

Example: The ocean was calm today.
      subject: *The ocean*   **predicate:** *was calm today*

1. Several sea gulls flew overhead.
2. They landed on the water.
3. John Day fished from the wharf.
4. He cast his line into the sea.
5. Fish swam below.
6. A large fish tugged on John's line.
7. The excited boy pulled in his catch.

LANGUAGE AND USAGE

Reminder

> ▸ The **complete subject** includes all the words that tell *whom* or *what* the sentence is about.
>
> ▸ The **complete predicate** includes all the words that tell what the subject *does* or *is*.
>
> | Complete Subjects | Complete Predicates |
> |---|---|
> | Many children | like cartoons. |
> | Cartoons | are exciting for people of all ages. |

Independent Practice

Tell what is underlined in each sentence. Write *CS* for the complete subject or *CP* for the complete predicate.

Example: Many people <u>listen to the radio</u>.   **CP**

1. People <u>enjoy different kinds of shows</u>.   CP
2. <u>Television</u> offers many kinds of programs.   CS
3. <u>Many adults</u> like evening comedy shows.   CS
4. Movies <u>are very popular</u>.   CP
5. <u>Walt Disney</u> made many fine films.   CS
6. <u>Children of all ages</u> enjoy his adventure films.   CS
7. My family <u>attends concerts</u>.   CP
8. <u>The sound of violins</u> is beautiful.   CS
9. <u>Grandmother</u> goes to many plays in the city.   CS
10. *My Fair Lady* <u>is her favorite musical</u>.   CP
11. The circus <u>is made up of many acts</u>.   CP
12. A happy audience <u>cheers</u>.   CP
13. <u>Other people</u> attend operas.   CS
14. <u>They</u> listen to the trained voices of the singers.   CS
15. Good shows <u>are an important part of life</u>.   CP

LANGUAGE AND USAGE

Simple Subjects

You have learned that the complete subject includes all the words that tell whom or what the sentence is about. In every complete subject, there is one main word. Sometimes this main word is a name. This main word or name tells exactly whom or what the sentence is about. It is called the **simple subject**. Sometimes the complete subject and the simple subject are the same. The simple subjects below are in bold type.

Complete Subjects	Complete Predicates
Many **people**	watch ball games at the park.
Martin Johnson	slides into third base.
He	pitched five innings.
The **palm** of his glove	is torn.

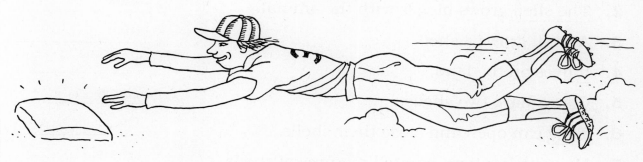

GUIDED PRACTICE The complete subject of each sentence is underlined. What is the simple subject?

Example: James Naismith invented basketball in 1891.
James Naismith

1. He was a teacher in Springfield, Massachusetts.
2. The head of the school wanted an indoor winter game.
3. Naismith tacked peach baskets to the walls of the gym.
4. The first players used soccer balls.
5. Each team had nine players.

REMINDER

> ▸ The **simple subject** is the main word or words in the complete subject. Sometimes the complete subject and the simple subject are the same.
>
> **Subjects**   **Predicates**
> Beautiful **seashells**   come from the ocean.
> **Mr. Roe**   collects all kinds of shells.

INDEPENDENT PRACTICE

The complete subject of each sentence is underlined. Write the simple subject.

Example: Most creatures with shells live in or near the water. — **creatures**

1. Animals with soft bodies live in shells. — **Animals**
2. The shell grows bigger with the animal. — **shell**
3. A snail grows a shell. — **snail**
4. It lives on the land. — **It**
5. Clams have two shells. — **Clams**
6. They can open and close their shells. — **They**
7. Many artists have painted pictures of shells. — **artists**
8. People collect shells as a hobby. — **People**
9. Ms. Roe strings shells to make necklaces. — **Ms. Roe**
10. She finds very few unbroken shells. — **She**
11. The tide brings the shells up onto the beach. — **tide**
12. Sunlight can change the color of shells. — **Sunlight**
13. The force of the ocean breaks many shells. — **force**
14. The broken pieces stay on the sand. — **pieces**
15. These treasures from the ocean are beautiful. — **treasures**

LANGUAGE AND USAGE

Simple Predicates

You know that the complete predicate includes all the words that tell what the subject does or is. In every complete predicate, there is one main word. This main word tells exactly what the subject does or is. It is called the **simple predicate**.

In each sentence below, the simple predicate is in a box.

Complete Subjects	Complete Predicates
Some students	go to space camp.
The camp	is in Alabama.
Campers	build rockets.
They	wear real space suits.

GUIDED PRACTICE The complete predicate of each sentence is underlined. What is the simple predicate?

Example: Students come from all over the country.
come

1. Campers are astronauts for a week.
2. They work in teams of ten.
3. The members name their teams after planets.
4. Some of the teams launch rockets into the air.
5. Other teams take a make-believe space flight.

LANGUAGE AND USAGE

REMINDER

> ▶ The **simple predicate** is the main word in the predicate.
>
> | Subjects | Predicates |
> |----------|------------|
> | Most dogs | **make** great pets. |
> | They | **are** good friends for children. |

INDEPENDENT PRACTICE

The complete predicate of each sentence is underlined. Write the simple predicate.

Example: People <u>teach dogs different kinds of tricks</u>.   teach

1. Puppies <u>catch newspapers in their mouths</u>.   catch
2. Other dogs <u>beg for food</u>.   beg
3. Many dogs <u>play quietly inside the house</u>.   play
4. Some pets <u>do important jobs</u>.   do
5. Guide dogs <u>help blind people</u>.   help
6. Watchdogs <u>guard people's homes</u>.   guard
7. These dogs <u>bark at strangers</u>.   bark
8. Dogs <u>depend on their owners for daily care</u>.   depend
9. These animals <u>need exercise every day</u>.   need
10. Meat and chicken <u>are good foods for them</u>.   are
11. Some children <u>feed vitamins to their dogs</u>.   feed
12. Some owners <u>enter their dogs in shows</u>.   enter
13. Judges <u>vote for the best dogs</u>.   vote
14. The most obedient dog <u>wins a blue ribbon</u>.   wins
15. The prettiest dog <u>gets a prize at some shows</u>.   gets
16. Dogs <u>make good pets indeed</u>!   make

LANGUAGE AND USAGE

Combining Sentences: Subjects and Predicates

There are ways to combine sentences to make your writing more interesting. Sometimes you can combine two sentences that have different subjects but the same predicate. Join the subjects with the connecting word *and* to make one **compound subject**.

Kim watched.
Ben watched. → Kim and Ben watched.

Sometimes you can combine two sentences that have different predicates but the same subject. Join the predicates with *and* to make one **compound predicate**.

They smiled.
They reported the news. → They smiled and reported the news.

GUIDED PRACTICE

How would you combine each pair of sentences into one sentence with a compound subject or a compound predicate?

Example: Carl visited Hollywood. Ann visited Hollywood. (subject)
Carl and Ann visited Hollywood.

1. Movies are filmed in Hollywood. TV shows are filmed in Hollywood. (subject) Movies and TV shows are filmed in Hollywood.

2. Ann visited a movie studio. Ann toured the sets. (predicate)
Ann visited a movie studio and toured the sets.

3. She liked the special effects. Her mother liked the special effects. (subject) She and her mother liked the special effects.

4. Carl went to a TV studio. Carl saw a show. (predicate)
Carl went to a TV studio and saw a show.

LANGUAGE AND USAGE

REMINDER

> ▶ You can combine two sentences with the same predicate. Join the subjects with *and* to make a **compound subject**.
>
> **Russ** liked to fish.
> **Cindy** liked to fish. ⟶ compound subject
> **Russ** *and* **Cindy** liked to fish.
>
> ▶ You can combine two sentences with the same subject. Join the predicates with *and* to make a **compound predicate**.
>
> They **walked** to the lake.
> They **rowed** to an island. ⟶ compound predicate
> They **walked** to the lake *and* **rowed** to an island.

INDEPENDENT PRACTICE

Write each pair of sentences as one sentence with a compound subject or a compound predicate. Join the underlined words with the connecting word *and*.

Example: My family went fishing. I went fishing.

My family and I went fishing.

1. We took out our gear. We baited our hooks. **We took out our gear and baited our hooks.**

2. Russ needed help with the line. Cindy needed help with the line. **Russ and Cindy needed help with the line.**

3. Dad caught some small fish. My sister caught some small fish. **Dad and my sister caught some small fish.**

4. Russ got the food basket. Russ passed out some fruit. **Russ got the food basket and passed out some fruit.**

5. Mom had a great time. Cindy had a great time. **Mom and Cindy had a great time.**

6. They caught the most fish. They pulled in the biggest trout. **They caught the most fish and pulled in the biggest trout.**

LANGUAGE AND USAGE

Correcting Run-on Sentences

When two sentences run into each other, they make a **run-on sentence**. Do not use run-on sentences in your writing.

A run-on sentence can be corrected by writing each complete thought as a separate sentence. Remember to use capital letters and end marks correctly.

> Incorrect: Our class visited a museum we saw whaling ships.
>
> Correct: Our class visited a museum. We saw whaling ships.

Guided Practice

Which of the following sentences are run-on sentences? Which sentences are correct? What are the two complete thoughts in each run-on sentence?

Example: Many cities have history museums there are many different kinds.   *run-on*
Many cities have history museums.
There are many different kinds.

1. History museums are fun they teach about the past.
2. A whole village can sometimes be a museum.
3. People dress in costumes people can ask them questions.

History museums are fun. They teach about the past.

People dress in costumes. Visitors can ask them questions.

LANGUAGE AND USAGE

REMINDER

> ▸ A **run-on sentence** has two complete thoughts that run into each other. Correct a run-on sentence by writing each complete thought as a separate sentence.
>
> **Incorrect:** They went on a trip the three of them took the train.
>
> **Correct:** They went on a trip. The three of them took the train.

INDEPENDENT PRACTICE

Correct each run-on sentence by writing it as two separate sentences.

Example: Cora and Theo visited Seaside Farm Mr. Li went with them.

Cora and Theo visited Seaside Farm.

Mr. Li went with them.

1. Seaside Farm is very old it's the oldest farm in the state.
   Seaside Farm is very old. It's the oldest farm in the state.

2. Seaside is two hundred years old it is a very large farm.
   Seaside is two hundred years old. It is a very large farm.

3. There are seventy chickens there they lay sixty eggs a day.
   There are seventy chickens there. They lay sixty eggs a day.

4. On some days there are hay rides many visitors enjoy them.
   On some days there are hay rides. Many visitors enjoy them.

5. People ride in a wagon full of hay everybody brings picnic lunches.
   People ride in a wagon full of hay. Everybody brings picnic lunches.

6. The owner lets visitors milk the cows this job isn't for everyone.
   The owner lets visitors milk the cows. This job isn't for everyone.

7. Farmers wake up early it takes all day to get the chores done.
   Farmers wake up early. It takes all day to get the chores done.

LANGUAGE AND USAGE

Common and Proper Nouns

There are two kinds of nouns. A noun that names any person, place, or thing is called a **common noun**. A noun that names a particular person, place, or thing is called a **proper noun**.

	Common Nouns	Proper Nouns
Persons	girl uncle queen	**M**arie **U**ncle **G**eorge **Q**ueen **E**lizabeth
Places	state country bay park	**K**ansas **C**anada **B**ay of **F**undy **G**lacier **N**ational **P**ark
Things	pet day holiday	**P**atches **S**aturday **F**ourth of **J**uly

When you write a proper noun, always begin it with a capital letter. If a proper noun is more than one word, capitalize the first letter of each important word.

Guided Practice

Find the common noun and the proper noun in each sentence. Which nouns should begin with capital letters?
CN underlined once; PN twice. PN should be cap.

Example: tanya is an explorer.
          common: *explorer* **proper:** *(cap.) Tanya*

1. Her kitten magellan is too!
2. Their trips to florida are always exciting.
3. Do the alligators in the everglades national park look scary?
4. The guides at cape canaveral are helpful.
5. Is there lost treasure in the gulf of mexico?

LANGUAGE AND USAGE

REMINDER

> ▸ A **common noun** names any person, place, or thing.
> ▸ A **proper noun** names a particular person, place, or thing.
> ▸ Capitalize each important word in a proper noun.
>
>   proper   common   common
>   noun   noun   noun
> The **Sahara Desert** is the largest **desert** in the **world**.

INDEPENDENT PRACTICE

Write *C* for each underlined common noun and *P* for each underlined proper noun. Then write the proper nouns correctly.

Example: At its longest, the *sahara* is 3200 miles.   **P — Sahara**

1. This desert is located in northern africa.   P—Africa
2. It stretches from the Atlantic Ocean to the red sea.   P—Red Sea
3. Only about four inches of rain fall there each year.   C
4. Very few people live there.   C
5. Most of the sahara is made up of rock, not sand.   P—Sahara
6. Some of the sand dunes rise over two hundred feet.   C
7. The namib desert is in southeastern Africa.   P—Namib Desert
8. It is along the atlantic ocean.   P—Atlantic Ocean
9. The orange river is south of the desert.   P—Orange River
10. Two more large rivers lie north of the desert.   C
11. The cape of good hope is also in Africa.   P—Cape of Good Hope
12. It is located at the southern tip of Africa.   C
13. An explorer first sailed around the Cape in 1487.   C
14. His name was dias.   P—Dias

LANGUAGE AND USAGE

Singular and Plural Nouns

A noun can name one or more than one. A noun that names only one person, place, or thing is called a **singular noun**. A noun that names more than one is called a **plural noun**.

Singular Nouns
One **goat** is in the **barn**.
This **hen** laid one **egg**.

Plural Nouns
Many **goats** are in those **barns**.
These **hens** laid a dozen **eggs**.

How to Form Plurals

Rules	Singular	Plural
Add *s* to most singular nouns.	one boy one puddle a rose	two boy**s** both puddle**s** ten rose**s**
Add *es* to a singular noun that ends with *s*, *x*, *ch*, or *sh*.	one bus this box one bunch a wish	three bus**es** some box**es** six bunch**es** many wish**es**

Guided Practice

What is the plural form of each of the following singular nouns?

Example: peach   *peaches*

1. brush *es*
2. gift *s*
3. class *es*
4. patch *es*
5. prize *s*
6. circus *es*
7. inch *es*
8. fox *es*

LANGUAGE AND USAGE

REMINDER

▸ A **singular noun** names one person, place, or thing.
▸ A **plural noun** names more than one person, place, or thing.
▸ To form plural nouns, add *s* to most singular nouns.
▸ Add *es* to singular nouns that end with *s, x, ch,* or *sh.*

Singular: toy   store   glass   fo*x*   lun*ch*   di*sh*
Plural: toy**s**   store**s**   glass**es**   fox**es**   lunch**es**   dish**es**

INDEPENDENT PRACTICE

Underline each singular noun. Write each plural noun.

Example: I have worked on my project for weeks. — weeks

1. One girl is building a house from tiny boxes. — boxes
2. A boy will hang balloons from the ceiling. — balloons
3. That picture was made with patches of cloth. — patches
4. One student is busy with paints and brushes. — paints, brushes
5. His picture is full of bright colors. — colors
6. Our teacher is making roses from paper. — roses
7. Bunches of flowers are on the table. — Bunches, flowers
8. Chairs will be placed in a circle on the floor. — Chairs
9. Each parent will bring snacks or drinks. — snacks, drinks
10. Sandwiches will be served on a large tray. — Sandwiches
11. My mother will bring glasses. — glasses
12. Invitations were sent to other classes. — Invitations, classes
13. Three teachers and their students are coming. — teachers, students
14. Some mothers and fathers will come too. — mothers, fathers
15. All the guests will enjoy our show. — guests

LANGUAGE AND USAGE

Nouns Ending with y

You have already learned some rules for making nouns plural. Here are two special rules for making the plural forms of nouns that end with *y*.

How to Form Plurals

Rules	Singular	Plural
If the noun ends with a vowel and *y*, add *s*.	one toy a monkey	many toy**s** five monkey**s**
If the noun ends with a consonant and *y*, change the *y* to *i* and add *es*.	one family this city a baby	some famil**ies** six cit**ies** two bab**ies**

Guided Practice

What is the plural form of each noun?

Example: pony *ponies*

1. berry — berries
2. holiday — holidays
3. turkey — turkeys
4. boy — boys
5. blue jay — blue jays
6. party — parties
7. lady — ladies
8. donkey — donkeys
9. puppy — puppies
10. sky — skies
11. hobby — hobbies
12. key — keys

LANGUAGE AND USAGE

REMINDER

▸ Some nouns end with a vowel and *y*. Add *s* to make these nouns plural.

▸ Some nouns end with a consonant and *y*. Change the *y* to *i* and add *es* to make these nouns plural.

Singular Nouns: one **valley**   one **country**
Plural Nouns: four **valleys**   two **countries**

INDEPENDENT PRACTICE

Write the plural form of each singular noun.

Example: penny   pennies

1. city — cities
2. subway — subways
3. alley — alleys
4. factory — factories
5. chimney — chimneys
6. company — companies
7. entry — entries
8. lobby — lobbies
9. story — stories
10. sky — skies
11. library — libraries
12. display — displays
13. highway — highways
14. journey — journeys
15. family — families
16. holiday — holidays
17. birthday — birthdays
18. toy — toys
19. puppy — puppies
20. day — days
21. play — plays
22. hobby — hobbies
23. party — parties
24. boy — boys
25. grocery — groceries
26. key — keys
27. supply — supplies
28. way — ways
29. monkey — monkeys
30. berry — berries
31. lady — ladies
32. daisy — daisies
33. majesty — majesties
34. bay — bays

LANGUAGE AND USAGE

More Plural Nouns

You know that you add *s* or *es* to form the plurals of most nouns. There are some nouns, however, that have special plural forms. Since these words follow no spelling pattern, you must remember them.

Singular and Plural Nouns			
Singular	**Plural**	**Singular**	**Plural**
one child	two child**ren**	each tooth	five t**ee**th
a man	many m**en**	one goose	both g**ee**se
this woman	three wom**en**	an ox	nine ox**en**
that foot	these f**ee**t	a mouse	some m**ic**e

Other nouns are the same in both singular and plural forms.

SINGULAR NOUNS
One **deer** nibbled the bark.
Did you see a **moose**?
I have a pet **sheep**.

PLURAL NOUNS
Several **deer** ate quietly.
Two **moose** crossed a stream.
These **sheep** have soft wool.

Guided Practice

Complete each sentence with the plural form of the underlined noun.

Example: One child helped both smaller __children__ tie their sneakers.

1. That man sang while two other __men__ played guitars.
2. This sheep is my pet, and those __sheep__ belong to Fred.
3. Pat hopped on one foot and then jumped with both __feet__.
4. One goose flew by, and three __geese__ swam in the pond.
5. Rex had a loose tooth, but his other __teeth__ were fine.
6. Ana saw one moose in Maine and four __moose__ in Canada.

LANGUAGE AND USAGE

REMINDER

▶ Some nouns have special plural forms.
▶ Some nouns have the same singular and plural forms.

Singular Nouns:	child	man	woman	foot	tooth	ox
Plural Nouns:	children	men	women	feet	teeth	oxen
Singular Nouns:	goose	mouse	deer	moose	sheep	
Plural Nouns:	geese	mice	deer	moose	sheep	

INDEPENDENT PRACTICE

Write each underlined noun. Label it *S* for singular or *P* for plural.

Example: The <u>children</u> saw an ox at the farm.   **children — P**

1. Some <u>men</u> hitched an ox to a wagon.   men — P
2. There were two more <u>oxen</u> in the barn.   oxen — P
3. Each ox had many strong white <u>teeth</u>.   teeth — P
4. A <u>woman</u> showed the children some baby sheep.   woman — S
5. Two men were cutting the wool from a big <u>sheep</u>.   sheep — S
6. One child saw some <u>mice</u> in the barn.   mice — P
7. A little <u>mouse</u> was near her foot.   mouse — S
8. She moved her <u>foot</u> out of the way.   foot — S
9. Several women were feeding some <u>geese</u>.   geese — P
10. A <u>goose</u> has feet like a duck's.   goose — S
11. A farmer saw some <u>deer</u> eating the corn.   deer — P
12. Can a moose eat as much as ten <u>sheep</u>?   sheep — P
13. Did you see three <u>moose</u> in the field?   moose — P
14. Each moose was even bigger than an <u>ox</u>.   ox — S
15. Every <u>child</u> had a good time at the farm.   child — S

LANGUAGE AND USAGE

Action Verbs

You know that every sentence has a subject and a predicate. The main word in the predicate is the verb. A **verb** is a word that can show action. When a verb tells what people or things do, it is called an **action verb**.

Subjects	Predicates
Rita and Eric	⬚dig⬚ slowly and carefully.
The students	⬚helped⬚ the scientists.
Rita	⬚uncovered⬚ some pottery.
The pieces of pottery	⬚provide⬚ clues about the past.

GUIDED PRACTICE

What is the action verb in each of the following sentences?

Example: Rita cleaned the pieces of pottery. *cleaned*

1. Eric <u>found</u> some old tools.
2. Raul <u>made</u> a map of the site.
3. Two students <u>stand</u> in the water.
4. They <u>hold</u> a tub with a screen in the bottom.
5. Water <u>fills</u> the tub.
6. Raul <u>pours</u> dirt into the tub.
7. Light objects <u>float</u> in the water.
8. Dirt <u>goes</u> through the screen.
9. The students <u>take</u> careful notes.
10. They <u>attach</u> labels to the objects.
11. The scientists <u>take</u> the objects to their lab.
12. They <u>learn</u> many things about early people.

LANGUAGE AND USAGE

REMINDER

> ▶ An **action verb** is a word that tells what people or things do.
>
>   verb
> The ship **sailed** away weeks ago.
>   predicate
>
>   verb
> Bill **looks** for the ship every day.
>   predicate

INDEPENDENT PRACTICE

Write the action verb in each underlined predicate.

Example: Bill stood on the dock.   **stood**

1. Bill watched the workers on the dock.   watched
2. The workers loaded boxes onto a large ship.   loaded
3. Bill's father works on this ship.   works
4. Some people boarded the ship.   boarded
5. The crew checked their names on a list.   checked
6. The captain blew the whistle.   blew
7. Bill heard three loud blasts.   heard
8. Tugboats towed the ship into deep water.   towed
9. The ship left the harbor after that.   left
10. Bill watches it for a long time.   watches
11. He walks home slowly.   walks
12. He misses his father already.   misses
13. Bill's father sends letters from far away.   sends
14. Bill goes to the docks again.   goes
15. The big ship arrives home at last!   arrives

LANGUAGE AND USAGE

Main Verbs and Helping Verbs

A verb may be more than one word. The **main verb** is the most important verb. The **helping verb** comes before it.

Some Common Helping Verbs		
am	was	has
is	were	have
are	will	had

The main verbs below are in bold. Helping verbs are in italics.

> Alfredo *is* **training** for the Olympics.
> He *has* **run** five miles each day.
> His coach *will* **help** him next week.

GUIDED PRACTICE

Find each helping verb and main verb.

Example: Sara had entered the summer Olympics.
helping verb: had   **main verb:** entered
HV is underlined once; MV, twice.

1. She was racing in a wheelchair race.
2. Sara had joined the Wheelchair Athlete Club.
3. The racers were using special racing wheelchairs.
4. They are training several times a week.
5. They have lifted weights too.
6. Sara has raced for several years.
7. She will race many more times.
8. She is practicing for next year's Olympics.

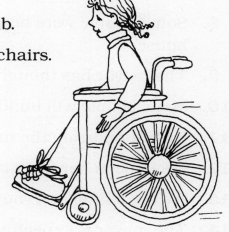

REMINDER

> ▸ A verb may be more than one word.
> ▸ The **main verb** is the most important verb.
> ▸ The **helping verb** comes before the main verb.
>
>   helping main   helping main
>   verb   verb   verb   verb
> I **am going** with my brother.   We **will walk** together.

INDEPENDENT PRACTICE

Look at each underlined verb. Write *helping* if it is the helping verb. Write *main* if it is the main verb.

Example: My brother Pablo <u>is</u> taking me to the park. — helping

1. We <u>will</u> see an old tree house. — helping
2. Pablo was <u>telling</u> me about it yesterday. — main
3. Pablo has <u>seen</u> it many times. — main
4. We <u>were</u> reading about it in the newspaper. — helping
5. The tree house <u>has</u> stood for years. — helping
6. Last year the wind <u>had</u> blown off the roof. — helping
7. The rain has <u>hurt</u> the house too. — main
8. Some people were <u>hoping</u> to save the old tree house. — main
9. The mayor has <u>thought</u> of a plan. — main
10. Many people <u>will</u> build a new tree house together. — helping
11. I am <u>supporting</u> the mayor's plan. — main
12. I have <u>seen</u> her work in our parks. — main
13. We <u>are</u> telling everyone in town the news. — helping
14. The mayor <u>is</u> expecting success. — helping
15. She has <u>worked</u> hard. — main

LANGUAGE AND USAGE

Present, Past, and Future

A verb tells when something happens. The **tense** of a verb lets you know whether something happens in the present, in the past, or in the future.

Present	**Past**	**Future**
hunt, hunts	hunted	will hunt
fly, flies	flew	will fly

Verb Tenses

A verb in the **present tense** shows action that is happening now.	Bats **hunt** at night. Now the bat **flies.**
A verb in the **past tense** shows action that has already happened. Many verbs in the past tense end with *-ed*.	It **hunted** last night. The bats **flew.**
A verb in the **future tense** shows action that will happen. Verbs in the future tense use the helping verb *will*.	They **will hunt** tonight. The bat **will fly.**

GUIDED PRACTICE What is the verb in each sentence? Is it in the present tense, the past tense, or the future tense?

Example: Sherry likes many kinds of animals.   *likes — present*

1. We will go to the library tomorrow.   fut.
2. I finished a book about birds yesterday.   past
3. Sherry will look for a book about bats.   fut.
4. She collects facts about animals.   pres.
5. The librarian always helps us.   pres.

LANGUAGE AND USAGE

REMINDER

> ▶ A **present tense** verb shows action that is happening now.
> ▶ A **past tense** verb shows action that has already happened.
> ▶ A **future tense** verb shows action that will happen.
>
> **Present Tense:** The Kamals **live** here now.
> **Past Tense:** The Kamals **lived** here last year.
> **Future Tense:** The Kamals **will live** here for many years.

INDEPENDENT PRACTICE

Write *present*, *past*, or *future* for each underlined verb.

Example: The Kamals <u>will leave</u> early in the morning. **future**

1. Kate Kamal's family <u>moved</u> to Los Angeles. — past
2. Mrs. Kamal <u>works</u> in Long Beach. — present
3. She <u>rides</u> a long way to her job every day. — present
4. Soon the Kamals <u>will move</u> to Long Beach. — future
5. Then Mrs. Kamal <u>will travel</u> only a few miles to work. — future
6. The Kamals <u>prepare</u> everything for the move. — present
7. Last week Mr. Kamal <u>found</u> many empty boxes. — past
8. He <u>saved</u> them for moving day. — past
9. Now the Kamals <u>pack</u> books and records into boxes. — present
10. The Kamals <u>will pack</u> the pots and pans next. — future
11. Tonight they <u>will eat</u> dinner at a restaurant. — future
12. Kate <u>gave</u> her new address to all her friends. — past
13. She <u>will give</u> us her new phone number soon. — future
14. Everyone <u>wished</u> Kate good luck. — past
15. Now the Kamals <u>wave</u> good-by to us. — present

LANGUAGE AND USAGE

Making Subjects and Verbs Agree

A verb in the present tense must **agree** with the subject of the sentence. This means that the subject and the verb must work together. They must both be singular or both be plural.

	Subject-Verb Agreement
Singular subjects	When the subject is a singular noun or *he*, *she*, or *it*, add *s* to the verb. A computer **helps** people. It **solves** problems.
Plural subjects	When the subject is a plural noun or *I*, *we*, *you*, or *they*, do not add *s* to the verb. Computers **help** people. They **solve** problems.

Guided Practice

Which verb correctly completes each sentence?

Example: Felita (own, owns) a small computer.
owns

1. She (use, <u>uses</u>) it to do her homework.
2. The computer (help, <u>helps</u>) her parents too.
3. Her brothers (<u>play</u>, plays) games on it.
4. Many companies also (<u>buy</u>, buys) computers.
5. Computers (<u>work</u>, works) very rapidly.
6. They (<u>store</u>, stores) a great deal of information.
7. I (<u>want</u>, wants) to learn more about computers.

LANGUAGE AND USAGE

REMINDER

> ▸ A present tense verb must **agree** with the subject of the sentence.
>
> ▸ Add *s* to the verb when the subject is a singular noun or *he, she,* or *it.*
>
> ▸ Do not add *s* to the verb when the subject is a plural noun or *I, we, you,* or *they.*
>
> The <u>cat</u> **plays**.   The <u>cats</u> **play**.
> My <u>puppy</u> **barks**.   Both <u>puppies</u> **bark**.
> Now <u>it</u> **hides**.   Now <u>they</u> **hide**.

INDEPENDENT PRACTICE

Write the verb that agrees with the underlined subject.

Example: My <u>dog</u> (follow, follows) me everywhere.   **follows**

1. <u>Razz</u> (hide, hides) in the tall grass.   **hides**
2. <u>We</u> (run, runs) together in the field.   **run**
3. <u>Dogs</u> (need, needs) daily exercise.   **need**
4. My <u>brother</u> (tell, tells) me about dog care.   **tells**
5. <u>He</u> (find, finds) a box for Razz.   **finds**
6. An old <u>towel</u> (keep, keeps) the box warm.   **keeps**
7. <u>Razz</u> (curl, curls) up in the bed.   **curls**
8. On cold nights, <u>Razz</u> (sleep, sleeps) on my bed.   **sleeps**
9. <u>We</u> (clean, cleans) Razz's fur every morning.   **clean**
10. <u>You</u> (comb, combs) a pet's fur regularly.   **comb**
11. My <u>sisters</u> (enjoy, enjoys) cats more than dogs.   **enjoy**
12. <u>They</u> (say, says) dogs are too noisy.   **say**
13. <u>I</u> (like, likes) dogs the best anyway.   **like**
14. <u>You</u> (care, cares) so much for your pet.   **care**

LANGUAGE AND USAGE

Irregular Verbs

Verbs that do not add -ed to show past action are called **irregular verbs**.

I **eat** now.   I **ate** earlier.   I **have eaten** already.

Because irregular verbs do not follow a regular pattern, you must remember their spellings.

IRREGULAR VERBS

Present	Past	Past with Helping Verb
begin	began	(has, have, had) begun
break	broke	(has, have, had) broken
bring	brought	(has, have, had) brought
come	came	(has, have, had) come
drive	drove	(has, have, had) driven
eat	ate	(has, have, had) eaten
give	gave	(has, have, had) given
grow	grew	(has, have, had) grown
know	knew	(has, have, had) known
say	said	(has, have, had) said
sing	sang	(has, have, had) sung
tell	told	(has, have, had) told
throw	threw	(has, have, had) thrown

GUIDED PRACTICE What are the past tense and the past with a helping verb for each irregular verb below?

Example: sing   *sang*   *sung*

1. eat — ate, eaten
2. bring — brought, brought
3. give — gave, given
4. know — knew, known
5. come — came, come
6. break — broke, broken

LANGUAGE AND USAGE

REMINDER

> ▸ Verbs that do not add *-ed* to show past action are called **irregular verbs**. You must remember their spellings.
>
> | **Present** | **Past** | **Past with Helping Verbs** |
> |---|---|---|
> | take | took | (has, have, had) taken |
> | wear | wore | (has, have, had) worn |
> | make | made | (has, have, had) made |

INDEPENDENT PRACTICE

Write the correct form of the verb in ( ) to complete each sentence.

Example: I had __eaten__ a cherry. **(ate, eaten)**

1. The pit of the cherry had __broken__ two teeth. **(broke, broken)**
2. The dentist had __known__ what to do. **(knew, known)**
3. He had __taken__ a picture of the teeth. **(took, taken)**
4. Dr. Levy __said__ the teeth were fine. **(say, said)**
5. Then he __took__ a piece of warm clay from a bowl. **(took, taken)**
6. Dr. Levy __made__ a mold of the teeth out of clay. **(make, made)**
7. Since then he has __begun__ the work. **(began, begun)**
8. The dentist has __made__ two caps for my teeth. **(make, made)**
9. He __began__ by gluing them to my chipped teeth. **(began, begun)**
10. Then he __told__ me to look in the mirror. **(tell, told)**
11. I __gave__ him a big smile. **(gave, given)**

LANGUAGE AND USAGE

The Special Verb be

The verb *be* has special forms for different subjects.

SUBJECTS	PRESENT	PAST
I	am	was
you	are	were
he, she, it	is	was
singular noun *(Lucia)*	is	was
we	are	were
they	are	were
plural noun *(stories)*	are	were

The verb *be* does not show action. It tells what someone or something is or is like.

I **am** a reporter.   You **are** a photographer.
That story **was** long.   Those cartoons **were** funny.

Guided Practice

Which verb correctly completes each sentence? Is it in the present tense or the past tense?

Example: Lucia (is, are) a reporter.   *is — present*

1. I (<u>was</u>, were) a sportswriter last year.   past
2. Two stories (was, <u>were</u>) about basketball.   past
3. You (is, <u>are</u>) in one of my stories.   pres.
4. My best story (<u>is</u>, are) about bikes.   pres.
5. We (was, <u>were</u>) winners every time!   past

LANGUAGE AND USAGE

REMINDER

> ▸ The special verb *be* does not show action. It tells what someone or something is or is like.
>
> ▸ Use the form of the verb *be* that agrees with the subject of the sentence.
>
> | Subject | Present | Past | Sentences |
> |---|---|---|---|
> | I | am | was | I **was** near the tree. |
> | he, she, it | is | was | He **was** surprised. |
> | singular noun | is | was | Silvio **is** here now. |
> | we, you, they | are | were | We **are** good friends. |
> | plural noun | are | were | Trees **were** around us. |

INDEPENDENT PRACTICE

Write the verb in ( ) that agrees with each underlined subject.

Example: This tree ____is____ a maple tree.
(**is, are**)

1. The <u>branches</u> __are__ close to the ground. (**is, are**)
2. <u>They</u> __are__ thick and strong. (**is, are**)
3. <u>I</u> __am__ up in the tree now. (**am, is**)
4. This <u>tree</u> __was__ my sister's favorite. (**was, were**)
5. My <u>sister</u> __is__ a good climber. (**is, are**)
6. <u>She</u> __is__ away at school now. (**am, is**)
7. Many <u>trees</u> __were__ here before the storm. (**was, were**)
8. The <u>storm</u> __was__ a very bad one. (**was, were**)
9. <u>It</u> __was__ the worst storm in years. (**was, were**)
10. <u>I</u> __was__ afraid of the strong winds. (**was, were**)

LANGUAGE AND USAGE

What Is an Adjective?

You know that a noun is a word that names a person, a place, or a thing. In your writing, you may sometimes want to describe or give more information about a person, a place, or a thing. One way to do this is to use adjectives.

An **adjective** is a word that describes or gives more information about a noun. An adjective can tell you *what kind* or *how many*. It usually comes before the noun it describes.

What Kind

We have a large dog.

The dog has a curly coat.

How Many

Two dogs played in the yard.

Many dogs like children.

You can use more than one adjective to describe a noun.

We have a large, friendly dog.

The dog has five tiny puppies.

Guided Practice

Find the adjectives that describe the underlined nouns. Does each adjective tell what kind, or does it tell how many? **Adjs. that tell what kind are underlined once; how many, twice.**

Example: Many dogs can learn to do useful work.
Many — how many   useful — what kind

1. Early people found that dogs made good hunters.
2. Strong sheepdogs help farmers with large herds of sheep.
3. One famous dog rescued forty lost people in the mountains.
4. Blind people use dogs to guide them through busy streets.
5. Some smart dogs learn to help deaf people.

LANGUAGE AND USAGE

REMINDER

> ▸ An **adjective** is a word that describes a noun.
> ▸ An adjective can tell *what kind* or *how many*.
>
> **What Kind:** We wanted to make **hot** soup.
>
> We like soup on a **cold, rainy** day.
>
> **How Many:** We made **ten** bowls of soup.
>
> There are **many** people in my family.

INDEPENDENT PRACTICE

Write the adjective or adjectives that describe each underlined noun.

Example: Soup is a healthy food.   healthy

1. We cooked homemade soup.   homemade
2. We had to buy several things for the soup.   several
3. We went to the new market on Brown Street.   new
4. I asked for three pounds of chicken.   three
5. I bought a few carrots too.   few
6. Eli got two fresh peppers.   two, fresh
7. We also needed some small onions.   some, small
8. Everything fit into one large bag.   one, large
9. At home we took out the big, heavy pot.   big, heavy
10. There are many ways to make soup.   many
11. We put in white rice and a dash of pepper.   white
12. A dash of pepper is a tiny bit.   tiny
13. Then we cooked it on a low flame.   low
14. I like soup that is a golden color.   golden
15. Everyone enjoyed the delicious, hot soup.   delicious, hot

LANGUAGE AND USAGE

Adjectives After be

An adjective can come after the word it describes. It usually follows a form of the verb *be*.

> The project is ready.
>
> I am excited.

You know that adjectives can describe nouns. They can also describe words like *I*, *it*, and *we*, which take the place of nouns.

GUIDED PRACTICE

Find the adjective in each sentence. What word does it describe? **Adj. is underlined once; word, twice.**

Example: The weather is beautiful.
beautiful   weather

1. The day is perfect.
2. The fair is exciting.
3. We were eager.
4. The bread is tasty.
5. It was difficult to make.
6. The rides are popular.
7. Paula is proud.
8. The chicken is fat.
9. The eggs are large.
10. Peter is afraid.
11. The sky is dark now.
12. I am tired.

LANGUAGE AND USAGE

REMINDER

> ▸ An adjective can follow the word it describes. It usually follows a form of the verb *be*.
>
> Science is **interesting**. They were **easy**.
> Experiments are **useful**. It was **fun** too.

INDEPENDENT PRACTICE

Write the adjective that follows each underlined word.

Example: The class was busy in science.   **busy**

1. The class was curious about eggs.   curious
2. Carmen is familiar with chickens.   familiar
3. Carmen's hens are beautiful.   beautiful
4. The eggs are brown.   brown
5. Carmen is proud of the eggs.   proud
6. An experiment was possible.   possible
7. Ms. Amato was helpful with the experiment.   helpful
8. We were excited about it.   excited
9. The machine was large.   large
10. The eggs were warm in the machine.   warm
11. Soon the eggs were open.   open
12. The chicks are tiny now.   tiny
13. They are weak too.   weak
14. Maria is gentle with them.   gentle
15. She is careful about the food.   careful
16. I am happy about the experiment.   happy

LANGUAGE AND USAGE

Using a, an, and the

The words *a*, *an*, and *the* are special adjectives called **articles**. Learn these rules for using articles.

> **With singular nouns:**
> Use *a* if the next word begins with a consonant.   **a f**lower
> Use *an* if the next word begins with a vowel.   **an i**ris
> Use *the* if the noun names a particular person, place, or thing.   **the** garden
>
> **With plural nouns:**
> Use *the*.   **the** flowers
>   **the** irises

GUIDED PRACTICE

Which article or articles could be used before each word?

Example: contest   a   the

1. award — an, the
2. orchid — an, the
3. students — the
4. prize — a, the
5. roses — the
6. bushes — the
7. evergreen — an, the
8. seeds — the
9. area — an, the
10. weed — a, the

LANGUAGE AND USAGE

REMINDER

> ▸ *A*, *an*, and *the* are special adjectives called **articles.**
> ▸ Use *a* before a word that begins with a consonant sound.
> ▸ Use *an* before a word that begins with a vowel sound.
>
> **With Singular Nouns:**   a coin   an old coin   the coin
> **With Plural Nouns:**   the coins   the old coins

INDEPENDENT PRACTICE

Write the correct article in ( ) to complete each sentence. The underlined letters are clues.

Example: Chang spent last summer on _____**an**_____ island. **(a, an)**

1. One day Chang took __a__ long walk. **(a, an)**
2. __The__ sun was bright that day. **(An, The)**
3. Chang passed __a__ white fence. **(a, an)**
4. He walked beyond __the__ small cottages. **(a, the)**
5. Soon he reached __the__ water. **(an, the)**
6. __The__ high waves were splashing. **(A, The)**
7. There he saw __a__ sand castle. **(a, an)**
8. It was __an__ enormous castle. **(a, an)**
9. Did __a__ child build it? **(a, an)**
10. Did __an__ adult do this careful work? **(a, an)**
11. Chang looked at __the__ tall towers. **(a, the)**
12. He peeked into __an__ opening. **(a, an)**
13. __An__ old coin lay inside the castle. **(A, An)**
14. Who could have left such __a__ strange coin? **(a, an)**
15. Chang returned __the__ next day. **(an, the)**
16. __The__ old coin and the castle were gone. **(A, The)**

LANGUAGE AND USAGE

Making Comparisons

Sometimes you may want to tell how things are alike or how they are different. You can use adjectives to compare.

You usually add *-er* to an adjective to compare two persons, places, or things, and *-est* to compare three or more.

Mindy took a **long** trip. (one trip described)
Lou's trip was **longer** than hers. (two trips compared)
I took the **longest** trip of all. (three or more compared)

Rules for Adding *-er* and *-est*

1. **If the adjective ends with e:**
   Drop the *e* before adding the ending.

   wid**e** + -er = **wider**
   wid**e** + -est = **widest**

2. **If the adjective ends with a single vowel and a consonant:**
   Double the consonant and add the ending.

   th**in** + -er = **thinner**
   th**in** + -est = **thinnest**

3. **If the adjective ends with a consonant and *y*:**
   Change the *y* to *i* before adding the ending.

   ti**ny** + -er = **tinier**
   ti**ny** + -est = **tiniest**

Guided Practice

What adjective completes each sentence?

1. Hawaii is <u>newer</u> than Alaska. (new)
2. However, every other state is <u>older</u> than Alaska. (old)
3. Alaska is the <u>biggest</u> of all the states. (big)

LANGUAGE AND USAGE

REMINDER

▶ Add *-er* to most adjectives to compare two persons, places, or things.

▶ Add *-est* to most adjectives to compare three or more persons, places, or things.

Adjective	Comparing Two	Comparing Three or More
tall	tall**er**	tall**est**
gentl*e*	gentl**er**	gentl**est**
fa*t*	fa**tter**	fa**ttest**
eas*y*	eas**ier**	eas**iest**

INDEPENDENT PRACTICE

Write the correct adjective in ( ) to complete each sentence.

Example: This elephant is ___taller___ than that one. **(taller, tallest)**

1. Elephants are the ___largest___ of all land animals. **(larger, largest)**

2. Elephants are one of the ___heaviest___ of all animals at birth. **(heavier, heaviest)**

3. Do elephants have the ___biggest___ brains of any land animals? **(bigger, biggest)**

4. African elephants are ___bigger___ than Indian elephants. **(bigger, biggest)**

5. Indian elephants have ___smaller___ ears than African elephants. **(smaller, smallest)**

6. Indian elephants have ___smoother___ skin than African elephants. **(smoother, smoothest)**

7. They also have ___shorter___ tusks than African elephants do. **(shorter, shortest)**

LANGUAGE AND USAGE

Comparing with more and most

You know that you add *-er* or *-est* to some adjectives when you want to compare. With long adjectives, use the words *more* and *most* to compare persons, places, or things. Use the word *more* to compare two. Use *most* to compare three or more.

> Tiger is a **playful** cat. (one cat described)
> Ginger is a **more playful** cat than Tiger. (two cats compared)
> Ike is the **most playful** cat of all. (three or more compared)
>
> Never add *-er* and *more* or *-est* and *most* to the same adjective.
>
> **Incorrect:** Tiger is more smarter than Ginger.
> Tiger is the most intelligentest cat.
> **Correct:** Tiger is **smarter** than Ginger.
> Tiger is the **most intelligent** cat.

GUIDED PRACTICE

What adjective completes each sentence correctly?

Example: Cats are among the __most common__ of all pets. (common)

1. Only the dog is __more popular__ than the cat. (popular)
2. One of the __most popular__ of all breeds of cat is the Siamese. (popular)
3. Some people think that a Persian cat is __more beautiful__ than any other cat. (beautiful)

LANGUAGE AND USAGE

REMINDER

▸ Use *more* with long adjectives to compare two persons, places, or things.

▸ Use *most* with long adjectives to compare three or more persons, places, or things.

The buffalo is an **enormous** animal.
The elephant is **more enormous** than a buffalo.
The blue whale is the **most enormous** animal of all.

Incorrect: The giraffe is <u>more</u> <u>taller</u> than the buffalo.
Its eyes are its <u>most</u> <u>usefulest</u> tool.

Correct: The giraffe is <u>taller</u> than the buffalo.
Its eyes are its <u>most</u> <u>useful</u> tool.

INDEPENDENT PRACTICE

Write <u>more</u> or <u>most</u> to complete each sentence correctly.

Example: A giraffe is one of the __**most**__ interesting of all animals.

1. Is the giraffe the __**most**__ unusual creature in the zoo?
2. It certainly has the __**most**__ amazing neck of all animals.
3. It is __**more**__ difficult for a giraffe to eat low leaves than high ones.
4. The giraffe is one of the __**most**__ silent of all animals.
5. The giraffe's sense of sight is __**more**__ important than its hearing.
6. Sight is the __**most**__ important of the giraffe's five senses.
7. Sleeping giraffes are __**more**__ comfortable standing up than lying down.
8. A baby giraffe is one of the __**most**__ enormous of all animal babies.
9. The mother giraffe is __**more**__ careful with the young than the father.

LANGUAGE AND USAGE

What Is a Pronoun?

You know that nouns name a person, a place, or a thing. You do not have to keep repeating nouns in your writing. Instead, you can use words called **pronouns**. A **pronoun** takes the place of one or more nouns. Read these two paragraphs.

Sara asked Jack and Leah to go to the seashore with Sara. Sara, Jack, and Leah spoke to Ms. Lanski. Ms. Lanski gave Sara, Jack, and Leah a special book. The book was about sea life.

Sara asked Jack and Leah to go to the seashore with **her**. **They** spoke to Ms. Lanski. **She** gave **them** a special book. **It** was about sea life.

What pronoun takes the place of the noun *Sara* in the first sentence of the second paragraph? What pronoun replaces *Sara, Jack, and Leah* in the second sentence?

Like the nouns they replace, pronouns are singular or plural. Look at the lists below. Notice that the pronoun *you* can be either singular or plural.

Singular Pronouns: I, me, you, he, him, she, her, it
Plural Pronouns: we, us, you, they, them

Guided Practice

Which words in these sentences are pronouns? Is each pronoun singular or plural?

Example: Sara said, "Come with me to the seashore."
me singular

1. Leah carried a pail. She planned to collect shells. **S**
2. Jack took a notebook. Sara wanted him to take notes. **S**
3. As the children walked along, they looked carefully. **P**

LANGUAGE AND USAGE

REMINDER

> ▸ A **pronoun** is a word that can take the place of one or more nouns.
>
> **Singular Pronouns:** I, me, you, he, him, she, her, it
> **Plural Pronouns:** we, us, you, they, them
>
> | Nouns | Pronouns |
> |---|---|
> | <u>Dennis</u> went to the park. | **He** went to the park. |
> | Dennis watched the <u>squirrels</u>. | Dennis watched **them**. |

INDEPENDENT PRACTICE

Underline the pronoun in each sentence. Then write *singular* or *plural* for each pronoun.

Example: Dennis had a camera. <u>He</u> took some pictures.   **singular**

1. <u>He</u> went to the park with Maya.   singular
2. <u>She</u> brought a camera along too.   singular
3. At the park, <u>they</u> started taking pictures.   plural
4. A little squirrel raced past <u>them</u>.   plural
5. Maya snapped a picture of <u>it</u>.   singular
6. The squirrel ran past <u>her</u> and behind a tree.   singular
7. "Don't run away from <u>us</u>," Maya called.   plural
8. "<u>We</u> have to be quick," said Dennis.   plural
9. Then <u>he</u> heard two soft voices.   singular
10. Two little boys walked up to <u>him</u>.   singular
11. "Show <u>me</u> how a camera works," a boy said to Dennis.   singular
12. "Please take pictures of <u>us</u>," said the other boy.   plural
13. "Both of <u>you</u> smile at the camera," said Maya.   plural

LANGUAGE AND USAGE

Subject Pronouns

You have learned that a pronoun can take the place of a noun. Like a noun, a pronoun can be used as the subject of a sentence. Remember that the subject tells whom or what the sentence is about.

Nouns
Juan did a project on insects.
Lola worked with Juan.
Juan and Lola gave a report.
Vin and I enjoyed the report.

Pronouns
He did a project on insects.
She worked with Juan.
They gave a report.
We enjoyed the report.

Not all pronouns can be used as subjects. Only the **subject pronouns** *I, you, he, she, it, we,* and *they* can be used as the subjects of sentences.

Subject Pronouns	
Singular	**Plural**
I	we
you	you
he, she, it	they

GUIDED PRACTICE Which subject pronoun could take the place of the underlined word or words in each sentence?

Example: Matt said to Barb, "<u>Barb</u> found a ladybug." **You**

1. <u>Matt and I</u> know that ladybugs are helpful insects. **We**
2. <u>Matt</u> told Barb more about ladybugs. **He**
3. Ladybugs are not really bugs. <u>Ladybugs</u> are beetles. **They**
4. <u>This beetle</u> eats insects that destroy plants. **It**
5. <u>Barb</u> did not disturb the ladybug. **She**

LANGUAGE AND USAGE

131

REMINDER

▶ Use only subject pronouns as the subjects of sentences.

Singular Subject Pronouns: I, you, he, she, it
Plural Subject Pronouns: we, you, they

Nouns	Pronouns
<u>Tara</u> has an exciting new toy.	**She** has an exciting new toy.
<u>Tara and Rico</u> play with the toy.	**They** play with the toy.

INDEPENDENT PRACTICE

Write the subject pronoun in ( ) that can take the place of the underlined word or words.

Example: <u>Aunt Lori</u> bought great toys for Rico.   **She**
              (She, We)

1. <u>Rico</u> is only four years old.   **(He, You)**   He
2. <u>Aunt Lori</u> has found toys to help Rico learn.   **(She, I)**   She
3. <u>This toy</u> teaches children to tell time.   **(It, They)**   It
4. <u>Those toys</u> are for making music.   **(We, They)**   They
5. <u>Tara</u> showed Rico a brand-new toy.   **(You, She)**   She
6. "<u>Tara and I</u> will play with the toy," said Rico.   **(They, We)**   We
7. "<u>Rico</u> will play with this toy all day," said Rico.   **(You, I)**   I
8. <u>Tara</u> showed Aunt Lori a new puzzle.   **(She, They)**   She
9. <u>This puzzle</u> was a map of the fifty states.   **(It, They)**   It
10. "<u>My brother and I</u> also have paints," Tara said.   **(We, He)**   We
11. <u>Uncle Joe and Aunt Lori</u> had bought these paints. **(She, They)**   They
12. <u>Uncle Joe</u> told Tara about toys of long ago.   **(He, We)**   He
13. <u>The toys</u> were balls made of wood.   **(It, They)**   They
14. <u>Hoops</u> were also used as toys long ago.   **(You, They)**   They

LANGUAGE AND USAGE

Object Pronouns

You know that subject pronouns may be used as the subjects of sentences. The pronouns *me, you, him, her, it, us,* and *them* are called **object pronouns**. Object pronouns follow action verbs and words such as *to, with, for,* and *at.*

Nouns	Pronouns
Ms. Rossi fed **the horses**.	Ms. Rossi fed **them**.
Sal helped **Ms. Rossi**.	Sal helped **her**.
Sal showed a pony to **Ed and me**.	Sal showed a pony to **us**.
Then Sal gave **the pony** a carrot.	Then Sal gave **it** a carrot.

Object Pronouns

Singular	Plural
me	us
you	you
him, her, it	them

Never use the object pronouns *me, him, her, us,* and *them* as subjects. You can use the pronouns *you* and *it* as either subject or object pronouns.

Guided Practice

Which object pronoun could take the place of the underlined word or words in each sentence?

Example: Sally rides <u>horses</u> every day.   **them**

1. Ed said to Sally, "Please teach <u>Ed</u>."   **me**
2. Sally took Ed to the stable with <u>Sally</u>.   **her**
3. Sally told Ed, "I will teach <u>Ed</u> grooming first."   **you**
4. She handed a brush to <u>Ed</u>.   **him**
5. Then Sally and Ed brushed <u>the horse</u>.   **it**

LANGUAGE AND USAGE

Reminder

> ▶ Use **object pronouns** after action verbs and words such as *to*, *with*, *for*, and *at*.
>
> **Singular Object Pronouns:** me, you, him, her, it
> **Plural Object Pronouns:** us, you, them
>
> | **Nouns** | **Pronouns** |
> |---|---|
> | Firefighters spoke to the children. | Firefighters spoke to **them**. |
> | Chief Drake gave Andrea a book. | Chief Drake gave **her** a book. |

Independent Practice

Write the object pronoun in ( ) that can take the place of the underlined word or words.

Example: "We can help Andrea," said the chief. **(her, she)**   her

1. One day Andrea smelled <u>smoke</u> in the house. **(it, them)**   it

2. She knew what the firefighters had taught <u>Andrea</u>. **(her, she)**   her

3. She felt the door before she opened <u>the door</u>. **(them, it)**   it

4. The door did not feel hot to <u>Andrea</u>. **(her, she)**   her

5. Andrea phoned the <u>firefighters</u>. **(they, them)**   them

6. "Give <u>Chief Drake</u> the address," said Chief Drake. **(me, I)**   me

7. Andrea gave the address to <u>Chief Tom Drake</u>. **(he, him)**   him

8. The fire trucks were ready for <u>the firefighters</u>. **(they, them)**   them

9. Soon the firefighters arrived at <u>the house</u>. **(it, you)**   it

10. Andrea's mother was outside with <u>Andrea</u>. **(her, she)**   her

11. "Can you help <u>Andrea and me</u>?" asked Mrs. Katz. **(we, us)**   us

12. The firefighters hooked a hose to <u>the truck</u>. **(it, me)**   it

LANGUAGE AND USAGE

Using I and me

When you talk or write about yourself, you use the pronoun *I* or *me*. Do you ever have trouble deciding whether to use *I* or *me* with another noun or pronoun? For example, should you say *Kim and me study* or *Kim and I study*? One way to check is to say the sentence to yourself with only *I* or *me*.

Kim and I study.　　　　　　　　　　**I** study.
Mrs. Perez teaches **Kim and me**.　　Mrs. Perez teaches **me**.
Ali studies with **Kim and me**.　　　Ali studies with **me**.

Remember to use *I* as the subject of a sentence. Use *me* after action verbs and after words such as *to, with, for,* and *at*.

When you talk about yourself and another person, always name yourself last.

Incorrect:　I and Kim help Ali.　　Ali thanks me and Kim.
Correct:　　**Kim and I** help Ali.　Ali thanks **Kim and me**.

GUIDED PRACTICE

Which words complete each sentence correctly?

Example: (Jen and I, I and Jen) met Maria.
Jen and I

1. Maria invited (me and Jen, Jen and me) to her house.
2. (I and Jen, Jen and I) walked home with Maria.
3. Maria talked to (Jen and me, me and Jen) about Mexico.
4. Jen and (I, me) were very interested.
5. Maria helped prepare dinner for Jen and (I, me).
6. Jen and (I, me) ate with Maria's family.
7. The food tasted wonderful to Jen and (I, me).

LANGUAGE AND USAGE

REMINDER

> ▸ Use the pronoun *I* as the subject of a sentence.
> ▸ Use the pronoun *me* after action verbs and after words such as *to*, *with*, *for*, and *at*.
> ▸ When speaking of yourself and another person, always name yourself last.
>
> <u>Rami and</u> **I** talked about fishing.   **I** talked about fishing.
> Klaus listened to <u>Rami and</u> **me**.   Klaus listened to **me**.

INDEPENDENT PRACTICE

Write the sentence in each pair that is correct.

Example: Klaus and I like summer.   I and Klaus like summer.
Klaus and I like summer.

1. Klaus and me enjoy hot weather and sunny days.
   Klaus and I enjoy hot weather and sunny days.
   Klaus and I enjoy hot weather and sunny days.

2. I like to swim and fish on long summer afternoons.
   Me like to swim and fish on long summer afternoons.
   I like to swim and fish on long summer afternoons.

3. Klaus loves to go swimming and fishing with me.
   Klaus loves to go swimming and fishing with I.
   Klaus loves to go swimming and fishing with me.

4. Klaus and I fish for hours in the Ohio River.
   I and Klaus fish for hours in the Ohio River.
   Klaus and I fish for hours in the Ohio River.

5. My mother has told many fishing stories to Klaus and I.
   My mother has told many fishing stories to Klaus and me.
   My mother has told many fishing stories to Klaus and me.

LANGUAGE AND USAGE

Possessive Pronouns

You have learned that possessive nouns show ownership. You can use pronouns in place of possessive nouns. A pronoun that shows ownership is a **possessive pronoun**.

Possessive Nouns	**Possessive Pronouns**
Pamela feeds **Pamela's** pet.	Pamela feeds **her** pet.
She fills **the pet's** dish.	She fills **its** dish.
The boys' gerbil is playful.	**Their** gerbil is playful.

Possessive Pronouns

Singular	Plural
my	our
your	your
her, his, its	their

GUIDED PRACTICE

Which possessive pronoun should you use in place of the underlined word or words?

Example: Max and I help Mr. Lee at <u>Mr. Lee's</u> shop.   his

1. Max likes <u>Max's</u> job at the pet store.   his
2. He gives the puppies <u>the puppies'</u> food.   their
3. Agnes is saving <u>Agnes's</u> money for a pet.   her
4. She will buy the parakeet and <u>the parakeet's</u> cage.   its
5. Agnes, you and <u>Agnes's</u> sister will like the parakeet.   your

LANGUAGE AND USAGE

REMINDER

> ▸ A possessive pronoun may be used in place of a possessive noun.
>
> **Singular Possessive Pronouns:** my, your, her, his, its
> **Plural Possessive Pronouns:** our, your, their
>
> | **Possessive Nouns** | **Possessive Pronouns** |
> |---|---|
> | Nola is Mack's sister. | Nola is **his** sister. |
> | This is Nola and Mack's playhouse. | This is **their** playhouse. |

INDEPENDENT PRACTICE

Write each sentence. Use the possessive pronoun in ( ) that can take the place of the underlined word or words.

Example: I have seen Mack and Nola's new playhouse. **(his, their)**
I have seen their new playhouse.

1. Nola and Mack built Nola and Mack's new playhouse. **(its, their)**
   Nola and Mack built their new playhouse.

2. Nola is proud of Nola's curtains for the playhouse. **(her, its)**
   Nola is proud of her curtains for the playhouse.

3. The children visited Mack's friend Mr. Rey. **(their, his)**
   The children visited his friend Mr. Rey.

4. Mr. Rey uses Mr. Rey's tools to make small furniture. **(your, his)**
   Mr. Rey uses his tools to make small furniture.

5. He said, "Nola, this chair is for Nola's playhouse." **(my, your)**
   He said, "Nola, this chair is for your playhouse."

6. Nola said, "I will keep the chair away from Nola's dog." **(its, my)**
   Nola said, "I will keep the chair away from my dog."

7. "The dog has enough of the dog's own toys," said Mack. **(its, their)**
   "The dog has enough of its own toys," said Mack.

LANGUAGE AND USAGE

Pronouns and Homophones

You know that homophones are words that sound alike but have different spellings and meanings. Writers often confuse some contractions and their homophones because these words sound alike. Study the chart below. Learn the spelling and the meaning of each homophone.

Homophone	Meaning	Sentence
it's	it is	**It's** a beautiful bird!
its	belonging to it	Take **its** picture.
they're	they are	**They're** odd birds.
their	belonging to them	**Their** wings are big!
there	in that place	**There** is another.
you're	you are	**You're** very lucky.
your	belonging to you	Get **your** camera.

GUIDED PRACTICE

Which word would you use to complete each sentence correctly?

Example: I hear (you're, your) entering the photo contest. *you're*

1. Which of (you're, your) pictures will you enter?
2. (They're, There) all so good!
3. The puppies love having (they're, their) picture taken.
4. The picture (their, there) on your desk is interesting.
5. (It's, Its) colors are sharp and clear.
6. (It's, Its) hard to choose the best one!

LANGUAGE AND USAGE

REMINDER

▶ Do not confuse the contractions *it's*, *they're*, and *you're* with their homophones *its*, *their*, *there*, and *your*.

Homophones:

You're a good painter.   **Your** picture is pretty.
It's a picture of a bird.   **Its** colors are bright.
They're going to paint too.   **There** are **their** paints.

INDEPENDENT PRACTICE

Write the words in ( ) to complete the sentences in each pair. Begin each sentence with a capital letter.

Example: All the children brought ___their___ paints. **(they're, their)** ___They're___ going to paint masks.

1. "May I use ___your___ yellow paint?" asked Teresa. **(your, you're)**
   "___You're___ welcome to use it," said Nikos.

2. "___It's___ going to be a large mask," said Olive. **(its, it's)**
   "I am painting ___its___ eyes yellow," Teresa said.

3. "___They're___ the strangest eyes!" said Nikos. **(they're, there)**
   "Paint some black spots right ___there___," he added.

4. "Now ___its___ eyes look better," said Teresa. **(its, it's)**
   "___It's___ a wonderful mask!" exclaimed Olive.

5. "___Your___ mask looks like a tiger," said Nikos. **(you're, your)**
   "___You're___ right about that," replied Teresa.

6. "Only ___its___ whiskers are missing," said Olive.
   (it's, its) "___It's___ time to clean up," said Teresa.

LANGUAGE AND USAGE

What Is an Adverb?

You know that an adjective is a word that describes a noun or a pronoun. Another kind of describing word is called an adverb. An **adverb** can describe a verb.

Adverbs give us more information about an action verb or a form of the verb *be*. They tell *how, when,* or *where*. Adverbs can come before or after the verbs they describe.

How: Maggie typed the letter carefully.
When: Then I sealed the envelope.
Where: All the stamps were upstairs.

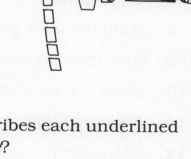

Study the lists below. They show adverbs that you use often in your writing. Most adverbs telling *how* end with *-ly*.

How	When	Where
angrily	always	downtown
carefully	finally	inside
fast	often	off
loudly	once	out
quickly	sometimes	there
sadly	then	upstairs

GUIDED PRACTICE   Find the adverb that describes each underlined verb. Does the adverb tell *how, when,* or *where*?

1. The mail carrier finally arrived.   when
2. We ran out to meet her.   where
3. Maggie clapped her hands excitedly.   how
4. I quickly opened the gold envelope.   how

LANGUAGE AND USAGE

REMINDER

> ▶ An **adverb** is a word that describes a verb.
> ▶ An adverb can tell *how, when,* or *where*.
>
> **How:** The baby lion's eyes opened **slowly**.
> **When:** **Then** it yawned.
> **Where:** The baby lion looked **around**.

INDEPENDENT PRACTICE

Write the adverb that describes each underlined verb.

Example: A mother lion carries her cubs gently.   <u>gently</u>

1. A mother lion sometimes leaves her cubs.   sometimes
2. She looks everywhere for food.   everywhere
3. The cubs stay inside.   inside
4. Patiently they wait for their mother.   Patiently
5. They play with each other happily.   happily
6. They sleep peacefully.   peacefully
7. The mother lion finally returns.   finally
8. Soon the cubs grow big and strong.   Soon
9. Now they travel with their group.   Now
10. Lions always travel as a group.   always
11. A young lion learns quickly from the other lions.   quickly
12. The older lions carefully guard the younger ones.   carefully
13. The big lions roar loudly.   loudly
14. Strange lions stay away.   away
15. The lions see a zebra there.   there
16. Then they chase it.   Then

LANGUAGE AND USAGE

REMINDER

> ▸ Add -er or -est to short adverbs to compare actions.
> ▸ For adverbs that end with -ly, use *more* or *most* to compare actions.
> ▸ Never use -er with *more*. Never use -est with *most*.

Adverb	Comparing Two	Comparing Three or More
fast	fast**er**	fast**est**
swiftly	**more** swiftly	**most** swiftly

INDEPENDENT PRACTICE

Write each sentence. Use the correct form of the adverb in ( ).

Example: Amy arrived at the pool (later, latest) than Ruth did.
Amy arrived at the pool later than Ruth did.

1. Amy works (harder, hardest) of all the divers in her class.
   Amy works hardest of all the divers in her class.

2. Amy learned the swan dive (more quickly, most quickly) than Ruth did.
   Amy learned the swan dive more quickly than Ruth did.

3. The back flip took (longer, longest) of all to learn.
   The back flip took longest of all to learn.

4. Ben dives (more skillfully, most skillfully) than Amy's last teacher.
   Ben dives more skillfully than Amy's last teacher.

5. He works (more closely, most closely) of all with Amy.
   He works most closely of all with Amy.

6. Does Ben keep his legs (straighter, straightest) than Amy does?
   Does Ben keep his legs straighter than Amy does?

7. Amy tries to spring (higher, more higher) than Ben does.
   Amy tries to spring higher than Ben does.

LANGUAGE AND USAGE

Comparing with Adverbs

You have already learned how adjectives are used to compare people, places, and things. You can also use adverbs to make comparisons. Add *-er* to short adverbs to compare two actions. Add *-est* to compare three or more actions.

> Bill skis **fast**. (one action)
> Louise skis **faster** than Bill does. (two actions)
> Kara skis **fastest** of the three. (three or more actions)
>
> For most adverbs that end with *-ly*, use *more* to compare two actions. Use *most* to compare three or more actions.
>
> Dee swam **gracefully**. (one action)
> Did Kato swim **more gracefully** than Dee? (two actions)
> Ty swam **most gracefully** of all. (three or more actions)
>
> Do not use *-er* or *-est* to compare adverbs that end with *-ly*. Never use *-er* with *more*. Never use *-est* with *most*.
>
> **Incorrect:** Faith skates smoothlier than Don.
> Don fell most hardest of all the skaters.
>
> **Correct:** Faith skates **more smoothly** than Don.
> Don fell **hardest** of all the skaters.

GUIDED PRACTICE What form of the adverb in ( ) correctly completes each sentence?

1. Today we practiced __longer__ than we did yesterday. (long)
2. Of all the team members, Ruth skated __most skillfully__. (skillfully)
3. Does Leslie skate __more quickly__ than Shawn? (quickly)
4. Andrew jumps __highest__ of us all. (high)
5. Of everyone on the team, Tara tries __hardest__. (hard)

LANGUAGE AND USAGE

Using good and well

Sometimes it may be hard to decide whether to use *good* or *well*. How can you make sure that you use these words correctly? Remember, *good* is an adjective that describes nouns. *Well* is an adverb that describes verbs.

Adjective	Adverb
Marcia is a good pilot.	She flies well.
This suit is good.	I choose my suits well.

Guided Practice

Which word is correct?

Example: Kipp's trips are all (good, well).   *good*

1. He plans (good, <u>well</u>) for his adventures.
2. Kipp's guidebook is (<u>good</u>, well).
3. His road maps are (<u>good</u>, well) too.
4. He has learned to read maps (good, <u>well</u>).
5. Kipp speaks several languages (good, <u>well</u>).
6. Talking to people helps him learn (good, <u>well</u>) about another country.
7. Kipp is (<u>good</u>, well) at taking pictures.
8. Photos help him remember his trips (good, <u>well</u>).
9. Kipp describes his travels (good, <u>well</u>).
10. Everyone listens (good, <u>well</u>) to his stories.
11. The presents that he brings to his family are always (<u>good</u>, well).
12. They think that Kipp's trips are (<u>good</u>, well)!

LANGUAGE AND USAGE

REMINDER

▸ Use the adjective *good* to describe nouns.
▸ Use the adverb *well* to describe verbs.

Adjective
Audrey chooses **good** colors.
This color is **good**.

Adverb
Audrey paints **well**.
Audrey chooses colors **well**.

INDEPENDENT PRACTICE

Write *good* or *well* to complete each sentence correctly. The underlined nouns and verbs are clues.

Example: Audrey had a ___**good**___ idea.

1. Audrey wanted to give her room a ___good___ paint job.
2. She went to a ___good___ store for paint.
3. She knew the store owner ___well___.
4. He showed Audrey a ___good___ way of choosing colors.
5. His color wheel showed colors that match ___well___.
6. Red goes ___well___ with yellow.
7. Blue is ___good___ for a cool look.
8. Light colors are ___good___ for making rooms seem larger.
9. A ___good___ shade of pink makes a room seem warmer.
10. Audrey planned her paint job ___well___.
11. She thought gray would be ___good___ for the windows and the door.
12. Light pink would be a ___good___ color for the walls.
13. Audrey painted her room ___well___.
14. Before painting, she covered the furniture ___well___.
15. After painting, she cleaned the paintbrushes ___well___.

LANGUAGE AND USAGE

Negatives

Sometimes when you write sentences, you use the word *no* or words that mean "no." A word that makes a sentence mean "no" is a **negative**. These sentences have negatives.

No one picked the beans.   I **didn't** water the garden.

The words *no, no one, nobody, none, nothing, nowhere,* and *never* are negatives. The word *not* and contractions made with *not* are also negatives. Never use two negatives together in a sentence.

Incorrect
There weren't no trees.

I won't never rake leaves!

Correct
There **weren't** any trees.
There were **no** trees.

I **won't** ever rake leaves!
I will **never** rake leaves!

Notice that there may be more than one correct way to write a sentence with a negative.

GUIDED PRACTICE

Which word in ( ) is correct?

Example: Eli doesn't want (any, no) leaves on the ground.   *any*

1. He can't go (nowhere, **anywhere**) until he has finished raking.
2. He never likes (**anything**, nothing) about yard work.
3. No one (never, **ever**) has time to help him.
4. Luckily there (**are**, aren't) no leaves left on the trees.
5. There won't be (no, **any**) more leaves to rake until next fall!

LANGUAGE AND USAGE

REMINDER

> ▸ A **negative** is a word that means "no."
> ▸ Do not use two negative words together in a sentence.
>
>   I had **never** heard about Helen Keller.
>   **No one** had told me about her.
>
> **Incorrect:**   I **never** knew **nothing** about Helen Keller.
> **Correct:**   I **never** knew **anything** about Helen Keller.
>   I knew **nothing** about Helen Keller.

INDEPENDENT PRACTICE

Write the correct word in ( ) to complete each sentence.

Example: Helen Keller's parents didn't have (any, no) help.   **any**

1. Helen Keller couldn't see or hear (nothing, anything).   **anything**
2. Her parents had never given her (any, no) training.   **any**
3. No one had (never, ever) helped their daughter.   **ever**
4. Helen couldn't go (nowhere, anywhere) by herself.   **anywhere**
5. Her parents couldn't teach her (anything, nothing).   **anything**
6. She hadn't (no, any) teacher until Annie Sullivan came.   **any**
7. Annie had never taught (anyone, no one) as bright.   **anyone**
8. At first Helen didn't want (none, any) of Annie's lessons.   **any**
9. Helen wouldn't do (nothing, anything) Annie wanted.   **anything**
10. Annie wouldn't (ever, never) let Helen misbehave.   **ever**
11. Helen hadn't met (anybody, nobody) as patient as Annie.   **anybody**

LANGUAGE AND USAGE

CAPITALIZATION, PUNCTUATION, AND USAGE GUIDE

ABBREVIATIONS

Abbreviations are shortened forms of words. Most abbreviations begin with a capital letter and end with a period.

Titles	Mr. *(Mister)* Mr. Juan Albino Mrs. *(Mistress)* Mrs. Frances Wong Ms. Leslie Clark	Sr. *(Senior)* John Helt, Sr. Jr. *(Junior)* John Helt, Jr. Dr. *(Doctor)* Dr. Janice Dodds
Words used in addresses	St. *(Street)* Rd. *(Road)*	Blvd. *(Boulevard)* Ave. *(Avenue)*
Days of the week	Sun. *(Sunday)* Mon. *(Monday)* Tues. *(Tuesday)* Wed. *(Wednesday)*	Thurs. *(Thursday)* Fri. *(Friday)* Sat. *(Saturday)*
Months of the year	Jan. *(January)* Feb. *(February)* Mar. *(March)*	Apr. *(April)* Aug. *(August)* Sept. *(September)*	Oct. *(October)* Nov. *(November)* Dec. *(December)*

Note: May, June, and July are not abbreviated.

States — The United States Postal Service uses two capital letters and no period in each of its state abbreviations.

AL *(Alabama)*	HI *(Hawaii)*	MA *(Massachusetts)*
AK *(Alaska)*	ID *(Idaho)*	MI *(Michigan)*
AZ *(Arizona)*	IL *(Illinois)*	MN *(Minnesota)*
AR *(Arkansas)*	IN *(Indiana)*	MS *(Mississippi)*
CA *(California)*	IA *(Iowa)*	MO *(Missouri)*
CO *(Colorado)*	KS *(Kansas)*	MT *(Montana)*
CT *(Connecticut)*	KY *(Kentucky)*	NE *(Nebraska)*
DE *(Delaware)*	LA *(Louisiana)*	NV *(Nevada)*
FL *(Florida)*	ME *(Maine)*	NH *(New Hampshire)*
GA *(Georgia)*	MD *(Maryland)*	NJ *(New Jersey)*

WRITER'S HANDBOOK

ABBREVIATIONS continued

States (continued)			
	NM *(New Mexico)*	PA *(Pennsylvania)*	VT *(Vermont)*
	NY *(New York)*	RI *(Rhode Island)*	VA *(Virginia)*
	NC *(North Carolina)*	SC *(South Carolina)*	WA *(Washington)*
	ND *(North Dakota)*	SD *(South Dakota)*	WV *(West Virginia)*
	OH *(Ohio)*	TN *(Tennessee)*	WI *(Wisconsin)*
	OK *(Oklahoma)*	TX *(Texas)*	WY *(Wyoming)*
	OR *(Oregon)*	UT *(Utah)*	

TITLES

Underlining Titles of books, newspapers, magazines, and TV series are underlined. The important words and the first and last words are capitalized.

<u>Life on the Mississippi</u>   <u>Newsweek</u>   <u>Nova</u>

Quotation Marks Put *quotation marks* (" ") around the titles of short stories, articles, songs, poems, and book chapters.

"The Necklace" *(short story)*
"Home on the Range" *(song)*

QUOTATIONS

Quotation Marks A *direct quotation* tells a speaker's exact words. Use *quotation marks* (" ") to set off a direct quotation from the rest of the sentence.

"Please put away your books now," said Mr. Emory.

Begin a quotation with a capital letter. When a quotation comes at the end of a sentence, use a comma to separate the quotation from the words that tell who is speaking. Put end marks inside the last quotation mark.

The driver announced, "This is the Summer Street bus."

Writing a conversation Begin a new paragraph each time a new person begins speaking.

   "Are your seats behind home plate or along the first-base line?" asked the voice on the phone.
   "I haven't bought any tickets yet," said Mr. Williams. "I was hoping that you would reserve three seats for me now."

CAPITALIZATION

Rules for capitalization

Capitalize the first word of every sentence.
What an unusual color the roses are!

Capitalize the pronoun *I*.
What should I do next?

Capitalize every important word in the names of particular people, pets, places, and things (proper nouns).
Rover   District of Columbia   Elm Street   Lincoln Memorial

Capitalize titles and initials that are parts of names.
Governor Bradford   Emily G. Hesse   Senator Smith

Capitalize family titles when they are used as names or as parts of names.
We visited Uncle Harry.   May we play now, Grandma?

Capitalize the names of months and days.
My birthday is on the last Monday in March.

Capitalize the names of groups.
Sutton Bicycle Club   National League

Capitalize the names of holidays.
Memorial Day   Fourth of July   Veterans Day

Capitalize the first and last words and all important words in the titles of books and newspapers.
From Earth to the Moon   The New York Times

Capitalize the first word in the greeting and the closing of a letter.
Dear Marcia,   Yours truly,

In an outline, each Roman numeral and capital letter is followed by a period. Capitalize the first word of each main topic and subtopic.
I. Types of libraries
   A. Large public library
   B. Bookmobile

WRITER'S HANDBOOK

151

CAPITALIZATION, PUNCTUATION, USAGE

PUNCTUATION

End marks	There are three end marks. A *period (.)* ends a statement or a command. A *question mark (?)* follows a question. An *exclamation point (!)* follows an exclamation.	

The scissors are on my desk.   *(statement)*
Look up the spelling of that word.   *(command)*
How is the word spelled?   *(question)*
This is your best poem so far!   *(exclamation)*

Apostrophe	**To form the possessive of a singular noun, add an apostrophe and *s* ('s).**

baby's   Russ's   grandmother's   family's

For a plural noun ending in *s*, add only an apostrophe (').

sisters'   families'   Smiths'   hound dogs'

For a plural noun that does not end in *s*, add an apostrophe and *s* ('s).

women's   mice's   children's

Use an apostrophe in contractions in place of dropped letters.

isn't *(is not)*   wasn't *(was not)*   I'm *(I am)*
can't *(cannot)*   we're *(we are)*   they've *(they have)*
won't *(will not)*   it's *(it is)*   they'll *(they will)*

Comma	**A *comma (,)* tells the reader to pause between the words that it separates.**

Use commas to separate items in a series. Put a comma after each item in the series except the last one.

Clyde asked if we had any apples, peaches, or grapes.

You can combine two short, related sentences to make one compound sentence. Use a comma and the connecting word *and*, *but*, or *or*.

Some students were at lunch, but others were studying.

Use commas to set off the words *yes, no,* and *well* when they are at the beginning of a sentence.

Well, it's just too cold out.   No, it isn't six yet.

WRITER'S HANDBOOK

PUNCTUATION continued

Comma (continued)

Use a comma or commas to set off the names of people who are spoken to directly.

Jean, help me fix this tire.   How was your trip, Grandpa?

Use a comma to separate the month and the day from the year.

Our nation was born on July 4, 1776.

Use a comma between the names of a city and a state.

Chicago, Illinois   Miami, Florida

Use a comma after the greeting in a friendly letter.

Dear Deena,   Dear Uncle Rudolph,

Use a comma after the closing in a letter.

Your nephew,   Sincerely yours,

PROBLEM WORDS

Words	Rules	Examples
a, an, the	These words are special adjectives called articles.	
a, an	Use *a* and *an* before singular nouns. Use *a* if a word begins with a consonant sound. Use *an* if a word begins with a vowel sound.	a banana an apple
the	Use *the* with both singular and plural nouns. Use *the* to point out particular persons, places, or things.	the apple the apples The books that I like are long.
are our	*Are* is a verb. *Our* is a possessive pronoun.	Are these gloves yours? This is our car.

WRITER'S HANDBOOK

PROBLEM WORDS continued

Words	Rules	Examples
doesn't	Use *doesn't* with singular nouns, *he, she,* and *it.*	Dad doesn't swim.
don't	Use *don't* with plural nouns, *I, you, we,* and *they.*	We don't swim.
good	Use the adjective *good* to describe nouns.	The weather looks good.
well	Use the adverb *well* to describe verbs.	She sings well.
its	*Its* is a possessive pronoun.	The dog wagged its tail.
it's	*It's* is a contraction of *it is.*	It's cold today.
let	*Let* means "to allow."	Please let me go swimming.
leave	*Leave* means "to go away from" or "to let stay."	I will leave soon. Leave it on my desk.
set	*Set* means "to put."	Set the vase on the table.
sit	*Sit* means "to rest or stay in one place."	Please sit in this chair.
their	*Their* is a possessive pronoun.	Their coats are on the bed.
there	*There* is an adverb. *There* means "in that place."	Is Carlos there? There is my book.
they're	*They're* is a contraction of *they are.*	They're going to the store.
two	*Two* is a number.	I bought two shirts.
to	*To* means "toward."	A squirrel ran to the tree.
too	*Too* means "more than enough" and "also."	I ate too many cherries. Can we go too?
your	*Your* is a possessive pronoun.	Are these your glasses?
you're	*You're* is a contraction of *you are.*	You're late again!

WRITER'S HANDBOOK

ADJECTIVE AND ADVERB USAGE

Use adjectives to describe nouns. Use adverbs to describe verbs.

This plant is tall. *(adj.)* It grew fast. *(adv.)*

Comparing

To compare two people, places, or things, add *-er* to many adjectives and adverbs.

This plant is taller than the other one. It grew faster.

To compare three or more people, places, or things, add *-est* to many adjectives or adverbs.

This plant is the tallest of the three. It grew fastest.

Double comparisons

Never combine *-er* with the word *more*. Do not combine *-est* with the word *most*.

She is a better (*not* more better) skier than he.
The third book is the longest (*not* most longest).

good, bad

When you use the adjectives *good* and *bad* to compare, you must change their form. Use *better* or *worse* to compare two. Use *best* or *worst* to compare three.

The weather today is worse than it was yesterday.
The forecast for tomorrow is the best one of the week.

more, most

With most long adjectives and with adverbs that end in *-ly*, use *more* to compare two people, places, things, or actions. Use *most* to compare three or more.

This song is more beautiful than the first one.
Of the five songs, this one was sung the most powerfully.

NEGATIVES

A negative word or negative contraction says "no" or "not." Do not use two negatives to express one negative idea.

INCORRECT: He can't see nothing.
CORRECT: He can't see anything.
CORRECT: He can see nothing.

WRITER'S HANDBOOK

PRONOUN USAGE

I and me	Use the pronoun *I* as the subject of a sentence. Use the pronoun *me* after action verbs and after words such as *to, with, for,* and *at*. When using *I* or *me* with other nouns or pronouns, name yourself last. Jan and I are going to the movies. She will telephone me. Beth and I will leave. Give the papers to Ron and me.
Subject and object pronouns	A pronoun used as a subject (*I, you, he, she, it, we,* or *they*) is called a subject pronoun. She did not disturb the grasshopper. A pronoun used as an object (*me, you, him, her, it, us,* or *them*) is called an object pronoun. Use object pronouns after action verbs and after words such as *to, with, for,* and *at*. The puppy likes us. Let's play with him.

VERB USAGE

Tenses	Avoid unnecessary shifts from one tense to another. The trains stopped, and everyone was (*not* is) surprised.
Irregular verbs	Irregular verbs do not add *-ed* or *-d* to show past action. Because irregular verbs do not follow a regular pattern, you must remember their spellings.

Present	Past	Past with helping verb	Present	Past	Past with helping verb
be	was	been	have	had	had
begin	began	begun	know	knew	known
break	broke	broken	make	made	made
bring	brought	brought	say	said	said
come	came	come	sing	sang	sung
drive	drove	driven	take	took	taken
eat	ate	eaten	tell	told	told
give	gave	given	throw	threw	thrown
grow	grew	grown	wear	wore	worn

WRITER'S HANDBOOK

Language and Usage Lessons

Overview

Language and Usage Lessons have been provided in the *Student Resource Book* as an optional resource for teachers who wish to integrate these skills into their reading/language arts curriculum. A Capitalization, Punctuation, and Usage Guide has also been provided as a useful handbook for students' own reference.

The Language and Usage Lessons provide opportunities for direct instruction in key language areas. Instruction, modeling, guided practice, and independent practice are provided for each skill. Students using *Houghton Mifflin Reading: The Literature Experience* will have a rich variety of reading, writing, listening, and speaking projects and activities; this section provides a useful support for those language arts areas.

The Language and Usage Lessons are organized into the following six major sections: The Sentence, Nouns, Verbs, Adjectives, Pronouns, and Adverbs.

Format of the Lessons

Each lesson is two pages in length. The first page provides instruction and guided practice. The second page provides a skill reminder and independent practice. Instruction, written to the student, can be used as a basis for a teacher-led discussion. Suggestions for modeling each lesson can be found on pages 158–162 of the *Student Resource Book Teacher's Annotated Edition*. Guided practice begins with an example and gives students an opportunity to practice the skill with the guidance and support of the teacher. Directions for the independent practice are clearly written and easy to follow. An example is provided to reinforce both the directions and the skill. Throughout, vocabulary has been kept at a level that allows students to focus on the skill.

How to Use the Lessons

The teacher's goals, style of teaching, and classroom organization will guide the use of this section. Because the lessons are grouped into skill categories and in many cases build sequentially, it is best to use them in the order presented. Some ideas for incorporating the lessons into your teaching plan follow.

1. You might decide to teach one of the language and usage skill categories with each of the seven themes in the student Anthology. For example, you might introduce the Capitalization, Punctuation, and Usage Guide during Theme 1 and encourage students to familiarize themselves with its contents so they can use it independently. You might then teach the six categories of Language and Usage Lessons sequentially, with the six remaining themes.

2. You might choose to use this section as a resource to supplement your own or a published instructional program in language, mechanics, and usage.

3. Annotations in the *Dinosauring* Journal Teacher's Edition provide suggestions for coordinating the activities in the student Journal with lessons in this section. You might find it helpful to familiarize yourself with this section and, knowing the contents, to use those suggestions to help students who indicate in their speech or writing a need for work in a particular area.

The flexibility of this section permits it to be used in a variety of ways. You, as a teacher, will know best how to use it.

Capitalization, Punctuation, and Usage Guide

This guide is a helpful handbook for student reference throughout the year. It can also be used for instruction. It includes the following sections, with rules and examples provided for each section: Abbreviations, Titles, Quotations, Capitalization, Punctuation, Problem Words, Adjective and Adverb Usage, Negatives, Pronoun Usage, and Verb Usage.

This guide will be a valuable resource to students in their writing. Annotations in the *Dinosauring* Journal Teacher's Edition note times when, as students are completing the activities in their Journals, it would be helpful to suggest that they refer to this guide.

Throughout the year you might want to remind students of the existence of this resource and to refer them to it whenever they indicate in their speech or writing a need for this support.

Modeling the Lessons

Page 85
Statements and Questions

Hold up a paper bag with a small object inside, such as a pencil. Tell students they can ask five questions to find out what it is. Write their questions on the board. (For example: What is it made of? It is made of wood and rubber.) Call attention to the sentences on the board. Ask students what is different about the end punctuation of the sentences. Help them see that some sentences end with periods and others end with question marks.

Page 87
Commands and Exclamations

Surprise students by dramatically slamming a book on a desk. Have them suggest sentences that describe their reactions. (For example: It scared me! I jumped in my seat!) Write these on the board. Explain that the exclamation point represents strong feelings in a sentence. Next, have students name things you often ask them to do. List these commands on the board. (For example: Pay attention. Open your books.) Point out that a sentence that tells someone to do something ends with a period.

Page 89
Complete Subjects and Complete Predicates

Draw two columns on the board, labeling one *Who or What* and the other *What Is or What Happens*. Remind students that a complete sentence has information for each column. Then have students suggest sentences, telling which part of the sentence belongs in which column. Write the suggestions in the appropriate columns. Tell students that in this lesson, they will learn more about these two sentence parts.

Page 91
Simple Subjects

Write this sentence on the board: *The ballplayer ran.* Have students name the complete subject (the ballplayer) and then expand the subject by adding words that describe the ballplayer more exactly. (For example: The tall, black-haired ballplayer ran.) Write students' sentences on the board, underlining the word *ballplayer* in each. Explain that no matter how long the complete subject is, there is always one main word that tells *who or what*.

Page 93
Simple Predicates

Write this sentence on the board: *The campers slept.* Remind students that a predicate includes all the words that tell what the subject of a sentence is or does. Ask students to identify the predicate (slept) and expand it by adding words that tell more about how or where the campers slept. (For example: The campers slept in their bunks in the cabin.) Write students' sentences on the board, underlining the word *slept*. Explain that a simple predicate is the main word that tells what the subject is or does.

Page 95
Combining Sentences: Subjects and Predicates

Point out to students that a compound sentence combines two simple sentences. Have students use a compound to name two things on their desk. (For example: A book is on my desk, and a pencil is on my desk.) Then ask them to think of another way to present the same information, using fewer words. Write this pattern on the board:

A _____ and a _____ are on my desk.

Have students use the pattern to change compound sentences into simple sentences with compound subjects.

Using the same method, direct students in writing sentences with compound predicates. Put this pattern on the board:

The skater _____ and _____ .

Ask students to supply the missing words.

Point out that a sentence can have both a compound subject and a compound predicate. You may also wish to explain that a compound subject or predicate can be connected with *or* instead of *and*.

Page 97
Correcting Run-on Sentences

Write the following on the board:

Li's family visited the Smithsonian Institution this museum has over 75 million items only a small number are displayed at one time.

Read the passage aloud without pausing after each complete thought. Help students see that ideas are confusing when they are run together in this way. Ask students to read the passage aloud, telling where they think one complete idea ends and another begins.

Page 99
Common and Proper Nouns

Write these words in a column on the board: *boy, girl, teacher, principal, school, town, park, river, store*. Point out that these words are nouns. Then ask volunteers to supply actual names of people and places that correspond to each noun on the board, and write these nouns in a second column. Point out that the nouns in the first column give the general name of a person, place, or thing while those in the second column name a particular place or thing.

Page 101
Singular and Plural Nouns

Write these word pairs on the board:

dollar   dollars
friend   friends

Ask a volunteer to tell which word in each pair he or she would rather have. Have the volunteer explain why dollars are preferable to a dollar and why friends are preferable to a friend. (The nouns *dollar* and *friend* stand for one and are called singular nouns. *Dollars* and *friends* stand for more than one and are called plural nouns.)

Tell students to use context clues to tell if a noun should be singular or plural. Words such as *many* and *all* precede plural nouns, while the word *a* means the noun following it is singular.

Page 103
Nouns Ending with *y*

Hold up a penny in one hand and a key in the other, and ask students to identify them. Write *penny* and *key* on the board. Next, hold up two pennies and two keys. Ask for the names and spellings of the plural nouns *pennies* and *keys*, and write them on the board. Have students look at the two plural forms and note that they are formed differently. Have volunteers suggest reasons for this. Explain that this lesson gives rules for forming plurals of nouns ending with *y*.

Page 105
More Plural Nouns

Ask students to listen to this poem, which makes fun of certain English plural nouns.

The plural of *man* is always *men*.
Is the plural of *pan* ever *pen*?
One is a *mouse*; two are *mice*.
One is a *house*. Are two ever *hice*?
I have a *foot*; I have two *feet*.
I have a *boot*. Have I two *beet*?
A *goose* and a *goose* equal *geese*,
But a *moose* and a *moose* are never —
   meese!

Have students name the made-up plural forms in the poem.

Page 107
Action Verbs

Write various action verbs — *jump, shave, cook, read* — on slips of paper. Then ask volunteers to pick a slip randomly and pantomime the action for the class. Have the class watch to determine the action being pantomimed. As students guess each action, write the verb on the board. Explain that these words are called action verbs. Then have students suggest other action verbs. Write these on the board.

Page 109
Main Verbs and Helping Verbs

Draw two columns on the board. In the first column, list the helping verbs *am, is, are, was,* and *were*. In the second, list the verbs *reading* and *playing*. Call on volunteers to combine a word in column 1 with a word in column 2. Ask them to use these two-word verbs in sentences. Repeat the process with *have, has,* and *had* in one column and *read* and *played* in another. Explain that the words in the first column are helping verbs because they help describe an action.

Page 111
Present, Past, and Future

Write the words *learn, learned,* and *will learn* on the board. Ask students for sentences with the word *learn* that tell what they are doing now. Guide them to give sentences with the present tense form. (For example: We learn about verbs.) Ask for sentences that use the word *learned* and that tell what students have learned in the past. Follow a similar procedure with the words *will learn* and future activities. Sum up by telling students that different forms of verbs tell when actions take place.

Explain that the word *tense* as used in the lesson simply means "time."

Page 113
Making Subjects and Verbs Agree

Ask two students to hop up and down as you write on the board:

The student__ hop__.

Ask what letter needs to be added to the sentence to describe the action. (an *s* to make the subject plural)

Next, ask one of the students to sit down. Remove the *s* from *students*. As the remaining student hops, ask the class what the sentence needs now. Elicit that an *s* must be added to *hop*. Add it, saying that the *s* ending makes the present tense verb singular to agree with the singular subject.

Page 115
Irregular Verbs

Explain that some verbs do not end in *ed* in the past tense. Then write these sentences on the board:

I sing today.

I _____ yesterday.

I have _____ for years.

Have students complete the sentences with the correct past tense forms for *sing*. (*sang, sung*) Point out that *sing* changes its spelling rather than adding *ed*. Have volunteers write three similar sentences for the irregular verbs *drive* and *tell*.

Page 117
The Special Verb *be*

Write the following on the board:

Mr. Wong teaches history.
Mr. Wong is a teacher.

Have students identify the verb in each sentence and tell which verb shows action. (*teaches*) Tell them that the other verb, *is*, does not show action, but, instead, tells what someone or something is like. Write *am, are, is, was,* and *were* on the board. Explain that these are all forms of the verb *be*. Have students make up sentences with these verbs as the simple predicate.

Page 119
What Is an Adjective?

Ask a volunteer to choose an object in the room. Without revealing its identity, the volunteer should give three words that describe the object. Write the words on the board. Have students use these describing words to guess the object. Once the object has been revealed, have students suggest other words to describe it. Write these additional words on the board. Tell students that describing words are called adjectives.

Tell students that the words *a, an,* and *the* are special kinds of adjectives that they will study soon.

Page 121
Adjectives After *be*

Write the following incomplete sentences on the board:

1. I am _____.
2. The _____ is _____.

For Sentence 1, ask volunteers to supply an adjective that they would use to describe themselves. For Sentence 2, ask volunteers to supply the name of an object for the first blank and an adjective that describes the object for the second blank. Indicate with an arrow that the adjective in each sentence comes after the noun or pronoun it describes.

Page 123
Using *a, an,* and *the*

Read these sentences aloud:

1. Do you want a banana?
2. Do you want the banana?

Ask students to tell how the meanings of the two sentences differ. (The word *the* refers to a particular banana while *a* means any banana at all.) Write this sentence on the board:

I want an apple and a banana.

Ask students why one fruit is preceded by *an* and one by *a*. (The word *an* is used before words beginning with vowel sounds. The word *a* is used before words beginning with consonant sounds.)

Page 125

Making Comparisons

Display three books that students use. Ask volunteers to provide three sentences that compare the books. Write the sentences on the board. (For example: The math book is thick. The reader is thicker. The social studies book is thickest of all.) Then ask students to compare the same books with other adjectives, such as *easy, hard,* or *heavy*. Point out the *-er* and *-est* endings. Help students conclude that the *-er* ending is used to compare two books while the *-est* ending is used to compare three or more.

Page 127

Comparing with *more* and *most*

Write these sentences on the board and ask volunteers to read each pair aloud:

May was enjoyabler than June.
May was more enjoyable than June.

My cat is the curiousest of all.
My cat is the most curious of all.

Ask students to identify the adjectives and tell which sounds better — *curiousest* or *most curious, enjoyabler* or *more enjoyable*. Tell students that since long adjectives sound awkward with *-er* and *-est* endings, we use the words *more* and *most* with such adjectives to make comparisons.

Tell students that some two-syllable adjectives compare with *-er* and *-est* endings, such as *quieter/quietest*, while others use *more* and *most*, as in *more helpful/most helpful*. Tell students to check their dictionaries if they are not sure about the correct form. Point out that adjectives with three or more syllables almost always compare with *more* and *most*.

Page 129

What Is a Pronoun?

Remind students that nouns name people, places, and things. Ask students to give examples of singular nouns and plural nouns. Write the examples on the board. Choose one of the examples and have students give three or four sentences about this noun. Write the sentences on the board as a paragraph without pronouns; that is, repeat the noun in every sentence. Call attention to the awkwardness of repeating the noun every time. Tell students that in this lesson they will be learning about words that replace nouns.

Page 131

Subject Pronouns

Write the following sentences on the board and ask volunteers to name the subject of each:

Sir Edmund Hillary climbed Mount Everest.
That mountain is the earth's highest mountain.

Then write the words *He* and *It* on the board. Have volunteers choose the pronoun that can replace the subject in each sentence and read the sentence with the pronoun. Have students supply two more sentences for which the words *He* and *It* can be used as the subjects.

Page 133

Object Pronouns

Help students name the subject pronouns they have studied. (*I, you, we, he, she, they, it*) Then help them list the remaining pronouns that are not subject pronouns. (*me, him, her, us, them*) Have them use these pronouns in sentences. Write sentences with object pronouns on the board. Ask students to comment on the position of these pronouns in the sentences. Explain that they do not serve as the subject of a sentence.

Page 135

Using *I* and *me*

Ask students to supply sentences about the activities they enjoy with friends. Write the sentences that contain *I* and *me* on the board. Remind students that *I* is a subject pronoun and *me* is an object pronoun. Point to any sentences in which *I* or *me* is used with another noun or pronoun. Help students to conclude that when *I* or *me* is used with another noun or pronoun, it is written or spoken last.

Stress that *me* should be used after words like *to, with, for,* and *at*. Explain that people sometimes mistakenly use *I* after these words because they think it sounds correct.

Page 137

Possessive Pronouns

Write the names of a few students on the

board. Ask each of these students to name a favorite possession. Write the item after each name to show the possessive, such as *Hilary's bicycle*. Remind students that possessive nouns show ownership.

Point out that certain pronouns can take the place of possessive nouns. Write *my, your, hers, his, its, our,* and *their* on the board. Have students rewrite the phrases, using possessive pronouns in place of the students' names, as in *her bicycle*.

Point out that *'s* is not added to possessive pronouns as it is to possessive nouns.

Page 139

Pronouns and Homophones

Write these sentences on the board:

It's my dog, and its name is Jo.
You're looking for your dogs.
They're there with their friends.

Ask students to identify the words in each sentence that sound alike. Then ask them to identify which words are contractions *(it's, you're, they're)* and which are possessive pronouns *(its, your, their)*. Point out that *there* is an adverb that means "in that place." Ask volunteers to use these pronouns and contractions in sentences.

Stress that a contraction uses an apostrophe while a possessive pronoun does not.

If students have trouble deciding when to use the contractions *it's, you're,* and *they're*, suggest they insert *it is, you are,* or *they are* in the sentence instead to see whether it makes sense.

Page 141

What Is an Adverb?

Write this sentence on the board:

Joe was singing.

Ask volunteers to suggest words that tell how Joe might have sung. (loudly, happily) Next, ask for single words that tell where he sang (upstairs, here) and when he sang (today, earlier). Proceed similarly with these sentences:

Jill was dancing.
Harry was eating.

Explain that words telling how, when, and where an action takes place are called adverbs.

Page 143

Comparing with Adverbs

Have students turn back to page 141, and have three students each read one paragraph from the page. After they have finished, write *loudly, fast,* and *clearly* on the board. Have students use these three adverbs to discuss and compare how the students read. (For example: Don read faster than Sue. Sue read the most clearly.) Write on the board the sentences that students suggest. Point out that adverbs can be used to compare two or more actions.

Stress that any adverb ending in *-ly*, even a short two-syllable adverb like *slowly*, should be used with *more* or *most* when making comparisons.

Page 145

Using *good* and *well*

On the board, write:

What a good drummer Dan is!
He also plays the piano well.

Ask students to name the word that *good* describes in the first sentence. (drummer) Then ask them to name the word in the second sentence that *well* describes. (plays) Help students see that *good* is an adjective because it describes a noun and that *well* is an adverb because it describes a verb. Ask students to use *good* and *well* in sentences that tell about their own skills and talents.

Page 147

Negatives

Shake your head to show "no," and ask students to tell what the gesture means. Write their suggestions on the board. (For example: No! Never! I can't. Don't do it! It's not right.) Then have students think of situations in which they have shaken their heads. Ask them to tell what they meant to say by doing this. Write their responses, underlining the negative word in each. Point out that the underlined words all mean "no." Explain that words that mean "no" are called negatives.

Remind students that a contraction is a shortened form of two words. Many contractions are formed from a helping verb and *not*. (For example: does + not = doesn't)

NO LONGER THE PROPERTY
OF THE
UNIVERSITY OF R. I. LIBRARY